RILEM State-of-the-Art Reports

RILEM STATE-OF-THE-ART REPORTS
Volume 41

RILEM, The International Union of Laboratories and Experts in Construction Materials, Systems and Structures, founded in 1947, is a non-governmental scientific association whose goal is to contribute to progress in the construction sciences, techniques and industries, essentially by means of the communication it fosters between research and practice. RILEM's focus is on construction materials and their use in building and civil engineering structures, covering all phases of the building process from manufacture to use and recycling of materials. More information on RILEM and its previous publications can be found on www.RILEM.net.

The RILEM State-of-the-Art Reports (STAR) are produced by the Technical Committees. They represent one of the most important outputs that RILEM generates – high level scientific and engineering reports that provide cutting edge knowledge in a given field. The work of the TCs is one of RILEM's key functions.

Members of a TC are experts in their field and give their time freely to share their expertise. As a result, the broader scientific community benefits greatly from RILEM's activities.

RILEM's stated objective is to disseminate this information as widely as possible to the scientific community. RILEM therefore considers the STAR reports of its TCs as of highest importance, and encourages their publication whenever possible.

The information in this and similar reports is mostly pre-normative in the sense that it provides the underlying scientific fundamentals on which standards and codes of practice are based. Without such a solid scientific basis, construction practice will be less than efficient or economical.

It is RILEM's hope that this information will be of wide use to the scientific community.

Indexed in SCOPUS, Google Scholar and SpringerLink.

Jianzhuang Xiao · Yamei Zhang · Amardeep Singh ·
Zengfeng Zhao
Editors

Structural Behaviour and Innovation of Recycled Aggregate Concrete

RILEM Technical Committee
273-RAC_Volume 1

rilem

Springer

Editors
Jianzhuang Xiao
College of Civil Engineering
Tongji University
Shanghai, China

Amardeep Singh
School of Civil and Architectural Engineering
Changzhou Institute of Technology
Changzhou, China

Yamei Zhang
School of Materials Science and Engineering
Southeast University
Nanjing, China

Zengfeng Zhao
College of Civil Engineering
Tongji University
Shanghai, China

ISSN 2213-204X ISSN 2213-2031 (electronic)
RILEM State-of-the-Art Reports
ISBN 978-3-031-84135-4 ISBN 978-3-031-84136-1 (eBook)
https://doi.org/10.1007/978-3-031-84136-1

This Springer imprint is published by the registered company Springer Nature Switzerland AG
The registered company address is: Gewerbestrasse 11, 6330 Cham, Switzerland

Members of TC 273-RAC

Sudhirkumar V. Barai, Indian Institute of Technology Kharagpur, India
Maria Chiara Bignozzi, University of Bologna, Italy
Donguk Choi, Hankyong National University, Republic of Korea
Luc Courard, Université de Liège, Belgium
Frank Dehn, Karlsruhe Institute of Technology (KIT), Germany
Miren Etxeberria, Universidad politécnica de Catalunya, Spain
Belen Gonzalez Fonteboa, Universidade da Coruna, Spain
Asad-Ur-Rehman Khan, NED University of Engineering and Technology, Pakistan
Alalea Kia, Imperial College London, UK
Eddie A. B. Koenders, Technical University of Darmstadt, Germany
Wengui Li, University of New South Wales, Australia
Fernando Martinez Abella, Fundación de Ingeniería Civil de Galícia—FICG, Spain
César Medina Martinez, Universidad Politecnica de Extramadura, Spain
Takafumi Noguchi, University of Tokyo, Japan
Kevin Paine, University of Bath, UK
Patrick Rougeau, CERIB, France
Nicolas Roussel, Université Gustave Eiffel, France
Mickael Saillio, Gustave Eiffel University, France
Gumersinda Seara-Paz, ETS Ingenieros de Caminos, canales y puertos-UDC, Spain
Surendra P. Shah, Northwestern University, USA
Mitsuhiro Shigeishi, Kumamoto University, Japan
Rui Silva, University of Minho, Portugal
Amardeep Singh, Changzhou Institute of Technology, China
Surinder P. Singh, Dr. B. R. Ambedkar National Institute of Technology, Jalandhar, Punjab, India
Nina Stirmer, University of Zagreb, Croatia
Assia Tegguer, Université Gustave Eiffel, France
Romildo D. Toledo Filho, COPPE/UFRJ, Brazil
Jianzhuang Xiao, Tongji Univeristy, China

Guang Ye, Delft University of Technology, The Netherlands
Yamei Zhang, Southeast University, China
Zengfeng Zhao, Tongji University, China
Yuxi Zhao, Zhejiang University, China

Contributors

Chi Sun Poon
Enric Vazquez
Enzo Martinelli
Francisco Agrela
François de Larrard
Gaifei Peng
Hakim Abdelgader
Hyun-Do Yun
Jiabin Li
Jian Yang
Jingwei Ying
Jin-Man Kim
John E. Bolander
Jorge de Brito
Kaijian Zhang
Khaled Sobhan
Khim Chye Gary Ong
Kil-Hee Kim
Luis Evangelista
Marco Pepe
Mukesh C. Limbachiya
Qiong Liu
Rafat Siddique
Sandeep Chaudhary

RILEM Publications

The following list is presenting the global offer of RILEM Publications, sorted by series. Each publication is available in printed version and/or in online version.

RILEM Proceedings (PRO)

PRO 1: Durability of High Performance Concrete (ISBN: 2-912143-03-9; e-ISBN: 2-351580-12-5; e-ISBN: 2351580125); *Ed. H. Sommer*

PRO 2: Chloride Penetration into Concrete (ISBN: 2-912143-00-04; e-ISBN: 2912143454); *Eds. L.-O. Nilsson and J.-P. Ollivier*

PRO 3: Evaluation and Strengthening of Existing Masonry Structures (ISBN: 2-912143-02-0; e-ISBN: 2351580141); *Eds. L. Binda and C. Modena*

PRO 4: Concrete: From Material to Structure (ISBN: 2-912143-04-7; e-ISBN: 2351580206); *Eds. J.-P. Bournazel and Y. Malier*

PRO 5: The Role of Admixtures in High Performance Concrete (ISBN: 2-912143-05-5; e-ISBN: 2351580214); *Eds. J. G. Cabrera and R. Rivera-Villarreal*

PRO 6: High Performance Fiber Reinforced Cement Composites—HPFRCC 3 (ISBN: 2-912143-06-3; e-ISBN: 2351580222); *Eds. H. W. Reinhardt and A. E. Naaman*

PRO 7: 1st International RILEM Symposium on Self-Compacting Concrete (ISBN: 2-912143-09-8; e-ISBN: 2912143721); *Eds. Å. Skarendahl and Ö. Petersson*

PRO 8: International RILEM Symposium on Timber Engineering (ISBN: 2-912143-10-1; e-ISBN: 2351580230); *Ed. L. Boström*

PRO 9: 2nd International RILEM Symposium on Adhesion between Polymers and Concrete ISAP '99 (ISBN: 2-912143-11-X; e-ISBN: 2351580249); *Eds. Y. Ohama and M. Puterman*

PRO 10: 3rd International RILEM Symposium on Durability of Building and Construction Sealants (ISBN: 2-912143-13-6; e-ISBN: 2351580257); *Ed. A. T. Wolf*

PRO 11: 4th International RILEM Conference on Reflective Cracking in Pavements (ISBN: 2-912143-14-4; e-ISBN: 2351580265); *Eds. A. O. Abd El Halim, D. A. Taylor and El H. H. Mohamed*

PRO 12: International RILEM Workshop on Historic Mortars: Characteristics and Tests (ISBN: 2-912143-15-2; e-ISBN: 2351580273); *Eds. P. Bartos, C. Groot and J. J. Hughes*

PRO 13: 2nd International RILEM Symposium on Hydration and Setting (ISBN: 2-912143-16-0; e-ISBN: 2351580281); *Ed. A. Nonat*

PRO 14: Integrated Life-Cycle Design of Materials and Structures—ILCDES 2000 (ISBN: 951-758-408-3; e-ISBN: 235158029X); (ISSN: 0356-9403); *Ed. S. Sarja*

PRO 15: Fifth RILEM Symposium on Fibre-Reinforced Concretes (FRC)—BEFIB'2000 (ISBN: 2-912143-18-7; e-ISBN: 291214373X); *Eds. P. Rossi and G. Chanvillard*

PRO 16: Life Prediction and Management of Concrete Structures (ISBN: 2-912143-19-5; e-ISBN: 2351580303); *Ed. D. Naus*

PRO 17: Shrinkage of Concrete—Shrinkage 2000 (ISBN: 2-912143-20-9; e-ISBN: 2351580311); *Eds. V. Baroghel-Bouny and P.-C. Aïtcin*

PRO 18: Measurement and Interpretation of the On-Site Corrosion Rate (ISBN: 2-912143-21-7; e-ISBN: 235158032X); *Eds. C. Andrade, C. Alonso, J. Fullea, J. Polimon and J. Rodriguez*

PRO 19: Testing and Modelling the Chloride Ingress into Concrete (ISBN: 2-912143-22-5; e-ISBN: 2351580338); *Eds. C. Andrade and J. Kropp*

PRO 20: 1st International RILEM Workshop on Microbial Impacts on Building Materials (CD 02) (e-ISBN 978-2-35158-013-4); *Ed. M. Ribas Silva*

PRO 21: International RILEM Symposium on Connections between Steel and Concrete (ISBN: 2-912143-25-X; e-ISBN: 2351580346); *Ed. R. Eligehausen*

PRO 22: International RILEM Symposium on Joints in Timber Structures (ISBN: 2-912143-28-4; e-ISBN: 2351580354); *Eds. S. Aicher and H.-W. Reinhardt*

PRO 23: International RILEM Conference on Early Age Cracking in Cementitious Systems (ISBN: 2-912143-29-2; e-ISBN: 2351580362); *Eds. K. Kovler and A. Bentur*

PRO 24: 2nd International RILEM Workshop on Frost Resistance of Concrete (ISBN: 2-912143-30-6; e-ISBN: 2351580370); *Eds. M. J. Setzer, R. Auberg and H.-J. Keck*

PRO 25: International RILEM Workshop on Frost Damage in Concrete (ISBN: 2-912143-31-4; e-ISBN: 2351580389); *Eds. D. J. Janssen, M. J. Setzer and M. B. Snyder*

PRO 26: International RILEM Workshop on On-Site Control and Evaluation of Masonry Structures (ISBN: 2-912143-34-9; e-ISBN: 2351580141); *Eds. L. Binda and R. C. de Vekey*

PRO 27: International RILEM Symposium on Building Joint Sealants (CD03; e-ISBN: 235158015X); *Ed. A. T. Wolf*

PRO 28: 6th International RILEM Symposium on Performance Testing and Evaluation of Bituminous Materials—PTEBM'03 (ISBN: 2-912143-35-7; e-ISBN: 978-2-912143-77-8); *Ed. M. N. Partl*

PRO 29: 2nd International RILEM Workshop on Life Prediction and Ageing Management of Concrete Structures (ISBN: 2-912143-36-5; e-ISBN: 2912143780); *Ed. D. J. Naus*

PRO 30: 4th International RILEM Workshop on High Performance Fiber Reinforced Cement Composites—HPFRCC 4 (ISBN: 2-912143-37-3; e-ISBN: 2912143799); *Eds. A. E. Naaman and H. W. Reinhardt*

PRO 31: International RILEM Workshop on Test and Design Methods for Steel Fibre Reinforced Concrete: Background and Experiences (ISBN: 2-912143-38-1; e-ISBN: 2351580168); *Eds. B. Schnütgen and L. Vandewalle*

PRO 32: International Conference on Advances in Concrete and Structures 2 vol. (ISBN (set): 2-912143-41-1; e-ISBN: 2351580176); *Eds. Ying-shu Yuan, Surendra P. Shah and Heng-lin Lü*

PRO 33: 3rd International Symposium on Self-Compacting Concrete (ISBN: 2-912143-42-X; e-ISBN: 2912143713); *Eds. Ó. Wallevik and I. Níelsson*

PRO 34: International RILEM Conference on Microbial Impact on Building Materials (ISBN: 2-912143-43-8; e-ISBN: 2351580184); *Ed. M. Ribas Silva*

PRO 35: International RILEM TC 186-ISA on Internal Sulfate Attack and Delayed Ettringite Formation (ISBN: 2-912143-44-6; e-ISBN: 2912143802); *Eds. K. Scrivener and J. Skalny*

PRO 36: International RILEM Symposium on Concrete Science and Engineering—A Tribute to Arnon Bentur (ISBN: 2-912143-46-2; e-ISBN: 2912143586); *Eds. K. Kovler, J. Marchand, S. Mindess and J. Weiss*

PRO 37: 5th International RILEM Conference on Cracking in Pavements—Mitigation, Risk Assessment and Prevention (ISBN: 2-912143-47-0; e-ISBN: 2912143764); *Eds. C. Petit, I. Al-Qadi and A. Millien*

PRO 38: 3rd International RILEM Workshop on Testing and Modelling the Chloride Ingress into Concrete (ISBN: 2-912143-48-9; e-ISBN: 2912143578); *Eds. C. Andrade and J. Kropp*

PRO 39: 6th International RILEM Symposium on Fibre-Reinforced Concretes—BEFIB 2004 (ISBN: 2-912143-51-9; e-ISBN: 2912143748); *Eds. M. Di Prisco, R. Felicetti and G. A. Plizzari*

PRO 40: International RILEM Conference on the Use of Recycled Materials in Buildings and Structures (ISBN: 2-912143-52-7; e-ISBN: 2912143756); *Eds. E. Vázquez, Ch. F. Hendriks and G. M. T. Janssen*

PRO 41: RILEM International Symposium on Environment-Conscious Materials and Systems for Sustainable Development (ISBN: 2-912143-55-1; e-ISBN: 2912143640); *Eds. N. Kashino and Y. Ohama*

PRO 42: SCC'2005—China: 1st International Symposium on Design, Performance and Use of Self-Consolidating Concrete (ISBN: 2-912143-61-6; e-ISBN: 2912143624); *Eds. Zhiwu Yu, Caijun Shi, Kamal Henri Khayat and Youjun Xie*

PRO 43: International RILEM Workshop on Bonded Concrete Overlays (e-ISBN: 2-912143-83-7); *Eds. J. L. Granju and J. Silfwerbrand*

PRO 44: 2nd International RILEM Workshop on Microbial Impacts on Building Materials (CD11) (e-ISBN: 2-912143-84-5); *Ed. M. Ribas Silva*

PRO 45: 2nd International Symposium on Nanotechnology in Construction, Bilbao (ISBN: 2-912143-87-X; e-ISBN: 2912143888); *Eds. Peter J. M. Bartos, Yolanda de Miguel and Antonio Porro*

PRO 46: ConcreteLife'06—International RILEM-JCI Seminar on Concrete Durability and Service Life Planning: Curing, Crack Control, Performance in Harsh Environments (ISBN: 2-912143-89-6; e-ISBN: 291214390X); *Ed. K. Kovler*

PRO 47: International RILEM Workshop on Performance Based Evaluation and Indicators for Concrete Durability (ISBN: 978-2-912143-95-2; e-ISBN: 9782912143969); *Eds. V. Baroghel-Bouny, C. Andrade, R. Torrent and K. Scrivener*

PRO 48: 1st International RILEM Symposium on Advances in Concrete through Science and Engineering (e-ISBN: 2-912143-92-6); *Eds. J. Weiss, K. Kovler, J. Marchand, and S. Mindess*

PRO 49: International RILEM Workshop on High Performance Fiber Reinforced Cementitious Composites in Structural Applications (ISBN: 2-912143-93-4; e-ISBN: 2912143942); *Eds. G. Fischer and V. C. Li*

PRO 50: 1st International RILEM Symposium on Textile Reinforced Concrete (ISBN: 2-912143-97-7; e-ISBN: 2351580087); *Eds. Josef Hegger, Wolfgang Brameshuber and Norbert Will*

PRO 51: 2nd International Symposium on Advances in Concrete through Science and Engineering (ISBN: 2-35158-003-6; e-ISBN: 2-35158-002-8); *Eds. J. Marchand, B. Bissonnette, R. Gagné, M. Jolin and F. Paradis*

PRO 52: Volume Changes of Hardening Concrete: Testing and Mitigation (ISBN: 2-35158-004-4; e-ISBN: 2-35158-005-2); *Eds. O. M. Jensen, P. Lura and K. Kovler*

PRO 53: High Performance Fiber Reinforced Cement Composites—HPFRCC5 (ISBN: 978-2-35158-046-2; e-ISBN: 978-2-35158-089-9); *Eds. H. W. Reinhardt and A. E. Naaman*

PRO 54: 5th International RILEM Symposium on Self-Compacting Concrete (ISBN: 978-2-35158-047-9; e-ISBN: 978-2-35158-088-2); *Eds. G. De Schutter and V. Boel*

PRO 55: International RILEM Symposium Photocatalysis, Environment and Construction Materials (ISBN: 978-2-35158-056-1; e-ISBN: 978-2-35158-057-8); *Eds. P. Baglioni and L. Cassar*

PRO 56: International RILEM Workshop on Integral Service Life Modelling of Concrete Structures (ISBN 978-2-35158-058-5; e-ISBN: 978-2-35158-090-5); *Eds. R. M. Ferreira, J. Gulikers and C. Andrade*

PRO 57: RILEM Workshop on Performance of cement-based materials in aggressive aqueous environments (e-ISBN: 978-2-35158-059-2); *Ed. N. De Belie*

PRO 58: International RILEM Symposium on Concrete Modelling—CONMOD'08 (ISBN: 978-2-35158-060-8; e-ISBN: 978-2-35158-076-9); *Eds. E. Schlangen and G. De Schutter*

PRO 59: International RILEM Conference on On Site Assessment of Concrete, Masonry and Timber Structures—SACoMaTiS 2008 (ISBN set: 978-2-35158-061-5; e-ISBN: 978-2-35158-075-2); *Eds. L. Binda, M. di Prisco and R. Felicetti*

PRO 60: Seventh RILEM International Symposium on Fibre Reinforced Concrete: Design and Applications—BEFIB 2008 (ISBN: 978-2-35158-064-6; e-ISBN: 978-2-35158-086-8); *Ed. R. Gettu*

PRO 61: 1st International Conference on Microstructure Related Durability of Cementitious Composites 2 vol., (ISBN: 978-2-35158-065-3; e-ISBN: 978-2-35158-084-4); *Eds. W. Sun, K. van Breugel, C. Miao, G. Ye and H. Chen*

PRO 62: NSF/ RILEM Workshop: In-situ Evaluation of Historic Wood and Masonry Structures (e-ISBN: 978-2-35158-068-4); *Eds. B. Kasal, R. Anthony and M. Drdácký*

PRO 63: Concrete in Aggressive Aqueous Environments: Performance, Testing and Modelling, 2 vol., (ISBN: 978-2-35158-071-4; e-ISBN: 978-2-35158-082-0); *Eds. M. G. Alexander and A. Bertron*

PRO 64: Long Term Performance of Cementitious Barriers and Reinforced Concrete in Nuclear Power Plants and Waste Management—NUCPERF 2009 (ISBN: 978-2-35158-072-1; e-ISBN: 978-2-35158-087-5); *Eds. V. L'Hostis, R. Gens and C. Gallé*

PRO 65: Design Performance and Use of Self-consolidating Concrete—SCC'2009 (ISBN: 978-2-35158-073-8; e-ISBN: 978-2-35158-093-6); *Eds. C. Shi, Z. Yu, K. H. Khayat and P. Yan*

PRO 66: 2nd International RILEM Workshop on Concrete Durability and Service Life Planning—ConcreteLife'09 (ISBN: 978-2-35158-074-5; ISBN: 978-2-35158-074-5); *Ed. K. Kovler*

PRO 67: Repairs Mortars for Historic Masonry (e-ISBN: 978-2-35158-083-7); *Ed. C. Groot*

PRO 68: Proceedings of the 3rd International RILEM Symposium on 'Rheology of Cement Suspensions such as Fresh Concrete (ISBN 978-2-35158-091-2; e-ISBN: 978-2-35158-092-9); *Eds. O. H. Wallevik, S. Kubens and S. Oesterheld*

PRO 69: 3rd International PhD Student Workshop on 'Modelling the Durability of Reinforced Concrete (ISBN: 978-2-35158-095-0); *Eds. R. M. Ferreira, J. Gulikers and C. Andrade*

PRO 70: 2nd International Conference on 'Service Life Design for Infrastructure' (ISBN set: 978-2-35158-096-7, e-ISBN: 978-2-35158-097-4); *Eds. K. van Breugel, G. Ye and Y. Yuan*

PRO 71: Advances in Civil Engineering Materials—The 50-year Teaching Anniversary of Prof. Sun Wei' (ISBN: 978-2-35158-098-1; e-ISBN: 978-2-35158-099-8); *Eds. C. Miao, G. Ye and H. Chen*

PRO 72: First International Conference on 'Advances in Chemically-Activated Materials—CAM'2010' (2010), 264 pp., ISBN: 978-2-35158-101-8; e-ISBN: 978-2-35158-115-5; *Eds. Caijun Shi and Xiaodong Shen*

PRO 73: 2nd International Conference on 'Waste Engineering and Management—ICWEM 2010' (2010), 894 pp., ISBN: 978-2-35158-102-5; e-ISBN: 978-2-35158-103-2, *Eds. J. Zh. Xiao, Y. Zhang, M. S. Cheung and R. Chu*

PRO 74: International RILEM Conference on 'Use of Superabsorbent Polymers and Other New Additives in Concrete' (2010) 374 pp., ISBN: 978-2-35158-104-9; e-ISBN: 978-2-35158-105-6; *Eds. O. M. Jensen, M. T. Hasholt, and S. Laustsen*

PRO 75: International Conference on 'Material Science—2nd ICTRC—Textile Reinforced Concrete—Theme 1' (2010) 436 pp., ISBN: 978-2-35158-106-3; e-ISBN: 978-2-35158-107-0; *Ed. W. Brameshuber*

PRO 76: International Conference on 'Material Science—HetMat—Modelling of Heterogeneous Materials—Theme 2' (2010) 255 pp., ISBN: 978-2-35158-108-7; e-ISBN: 978-2-35158-109-4; *Ed. W. Brameshuber*

PRO 77: International Conference on 'Material Science—AdIPoC—Additions Improving Properties of Concrete—Theme 3' (2010) 459 pp., ISBN: 978-2-35158-110-0; e-ISBN: 978-2-35158-111-7; *Ed. W. Brameshuber*

PRO 78: 2nd Historic Mortars Conference and RILEM TC 203-RHM Final Workshop—HMC2010 (2010) 1416 pp., e-ISBN: 978-2-35158-112-4; *Eds. J. Válek, C. Groot and J. J. Hughes*

PRO 79: International RILEM Conference on Advances in Construction Materials Through Science and Engineering (2011) 213 pp., ISBN: 978-2-35158-116-2, e-ISBN: 978-2-35158-117-9; *Eds. Christopher Leung and K. T. Wan*

PRO 80: 2nd International RILEM Conference on Concrete Spalling due to Fire Exposure (2011) 453 pp., ISBN: 978-2-35158-118-6; e-ISBN: 978-2-35158-119-3; *Eds. E.A.B. Koenders and F. Dehn*

PRO 81: 2nd International RILEM Conference on Strain Hardening Cementitious Composites (SHCC2-Rio) (2011) 451 pp., ISBN: 978-2-35158-120-9; e-ISBN: 978-2-35158-121-6; *Eds. R.D. Toledo Filho, F.A. Silva, E.A.B. Koenders and E.M.R. Fairbairn*

PRO 82: 2nd International RILEM Conference on Progress of Recycling in the Built Environment (2011) 507 pp., e-ISBN: 978-2-35158-122-3; *Eds. V.M. John, E. Vazquez, S.C. Angulo and C. Ulsen*

PRO 83: 2nd International Conference on Microstructural-related Durability of Cementitious Composites (2012) 250 pp., ISBN: 978-2-35158-129-2; e-ISBN: 978-2-35158-123-0; *Eds. G. Ye, K. van Breugel, W. Sun and C. Miao*

PRO 84: CONSEC13—Seventh International Conference on Concrete under Severe Conditions—Environment and Loading (2013) 1930 pp., ISBN: 978-2-35158-124-7; e-ISBN: 978-2- 35158-134-6; *Eds. Z.J. Li, W. Sun, C.W. Miao, K. Sakai, O.E. Gjorv and N. Banthia*

PRO 85: RILEM-JCI International Workshop on Crack Control of Mass Concrete and Related issues concerning Early-Age of Concrete Structures—ConCrack 3—Control of Cracking in Concrete Structures 3 (2012) 237 pp., ISBN: 978-2-35158-125-4; e-ISBN: 978-2-35158-126-1; *Eds. F. Toutlemonde and J.-M. Torrenti*

PRO 86: International Symposium on Life Cycle Assessment and Construction (2012) 414 pp., ISBN: 978-2-35158-127-8, e-ISBN: 978-2-35158-128-5; *Eds. A. Ventura and C. de la Roche*

PRO 87: UHPFRC 2013—RILEM-fib-AFGC International Symposium on Ultra-High Performance Fibre-Reinforced Concrete (2013), ISBN: 978-2-35158-130-8, e-ISBN: 978-2-35158-131-5; *Eds. F. Toutlemonde*

PRO 88: 8th RILEM International Symposium on Fibre Reinforced Concrete (2012) 344 pp., ISBN: 978-2-35158-132-2; e-ISBN: 978-2-35158-133-9; *Eds. Joaquim A.O. Barros*

PRO 89: RILEM International workshop on performance-based specification and control of concrete durability (2014) 678 pp., ISBN: 978-2-35158-135-3; e-ISBN: 978-2-35158-136-0; *Eds. D. Bjegović, H. Beushausen and M. Serdar*

PRO 90: 7th RILEM International Conference on Self-Compacting Concrete and of the 1st RILEM International Conference on Rheology and Processing of Construction Materials (2013) 396 pp., ISBN: 978-2-35158-137-7; e-ISBN: 978-2-35158-138-4; *Eds. Nicolas Roussel and Hela Bessaies-Bey*

PRO 91: CONMOD 2014—RILEM International Symposium on Concrete Modelling (2014), ISBN: 978-2-35158-139-1; e-ISBN: 978-2-35158-140-7; *Eds. Kefei Li, Peiyu Yan and Rongwei Yang*

PRO 92: CAM 2014—2nd International Conference on advances in chemically-activated materials (2014) 392 pp., ISBN: 978-2-35158-141-4; e-ISBN: 978-2-35158-142-1; *Eds. Caijun Shi and Xiadong Shen*

PRO 93: SCC 2014—3rd International Symposium on Design, Performance and Use of Self-Consolidating Concrete (2014) 438 pp., ISBN: 978-2-35158-143-8; e-ISBN: 978-2-35158-144-5; *Eds. Caijun Shi, Zhihua Ou and Kamal H. Khayat*

PRO 94 (online version): HPFRCC-7—7th RILEM conference on High performance fiber reinforced cement composites (2015), e-ISBN: 978-2-35158-146-9; *Eds. H.W. Reinhardt, G.J. Parra-Montesinos and H. Garrecht*

PRO 95: International RILEM Conference on Application of superabsorbent polymers and other new admixtures in concrete construction (2014), ISBN: 978-2-35158-147-6; e-ISBN: 978-2-35158-148-3; *Eds. Viktor Mechtcherine and Christof Schroefl*

PRO 96 (online version): XIII DBMC: XIII International Conference on Durability of Building Materials and Components (2015), e-ISBN: 978-2-35158-149-0; *Eds. M. Quattrone and V.M. John*

PRO 97: SHCC3—3rd International RILEM Conference on Strain Hardening Cementitious Composites (2014), ISBN: 978-2-35158-150-6; e-ISBN: 978-2-35158-151-3; *Eds. E. Schlangen, M.G. Sierra Beltran, M. Lukovic and G. Ye*

PRO 98: FERRO-11—11th International Symposium on Ferrocement and 3rd ICTRC—International Conference on Textile Reinforced Concrete (2015), ISBN: 978-2-35158-152-0; e-ISBN: 978-2-35158-153-7; *Ed. W. Brameshuber*

PRO 99 (online version): ICBBM 2015—1st International Conference on Bio-Based Building Materials (2015), e-ISBN: 978-2-35158-154-4; *Eds. S. Amziane and M. Sonebi*

PRO 100: SCC16—RILEM Self-Consolidating Concrete Conference (2016), ISBN: 978-2-35158-156-8; e-ISBN: 978-2-35158-157-5; *Ed. Kamal H. Kayat*

PRO 101 (online version): III Progress of Recycling in the Built Environment (2015), e-ISBN: 978-2-35158-158-2; *Eds I. Martins, C. Ulsen and S. C. Angulo*

PRO 102 (online version): RILEM Conference on Microorganisms-Cementitious Materials Interactions (2016), e-ISBN: 978-2-35158-160-5; *Eds. Alexandra Bertron, Henk Jonkers and Virginie Wiktor*

PRO 103 (online version): ACESC'16—Advances in Civil Engineering and Sustainable Construction (2016), e-ISBN: 978-2-35158-161-2; *Eds. T.Ch. Madhavi, G. Prabhakar, Santhosh Ram and P.M. Rameshwaran*

PRO 104 (online version): SSCS'2015—Numerical Modeling—Strategies for Sustainable Concrete Structures (2015), e-ISBN: 978-2-35158-162-9

PRO 105: 1st International Conference on UHPC Materials and Structures (2016), ISBN: 978-2-35158-164-3; e-ISBN: 978-2-35158-165-0

PRO 106: AFGC-ACI-fib-RILEM International Conference on Ultra-High-Performance Fibre-Reinforced Concrete—UHPFRC 2017 (2017), ISBN: 978-2-35158-166-7; e-ISBN: 978-2-35158-167-4; *Eds. François Toutlemonde and Jacques Resplendino*

PRO 107 (online version): XIV DBMC—14th International Conference on Durability of Building Materials and Components (2017), e-ISBN: 978-2-35158-159-9; *Eds. Geert De Schutter, Nele De Belie, Arnold Janssens and Nathan Van Den Bossche*

PRO 108: MSSCE 2016—Innovation of Teaching in Materials and Structures (2016), ISBN: 978-2-35158-178-0; e-ISBN: 978-2-35158-179-7; *Ed. Per Goltermann*

PRO 109 (2 volumes): MSSCE 2016—Service Life of Cement-Based Materials and Structures (2016), ISBN Vol. 1: 978-2-35158-170-4; Vol. 2: 978-2-35158-171-4; Set Vol. 1&2: 978-2-35158-172-8; e-ISBN : 978-2-35158-173-5; *Eds. Miguel Azenha, Ivan Gabrijel, Dirk Schlicke, Terje Kanstad and Ole Mejlhede Jensen*

PRO 110: MSSCE 2016—Historical Masonry (2016), ISBN: 978-2-35158-178-0; e-ISBN: 978-2-35158-179-7; *Eds. Inge Rörig-Dalgaard and Ioannis Ioannou*

PRO 111: MSSCE 2016—Electrochemistry in Civil Engineering (2016); ISBN: 978-2-35158-176-6; e-ISBN: 978-2-35158-177-3; *Ed. Lisbeth M. Ottosen*

PRO 112: MSSCE 2016—Moisture in Materials and Structures (2016), ISBN: 978-2-35158-178-0; e-ISBN: 978-2-35158-179-7; *Eds. Kurt Kielsgaard Hansen, Carsten Rode and Lars-Olof Nilsson*

PRO 113: MSSCE 2016—Concrete with Supplementary Cementitious Materials (2016), ISBN: 978-2-35158-178-0; e-ISBN: 978-2-35158-179-7; *Eds. Ole Mejlhede Jensen, Konstantin Kovler and Nele De Belie*

PRO 114: MSSCE 2016—Frost Action in Concrete (2016), ISBN: 978-2-35158-182-7; e-ISBN: 978-2-35158-183-4; *Eds. Marianne Tange Hasholt, Katja Fridh and R. Doug Hooton*

PRO 115: MSSCE 2016—Fresh Concrete (2016), ISBN: 978-2-35158-184-1; e-ISBN: 978-2-35158-185-8; *Eds. Lars N. Thrane, Claus Pade, Oldrich Svec and Nicolas Roussel*

PRO 116: BEFIB 2016—9th RILEM International Symposium on Fiber Reinforced Concrete (2016), ISBN: 978-2-35158-187-2; e-ISBN: 978-2-35158-186-5; *Eds. N. Banthia, M. di Prisco and S. Soleimani-Dashtaki*

PRO 117: 3rd International RILEM Conference on Microstructure Related Durability of Cementitious Composites (2016), ISBN: 978-2-35158-188-9; e-ISBN: 978-2-35158-189-6; *Eds. Changwen Miao, Wei Sun, Jiaping Liu, Huisu Chen, Guang Ye and Klaas van Breugel*

PRO 118 (4 volumes): International Conference on Advances in Construction Materials and Systems (2017), ISBN Set: 978-2-35158-190-2; Vol. 1: 978-2-35158-193-3; Vol. 2: 978-2-35158-194-0; Vol. 3: ISBN:978-2-35158-195-7; Vol. 4: ISBN:978-2-35158-196-4; e-ISBN: 978-2-35158-191-9; *Ed. Manu Santhanam*

PRO 119 (online version): ICBBM 2017—Second International RILEM Conference on Bio-based Building Materials, (2017), e-ISBN: 978-2-35158-192-6; *Ed. Sofiane Amziane*

PRO 120 (2 volumes): EAC-02—2nd International RILEM/COST Conference on Early Age Cracking and Serviceability in Cement-based Materials and Structures, (2017), Vol. 1: 978-2-35158-199-5, Vol. 2: 978-2-35158-200-8, Set: 978-2-35158-197-1, e-ISBN: 978-2-35158-198-8; *Eds. Stéphanie Staquet and Dimitrios Aggelis*

PRO 121 (2 volumes): SynerCrete18: Interdisciplinary Approaches for Cement-based Materials and Structural Concrete: Synergizing Expertise and Bridging Scales of Space and Time, (2018), Set: 978-2-35158-202-2, Vol.1: 978-2-35158-211-4, Vol.2: 978-2-35158-212-1, e-ISBN: 978-2-35158-203-9; *Eds. Miguel Azenha, Dirk Schlicke, Farid Benboudjema, Agnieszka Knoppik*

PRO 122: SCC'2018 China—Fourth International Symposium on Design, Performance and Use of Self-Consolidating Concrete, (2018), ISBN: 978-2-35158-204-6, e-ISBN: 978-2-35158-205-3; *Eds. C. Shi, Z. Zhang, K. H. Khayat*

PRO 123: Final Conference of RILEM TC 253-MCI: Microorganisms-Cementitious Materials Interactions (2018), Set: 978-2-35158-207-7, Vol.1: 978-2-35158-209-1, Vol.2: 978-2-35158-210-7, e-ISBN: 978-2-35158-206-0; *Ed. Alexandra Bertron*

PRO 124 (online version): Fourth International Conference Progress of Recycling in the Built Environment (2018), e-ISBN: 978-2-35158-208-4; *Eds. Isabel M. Martins, Carina Ulsen, Yury Villagran*

PRO 125 (online version): SLD4—4th International Conference on Service Life Design for Infrastructures (2018), e-ISBN: 978-2-35158-213-8; *Eds. Guang Ye, Yong Yuan, Claudia Romero Rodriguez, Hongzhi Zhang, Branko Savija*

PRO 126: Workshop on Concrete Modelling and Material Behaviour in honor of Professor Klaas van Breugel (2018), ISBN: 978-2-35158-214-5, e-ISBN: 978-2-35158-215-2; *Ed. Guang Ye*

PRO 127 (online version): CONMOD2018—Symposium on Concrete Modelling (2018), e-ISBN: 978-2-35158-216-9; *Eds. Erik Schlangen, Geert de Schutter, Branko Savija, Hongzhi Zhang, Claudia Romero Rodriguez*

PRO 128: SMSS2019—International Conference on Sustainable Materials, Systems and Structures (2019), ISBN: 978-2-35158-217-6, e-ISBN: 978-2-35158-218-3

PRO 129: 2nd International Conference on UHPC Materials and Structures (UHPC2018-China), ISBN: 978-2-35158-219-0, e-ISBN: 978-2-35158-220-6

PRO 130: 5th Historic Mortars Conference (2019), ISBN: 978-2-35158-221-3, e-ISBN: 978-2-35158-222-0; *Eds. José Ignacio Álvarez, José María Fernández, Íñigo Navarro, Adrián Durán, Rafael Sirera*

PRO 131 (online version): 3rd International Conference on Bio-Based Building Materials (ICBBM2019), e-ISBN: 978-2-35158-229-9; *Eds. Mohammed Sonebi, Sofiane Amziane, Jonathan Page*

PRO 132: IRWRMC'18—International RILEM Workshop on Rheological Measurements of Cement-based Materials (2018), ISBN: 978-2-35158-230-5, e-ISBN: 978-2-35158-231-2; *Eds. Chafika Djelal, Yannick Vanhove*

PRO 133 (online version): CO2STO2019—International Workshop CO2 Storage in Concrete (2019), e-ISBN: 978-2-35158-232-9; *Eds. Assia Djerbi, Othman Omikrine-Metalssi, Teddy Fen-Chong*

PRO 134: 3rd ACF/HNU International Conference on UHPC Materials and Structures—UHPC'2020, ISBN: 978-2-35158-233-6, e-ISBN: 978-2-35158-234-3; *Eds. Caijun Shi and Jiaping Liu*

PRO 135: Fourth International Conference on Chemically Activated Materials (CAM2021), ISBN: 978-2-35158-235-0, e-ISBN: 978-2-35158-236-7; *Eds. Caijun Shi and Xiang Hu*

PRO 136: Lime Based Materials for Repairing Historic Structures (2022), ISBN: 978-2-35158-240-4, e-ISBN: 978-2-35158-237-4; *Eds. Ioanna Papayianni and Jan Válek*

PRO 137 (online version): V International Conference Progress of Recycling in the Built Environment (2023), e-ISBN: 978-2-35158-238-1; *Eds. Ines Döring, Yury Villagrán Zaccardi, Anette Müller*

RILEM Reports (REP)

Report 19: Considerations for Use in Managing the Aging of Nuclear Power Plant Concrete Structures (ISBN: 2-912143-07-1); *Ed. D. J. Naus*

Report 20: Engineering and Transport Properties of the Interfacial Transition Zone in Cementitious Composites (ISBN: 2-912143-08-X); *Eds. M. G. Alexander, G. Arliguie, G. Ballivy, A. Bentur and J. Marchand*

Report 21: Durability of Building Sealants (ISBN: 2-912143-12-8); *Ed. A. T. Wolf*

Report 22: Sustainable Raw Materials—Construction and Demolition Waste (ISBN: 2-912143-17-9); *Eds. C. F. Hendriks and H. S. Pietersen*

Report 23: Self-Compacting Concrete state-of-the-art report (ISBN: 2-912143-23-3); *Eds. Å. Skarendahl and Ö. Petersson*

Report 24: Workability and Rheology of Fresh Concrete: Compendium of Tests (ISBN: 2-912143-32-2); *Eds. P. J. M. Bartos, M. Sonebi and A. K. Tamimi*

Report 25: Early Age Cracking in Cementitious Systems (ISBN: 2-912143-33-0); *Ed. A. Bentur*

Report 26: Towards Sustainable Roofing (Joint Committee CIB/RILEM) (CD 07) (e-ISBN 978-2-912143-65-5); *Eds. Thomas W. Hutchinson and Keith Roberts*

Report 27: Condition Assessment of Roofs (Joint Committee CIB/RILEM) (CD 08) (e-ISBN 978-2-912143-66-2); *Ed. CIB W 83/RILEM TC166-RMS*

Report 28: Final report of RILEM TC 167-COM 'Characterisation of Old Mortars with Respect to Their Repair (ISBN: 978-2-912143-56-3); *Eds. C. Groot, G. Ashall and J. Hughes*

Report 29: Pavement Performance Prediction and Evaluation (PPPE): Interlaboratory Tests (e-ISBN: 2-912143-68-3); *Eds. M. Partl and H. Piber*

Report 30: Final Report of RILEM TC 198-URM 'Use of Recycled Materials' (ISBN: 2-912143-82-9; e-ISBN: 2-912143-69-1); *Eds. Ch. F. Hendriks, G. M. T. Janssen and E. Vázquez*

Report 31: Final Report of RILEM TC 185-ATC 'Advanced testing of cement-based materials during setting and hardening' (ISBN: 2-912143-81-0; e-ISBN: 2-912143-70-5); *Eds. H. W. Reinhardt and C. U. Grosse*

Report 32: Probabilistic Assessment of Existing Structures. A JCSS publication (ISBN 2-912143-24-1); *Ed. D. Diamantidis*

Report 33: State-of-the-Art Report of RILEM Technical Committee TC 184-IFE 'Industrial Floors' (ISBN 2-35158-006-0); *Ed. P. Seidler*

Report 34: Report of RILEM Technical Committee TC 147-FMB 'Fracture mechanics applications to anchorage and bond' Tension of Reinforced Concrete Prisms—Round Robin Analysis and Tests on Bond (e-ISBN 2-912143-91-8); *Eds. L. Elfgren and K. Noghabai*

Report 35: Final Report of RILEM Technical Committee TC 188-CSC 'Casting of Self Compacting Concrete' (ISBN 2-35158-001-X; e-ISBN: 2-912143-98-5); *Eds. Å. Skarendahl and P. Billberg*

Report 36: State-of-the-Art Report of RILEM Technical Committee TC 201-TRC 'Textile Reinforced Concrete' (ISBN 2-912143-99-3); *Ed. W. Brameshuber*

Report 37: State-of-the-Art Report of RILEM Technical Committee TC 192-ECM 'Environment-conscious construction materials and systems' (ISBN: 978-2-35158-053-0); *Eds. N. Kashino, D. Van Gemert and K. Imamoto*

Report 38: State-of-the-Art Report of RILEM Technical Committee TC 205-DSC 'Durability of Self-Compacting Concrete' (ISBN: 978-2-35158-048-6); *Eds. G. De Schutter and K. Audenaert*

Report 39: Final Report of RILEM Technical Committee TC 187-SOC 'Experimental determination of the stress-crack opening curve for concrete in tension' (ISBN 978-2-35158-049-3); *Ed. J. Planas*

Report 40: State-of-the-Art Report of RILEM Technical Committee TC 189-NEC 'Non-Destructive Evaluation of the Penetrability and Thickness of the Concrete Cover' (ISBN 978-2-35158-054-7); *Eds. R. Torrent and L. Fernández Luco*

Report 41: State-of-the-Art Report of RILEM Technical Committee TC 196-ICC 'Internal Curing of Concrete' (ISBN 978-2-35158-009-7); *Eds. K. Kovler and O. M. Jensen*

Report 42: 'Acoustic Emission and Related Non-destructive Evaluation Techniques for Crack Detection and Damage Evaluation in Concrete'—Final Report of RILEM Technical Committee 212-ACD (e-ISBN: 978-2-35158-100-1); *Ed. M. Ohtsu*

Report 45: Repair Mortars for Historic Masonry—State-of-the-Art Report of RILEM Technical Committee TC 203-RHM (e-ISBN: 978-2-35158-163-6); *Eds. Paul Maurenbrecher and Caspar Groot*

Report 46: Surface delamination of concrete industrial ffioors and other durability related aspects guide—Report of RILEM Technical Committee TC 268-SIF (e-ISBN: 978-2-35158-201-5); *Ed. Valerie Pollet*

Contents

About the Editors

Prof. Dr. Jianzhuang Xiao (Committee Chair) has been doing fundamental research on the material characteristics and structural behaviours of recycled aggregate concrete for over 20 years. He has won one second-class National Awards of Scientific and Technology Progress in China and six first-class Province/Ministry Awards of Scientific and Technology Progress in China. He was awarded the Alexander von Humboldt Foundation Fellowship in Germany in 2004, and in 2024, he was honored with the Humboldt Research Award. He has 82 granted patents (1 in the US, 2 in Australia, 3 in South Africa and 76 in China). He has served as the chair of 23 academic conferences, both international and national. With a total of 563 scientific publications (512 peer-reviewed articles, 11 authored academic books and 40 conference papers). His H-index reaches 73, with 19,738 citations in WOS in January, 2025, and he has been consistently recognized as a World's Top 2% Scientists (2019–2024).

Prof. Dr. Yamei Zhang is a Professor in School of Materials Science and Engineering from Southeast University in Nanjing, China. She is also an affiliate professor in the department of civil engineering at the University of British Columbia, Canada, the director of Nanjing Institute for Intelligent Additive Manufacturing. She serves on several national and international academic committees which include the chair of the Committee of concrete 3D printing, the Chinese Ceramic Society; chair of Technical Board, Asia Concrete Federation; and National Convener of RILEM China. She is now an associate editor of the *Journal of Cement and Concrete Composites* and board members of several other international journals. Her research covers cementitious composites, 3D printing, and utilization of solid wastes. Her team has close collaboration with well-known industrial partners including Holcim, CR Cement, China Construction Eighth Engineering Division, and etc. Her work on the utilization of solid wastes in concrete 3D printing was reported by CCTV 9.

Dr. Amardeep Singh is a distinguished researcher specializing in low-carbon cementitious materials, the utilization of construction and demolition waste in concrete, and 3D concrete printing. He earned his Bachelor of Civil Engineering from DAV Institute of Engineering and Technology, India, in 2009, followed by a Master of Engineering from Dr. B. R. Ambedkar National Institute of Technology, India, in 2012, and a Ph.D. from Tongji University, China, in 2020. He currently holds a faculty position at Changzhou Institute of Technology and has been honored with the prestigious Double Doctorate of Innovation and Entrepreneurship title by the Jiangsu Provincial Government, in recognition of his contributions to scientific research and regional development. Currently, he is serving as managing editor of *Low-Carbon Materials* and *Green Construction*. He has authored 59 peer-reviewed publications, with 1448 citations, reflecting an H-index of 20 and was granted one China Patent. He currently involves in two RILEM Technical Committees 303-PFC, focusing on performance requirements and testing of fresh printable cement-based materials, and 304-ADC, which assesses additively manufactured concrete materials and structures.

Dr. Zengfeng Zhao research interests cover the valorization of recycled concrete aggregates and by-products in concrete in order to reduce their environmental impact and CO_2 emissions. He is currently a research professor at the College of Civil Engineering, Tongji University. He received his Ph.D. in civil engineering from University of Lille 1 in France in 2014. He was selected for the Marie Curie fellowship, the Shanghai Leader Oversea Program, and the Shanghai Pujiang Talents Program. He currently involves in eight Technical Committees of RILEM including 273-RAC, 281-CCC, 282-CCL, 303-PFC, 304-ADC, 309-MCP, 313-MMS, and CUC. He is also serving as the managing editor of *Low-Carbon Materials and Green Construction* and guest editor of *Buildings*. He has published over 50 fully reviewed journal papers and was granted three China Patents.

Abbreviations

ADR	Advanced Dry Recovery
AM	Adhered Mortar
ANN	Artificial Neural Networks
CDW	Construction and Demolition Waste
CF	Concrete Fines
CH	Calcium Hydrate
CM	Calcium Metasilicate
CPM	Compressible Packing Model
CT	Computed Tomography
DVR	Direct Volume Replacement
DWR	Direct Weight Replacement
EMV	Equivalent Mortar Volume
FA	Fly Ash
HPC	High Performance Concrete
HRM	Heating and Rubbing Method
ITZ	Interfacial Transition Zone
LCA	Life Cycle Assessment
MA	Mixed Aggregates
MICP	Microbial Induced Carbonation Precipitation
MIP	Mercury Intrusion Porosimetry
MSE	Mean Squared Error
NA	Natural Aggregates
NAC	Natural Aggregate Concrete
NCA	Natural Coarse Aggregate
NFA	Natural Fine Aggregate
NL	Nano-Limestone
NS	Nano-Silica
NVC	Normal Vibrated Concrete
OCA	Original Coarse Aggregate
ODD	Oven-Dry Density
OPC	Ordinary Portland Cement

PPC	Pozzolanic Portland Cement
PPM	Particle Packing Method
PVA	Polyvinyl Alcohol
RA	Recycled Aggregate
RAC	Recycled Aggregate Concrete
RBA	Recycled Brick Aggregates
RCA	Recycled Coarse Aggregate
RFA	Recycled Fine Aggregates
RMA	Recycled Mortar Aggregates
RP	Recycled Powder
SCC	Self-compacting Concrete
SCI	Science Citation Index
SCM	Supplemental Cementitious Materials
SEM	Scanning Electron Microscope
SF	Silica Fume
SSD	Saturated Surface Dry
TM	Triple Mixing Method
TSMA	Two-Stage Mixing Approach
VC	Voids Content

Properties of Recycled Aggregates and Related Specifications

Donguk Choi, Romildo Dias Toledo Filho, and Yamei Zhang

1 Introduction

According to the World Business Council for Sustainable Development (Jane and Grayson 2017), approximately 25 billion tons of concrete are produced each year globally. Because the world population is 7 billion, concrete is produced at a rate of about 3.5 tons per capita to be used to construct concrete buildings, pavement, tunnels, bridges, and other infrastructure. A concrete structure must be demolished when it reaches the end of its useful life or a socio-economic cause, resulting in waste. Construction generates waste, though the amount is less than that of demolition (Poon et al. 2001). Every year, approximately 1.3 billion tons of waste are generated in Europe, with construction and demolition waste (CDW) accounting for approximately 40%, or 510 million tons (Jane and Grayson 2017). According to Townsend et al. (2014), the United States produced approximately 480 million tons of CDW in 2012. The CDW comprises about 100 million tons of mixed waste, 310 million tons of bulk aggregate (mostly concrete), and 70 million tons of reclaimed asphalt pavement. The recycling industry was expected to recover more than 70% of waste (corresponding to a 35% recycling rate for mixed waste, 85% recycling rate for bulk aggregate, and 99% recycling rate for recovered asphalt pavement). Approximately 1.55 billion tons of CDW are produced in China (Li et al. 2016).

D. Choi
School of Architecture and Design Convergence, Hankyong National University, Anseong, South Korea

R. D. T. Filho
Civil Engineering Program, COPPE, Federal University of Rio de Janeiro, Rio de Janeiro, Brazil

Y. Zhang (✉)
School of Materials Science and Engineering, Southeast University, Nanjing, China
e-mail: ymzhang@seu.edu.cn

J. Xiao et al. (eds.), *Structural Behaviour and Innovation of Recycled Aggregate Concrete*, RILEM State-of-the-Art Reports 41,
https://doi.org/10.1007/978-3-031-84136-1_1

The composition of the CDW varies depending on the type of structure and region, but it typically includes soils, metals, concrete rubbles, bricks, asphalt, tiles, plastics, glass, and wood pieces, as well as some minor impurities such as gypsum (Zhao et al. 2020a, b; Courard et al. 2020). Concrete and masonry elements are frequently used to create recycled aggregate (RA). Poon et al. (2001) used Table 1 to investigate the type and amount of CDW in Hong Kong. Waste typically accounts for 10–20% of the total weight of building materials delivered to a construction site during building construction. The amount of waste generated by building demolition is 10–20 times the weight generated by construction. Waste separation through selective demolition is critical to reducing the amount of mixed waste, but it is dependent on the demolition method used, such as (a) explosion, (b) use of wrecking balls, (c) use of hydraulic crushers and pulverizers, and (d) top-down method. When the top-down demolition method was used, the amount of mixed waste was significantly reduced (in highly congested areas of Hong Kong). To reverse the construction process, the demolition process is typically carried out selectively.

The composition of CDW differs from a region to another. For example, in the Mediterranean countries such as Spain, Portugal, Italy or Greece, the architectural elements used in buildings often include ceramic elements combined with mortar and

Table 1 Composition of CDW in Hong Kong (Poon et al. 2001)

Constituent	Percent	
	Building demolition waste	Building waste
Asphalt	1.61	0.13
Concrete	19.99	9.27
Reinforced concrete	33.11	8.25
Dirt, soil, mud	11.91	30.55
Rock	6.83	9.74
Rubble	4.95	14.13
Wood	7.15	10.53
Bamboo	0.31	0.30
Block concrete	1.11	0.90
Brick	6.33	5.00
Glass	0.20	0.56
Other organics	1.30	3.05
Plastic pipe	0.61	1.13
Sand	1.44	1.70
Trees	0.00	0.12
Fixtures	0.04	0.03
Junk	0.07	0.24
Metal (ferrous)	3.41	4.36
Total	100.00	100.00

concrete, including some gypsum in interior faces. Therefore, a lot of ceramic particles are typically mixed with concrete. Agrela et al. 2011 investigated the constituents of 35 recycled coarse aggregates produced in thirteen different RA production plants in Spain with results summarized in Section 2.1 (Table 2).

2 RA Production Technology and Its Properties

2.1 Production Technology

Recycled coarse aggregate (RCA) has the following three distinctive characteristics which do not belong to its counterpart, natural aggregates (NA):

A Adhered mortar of old concrete exists on the surface of NA;
B Various organic and inorganic foreign matters are included in the source materials of RA which cannot be entirely removed; and
C Fines produced in the production process of RA are included as part of RA.

The plant for RA production incorporates various types of crushers, screens, devices used for removal of foreign matter, and transfer equipment (Hansen 1986). While the present production procedure of RA is more sophisticated than it was about 30 years ago, at the time of the second state-of-the-art report of RILEM, the fundamental aspects of RA production remain the same: RA production consists of crushing and sorting/screening. Figure 1 illustrates a production process of recycled coarse and fine aggregates, typical in South Korea (Sim and Park 2011; Kim et al. 2017). Concrete rubbles as part of CDW are first brought to the production plant. Large concrete debris has to be reduced in size to less than 1,000 mm. The reduced concrete debris is brought to a scalping screen where particles smaller than 80–100 mm are separated, and larger particles are put into a jaw crusher, further reducing the particle size to 150–200 mm. Above conveyor belts, magnetic separation of metallic objects, manual separation, air separation, and flotation of lightweight materials such as styrofoam, papers, and small pieces of wood remove foreign materials following the primary crushing. The concrete particles are then brought to the second crusher, such as a cone crusher, where the particles are further reduced to <75 mm. After screening the output from the second crusher, particles smaller than 10–13 mm are typically collected separately and later used as raw material for the secondary building products (bricks and blocks, etc.). Foreign matters are removed again using magnetic and air separation and flotation, while the output of this process is now suitable for use as RA for road sub-base. When higher quality RA for use in structural concrete is desired, the particles are further processed using the third crusher, an impact crusher typ., which produces the particles to less than 40 mm in size as RA. The RA, after screening, is divided into fractions of less than 5, 5–25, and 25–40 mm. The 5–25 mm fraction is coarse RA suitable for road base material and structural concrete. The under 5 mm fraction and 25–40 mm fraction are further processed to the fourth

Table 2 Constituents of RA in Spain (Agrela et al. 2011)

	Constituents of coarse RA				
	Concrete	Brick	Asphalt	Gypsum	Glass
CRA1	97.7	1.9	0.3	0	0
CRA2	93.7	3.4	2.9	0	0
CRA3	91.5	7.2	1.1	0	0.1
CRA4	91.1	4.1	4.6	0.1	0.1
CRA5	90.5	7.2	1.2	1	0
MixRA1	88.9	6	5	0.1	0
MixRA2	88.3	9.8	1.6	0.1	0.2
MixRA3	84.42	14.22	0.34	0.5	0.3
MixRA4	83.8	7.22	11.1	0	0
MixRA5	80.1	19	0.7	0	0
MixRA6	80	15	0	0	0.01
MixRA7	78.84	19.33	0.3	0.42	0.29
MixRA8	78.74	14.48	4.55	2.1	0
MixRA9	78.7	12.7	3.8	2.7	1.53
MixRA10	78.7	16.9	6.9	0.1	0
MixRA11	78.47	14.74	4.49	1.01	0.34
MixRA12	78	18.8	0.65	1.1	0.55
MixRA13	76.6	20.6	1	0.1	0.1
MixRA14	72.5	19	3.6	3.2	1.51
MixRA15	71.7	27.8	0	0.46	0
MixRA16	71.5	16.4	11.7	0.2	0
MixRA17	70.44	22.21	4.03	2.63	0.15
MixRA18	70.3	28.9	0.2	0.15	0.43
MixRA19	70.01	12.9	16.7	0	0
CerRA1	65.08	30.08	2.8	1.8	0.17
CerRA2	59.4	40	2.2	0	1.3
CerRA3	58	33	0	4	0.6
CerRA4	55.6	36.9	5.1	2.7	0
CerRA5	46.5	39.3	6.9	9.9	0.75
CerRA6	45.69	51.57	0.29	1.66	0.28
CerRA7	45.3	52.59	0.12	1.32	0.27
CerRA8	44.89	53.11	0.17	1.15	0.42
CerRA9	44.14	53.9	0.22	1.2	0.34
CerRA10	38	49.5	5.9	3	2.3
CerRA11	36.38	47.76	4.73	7.5	0.3
Medium value RCA	92.90	4.76	2.02	0.22	0.04

(continued)

Table 2 (continued)

	Constituents of coarse RA				
	Concrete	Brick	Asphalt	Gypsum	Glass
Medium value MixRA	77.90	16.63	4.03	0.78	0.28
Medium value CerRA	49.00	44.34	2.58	3.11	0.61

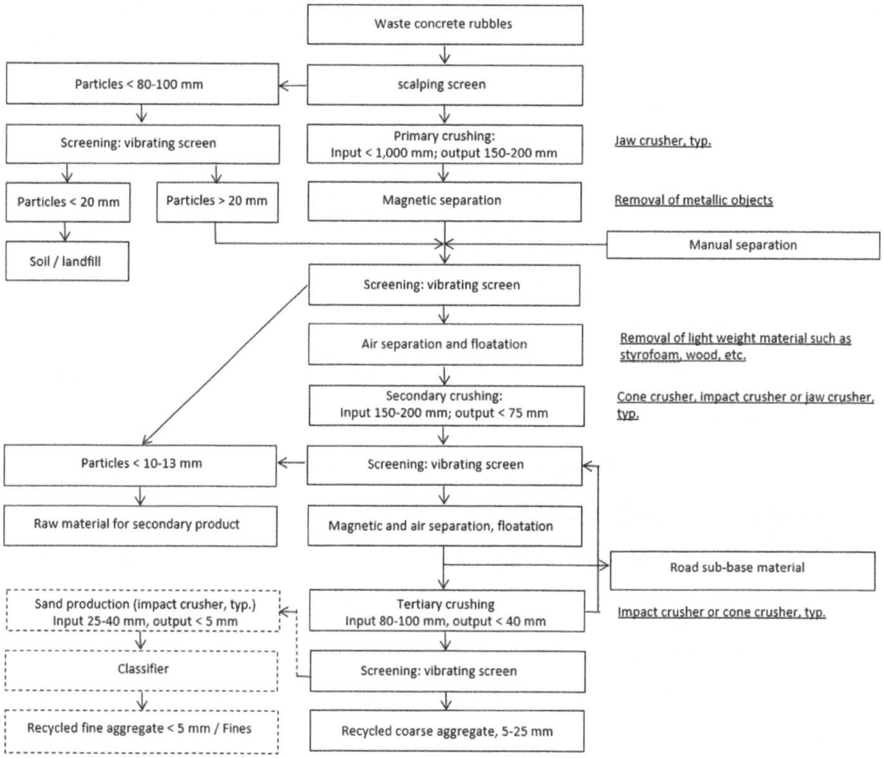

Fig. 1 Production process flowchart of recycled coarse and fine aggregates in a South Korea

crusher (impact crusher or roll crusher, typ.) to produce fine RA (the fine aggregates production process is shown using dashed boxes in Fig. 1).

2.1.1 Crushing

The jaw crusher is most often used as a primary crusher among the various crushers due to its capacity to handle large concrete rubbles, while the gyratory crusher can also be used for primary crushing (Kim et al. 2017). The crushing chamber of a jaw crusher consists of two metallic plates that face each other at an angle, a fixed jaw and a swing jaw, as shown in Table 3 and 4. The swing jaw oscillates back and forth relative to the fixed jaw, driven by an eccentric wheel. Source input material is crushed at the lower part by compression between the fixed jaw and the swing jaw. The particle size reduction depends on the size of the gap between the plates, which is adjustable. A cone crusher, which works on the same principle as a gyratory crusher, is suitable for use as the second crusher with its capacity to produce particles of enhanced shape. The crushing chamber of a cone crusher consists of a mantle (cone) and a concave. Particles reduced in size by the primary crusher are fed through the top of the cone crusher and flow over the mantle, rotating eccentrically into a concave and crushed in compression between the mantle and the concave. An impact crusher hurls particles to breaker plates by multiple blades or hammers connected to the high speed spinning motor. Particles fed from above the crushing chamber are subjected to impact force as they collide at high speed against the breaker plates and with each other. An impact crusher produces particles with enhanced shape and is often used for secondary or tertiary crushing. There are two types of impact crushers, vertical and horizontal type, depending on the direction of the rotor shaft, where the type and magnitude of the force on particles are slightly different depending on the type.

Table 3 Crusher types

Type	Cross-section	Type	Cross-section
Jaw crusher	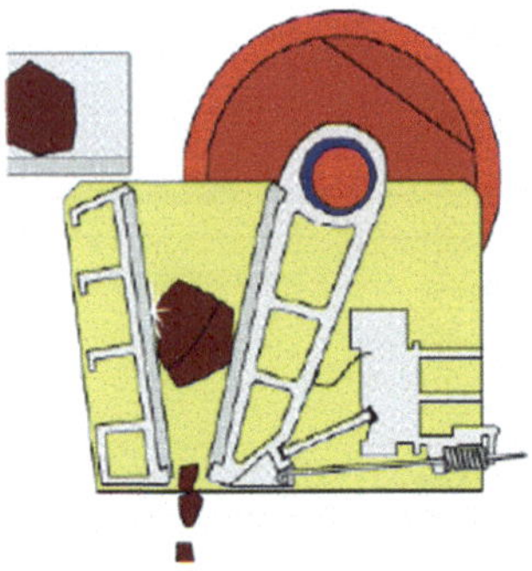	Vertical impact crusher	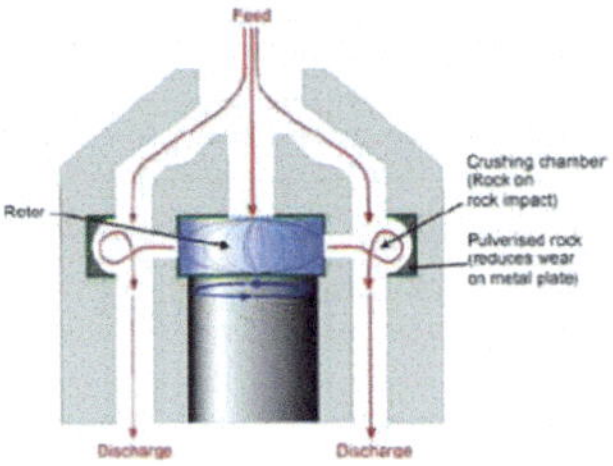
Cone crusher	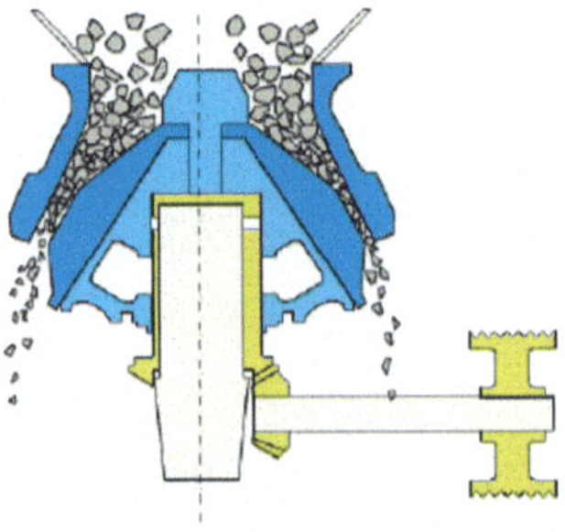	Horizontal impact crusher	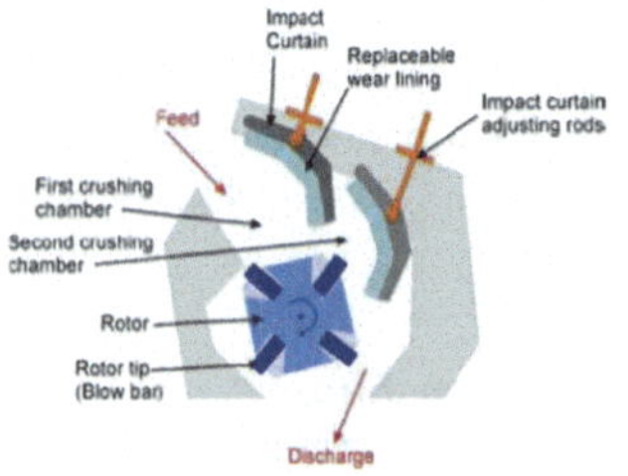

Table 4 Use and crushing mechanism of various crushers

Crusher type	Purpose of use	Size of input particles, type	Size of output particles, type	Crushing mechanism	Remarks
Jaw crusher	P, S	<1,000 mm	<150–200 mm	Compression	Reduction ratio = 5:1
Cone crusher	S, T	<150–200 mm	<40–80 mm	Compression	–
Impact crusher	S, T	<40 mm (vert. type) <150–200 (horiz. type)	<25 mm (vert. type) <40–80 mm (horiz. type)	Impact	–

Note P—primary crushing, S—secondary crushing, T—tertiary crushing; Reduction ratio = size of input material: the size of output material

2.1.2 Sorting and Screening

Various foreign matters other than concrete, such as soil, paper, cloth, plastics, wood, metals, brick, glass, etc., included in the input waste concrete rubbles should be sorted out, and the aggregates need to be screened according to different sizes during the RA production process. A grizzly screen or a trommel screen is often used to separate soil from waste concrete rubbles before or after the first stage of crushing. Steel can be effectively picked up using magnetic separation, but some metallic objects such as aluminum and stainless steel cannot be collected by magnetic separation and must be manually sorted out. Air separation can separate light materials such as styrofoam, vinyl, paper, lightweight wood, etc. Separation by flotation can be used to remove lightweight materials. Vibrating screens of various types and sizes are typically used for sieving. Fine fractions are challenging to be sieved and need to be collected using classifiers such as an air separator. A separate water treatment system needs to be added in wet processing using a spiral classifier or an inclined cyclone, further separating the used water into sludge and water. The dewatered sludge is typically sent to the landfill, while the treated water can be reused.

2.1.3 Possible Damage to RA during the Production Procedure

As the RA is produced using various crushers, microstructural damage to particles during the production process is possible. The microstructure of laboratory-produced RA was investigated using fluorescent microscopy and image analysis by Nagataki et al. (2004). Three laboratory-produced concretes with low (0.35), medium (0.45), and high (0.63) water-cement ratios were prepared as the source of RA using ordinary Portland cement, coarse sandstone aggregate, and river sand. The source concrete was crushed using a jaw crusher and an impact crusher (level 1) followed by mechanical grinding (level 2) or by double mechanical grinding (level 3), as shown in Fig. 2.

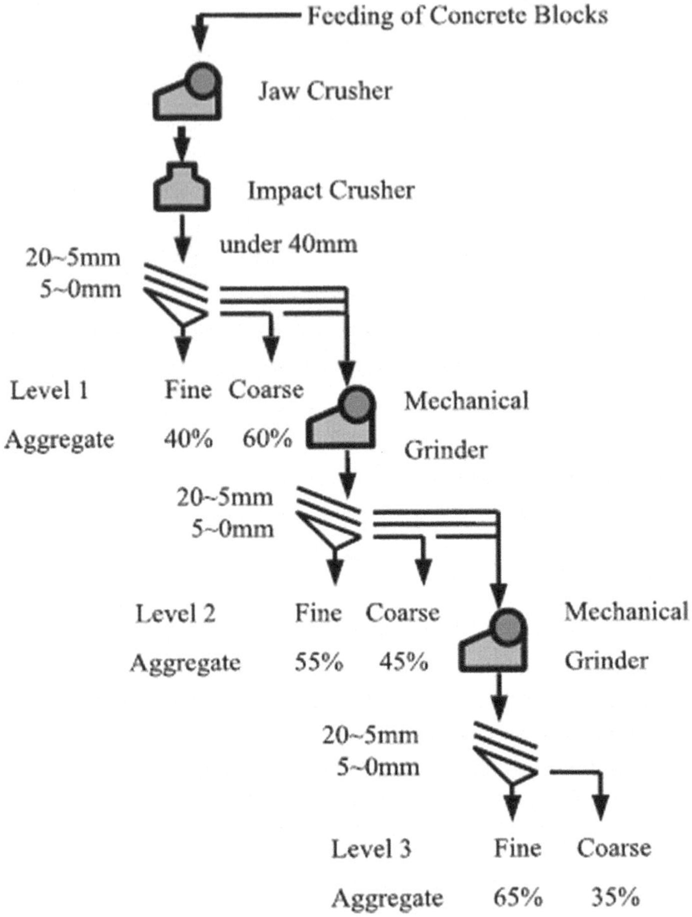

Fig. 2 Flow of the recycling process (Nagataki et al. 2004)

In the microstructural investigation on samples of RA as well as the sliced core samples of the source concrete used as a reference, possible cracking patterns were classified into three categories: original coarse aggregate (OCA) cracks, adhered mortar (AM) cracks, and interface (ITZ) cracks. It was found that the sedimentary sandstone particles initially had segmented fine fissures up to 30 μm long and coarse cavities larger than 50 μm in diameter that were friable. Level 1 crushing considerably reduced the density of the cracks in the OCA by eliminating the particles with micro defects and irregular voids. Processing up to level 3 reintroduced new cracks, but the amount was negligible. RA obtained at each stage did not show any loss of integrity while processing up to level 3 increased the physical performance of the concrete aggregates by reducing AM.

2.2 *Properties of RA*

2.2.1 Morphology

The crushing produces angular particles with sharp corners and rough textures (Mamlouk and Zaniewski 2018). Crushed angular and rough-textured aggregates such as RA may produce bulk materials with higher stability than rounded and smooth textured aggregates. At the same time, the angular and rough-textured aggregates may be more challenging to work into place than rounded aggregates since their shapes make it difficult for them to slide across each other. The shape index of coarse aggregate such as flat, elongated, or flat and elongated (or flaky) can be determined using ASTM D4791, EN 933-3, and EN 933-4. ASTM D3398 can test aggregate particle shape and texture, although this standard is often used to determine the influence of shape and texture of aggregates on asphalt concrete's compaction and strength characteristics (Jamkar and Rao 2004).

The particle shape on particle breakage (i.e., particle fracture/fragmentation) was investigated experimentally and analytically using basaltic waste rock, RA, and crushed blocks by Afshar et al. (2017). Depending on the degree of sphericity and aspect ratio, the aggregates were grouped as bulky (closer in shape to spheroid), elongated (closer to prolate spheroid) and flaky (comparable to oblate spheroid) in this study. The single-particle crushing test and particle assembly crushing test indicated that the shape effect on particle breakage is more dominant than the mineralogical or microstructural effect. Particles within the same shape category fractured at approximately the same load range regardless of the material type. The granular materials with a higher degree of sphericity (an aspect ratio close to 1) and a lower flakiness index experienced less particle breakage.

2.2.2 Adhered Mortar

Recycled aggregate's physical properties, such as absorption capacity, density, and abrasion resistance, are affected by the amount of mortar from old concrete attached to NA. Although there is no standardized procedure to determine the AM amount at present, various methods attempted by researchers can be categorized as follows (Hansen and Narud 1983; Tam and Tam 2008; Juan and Gutiérrez 2009; Padmini, et al. 2009; Zega et al. 2010; Sui and Mueller 2012; Ismail and Ramli 2013; Al-Bayati et al. 2016; Zhao et al. 2015, 2018):

- Production of a new concrete made with RA and colored cement;
- Mechanical grinding using devices such as Micro-Deval device;

  Pre-soaking treatment using an acid such as HCl and $C_2H_4O_2$;

- Salicylic acid and methanol dissolution method to measure the hardened cement paste content

- Heat treatment under various temperatures; and
- Combination of the above methods.

Hansen and Narud (1983) produced a high-strength (56 MPa), a medium strength (34 MPa), and a low strength (14 MPa) concrete from natural gravel by varying the water-cement ratio. Concretes were crushed using a jaw crusher at 38 days after the strength test and screened into three size fractions of 4–8 mm, 8–16 mm, and 16–32 mm for each strength class. A representative sample of each grade and size fraction of RA was mixed with red-colored cement and cast into cubes. After hardening, the cubes were cut into slices and polished. The volume percentage of old mortar attached to gravel particles in each grade and size fraction of RA were visually determined using a linear traverse method. The amount of mortar remaining attached to natural gravel particles in RA varied from around 30% for 16–32 mm fractions to more than 60% for the 4–8 mm fractions. Juan and Gutiérrez (2009), as a result of literature review, have found that the mortar content results are different depending on the test applied: 25—70% for treatment with a hydrochloric acid solution, 25–65% for production of new concrete, and 40—55% for thermal treatment. In the same study by Juan and Gutiérrez (2009), 15 samples of different RA were obtained after double crushing with impact crusher and separation with a vibrating screen. Strength information of original concretes was not disclosed. The samples were divided into 4–8 mm and 8–16 mm fractions. The thermal method was chosen. A RA sample was immersed in water for two hours, dried at 500 °C for two hours, and then immersed in cold water where the sudden cooling caused stress and cracks in the mortar, which was easy to be removed. It was necessary to remove the remaining mortar using a rubber hammer or by scratching. The 4–8 mm fraction had a mortar content from 33 to 55%, while, for the 8–16 mm fraction, it ranged from 23 to 44%.

Some researchers reported mortar much lower than those reported above (Ismail and Ramli 2013; Al-Bayati et al. 2016). In a study by Al-Bayati et al. 2016, the RA samples were obtained from crushing concrete returned to a ready-mixed concrete plant for various reasons. Coarse RA samples of 4.75–19 mm fraction were generated from original concretes, and RA was subject to various treatment methods: pre-soaking in an acidic solution composed of HCl and $C_2H_4O_2$ at a low concentration of 0.1 M for 24 h at room temperature while other RA samples were heated at four different temperatures: 250, 350, 500, and 750 °C for one hour. Both pre-soaked and heated samples were then placed in the Micro-Deval device for 15 minutes without metal balls in dry condition. The AM amount was determined after the samples were washed and sieved with a 4.75 mm sieve. The amount of AM ranged between 3.0 and 9.8%, except for the very high-temperature treatment at 750 °C, which resulted in 41% AM amount. It can be concluded that the amount of AM varies widely between 3 and 70% depending on the production technologies (e.g., several crushes) involved, as shown in Fig. 3 (Juan and Gutiérrez 2009; Silva et al. 2014). The AM amount is more significant in the fine fraction than it is in the coarse fraction. In the case

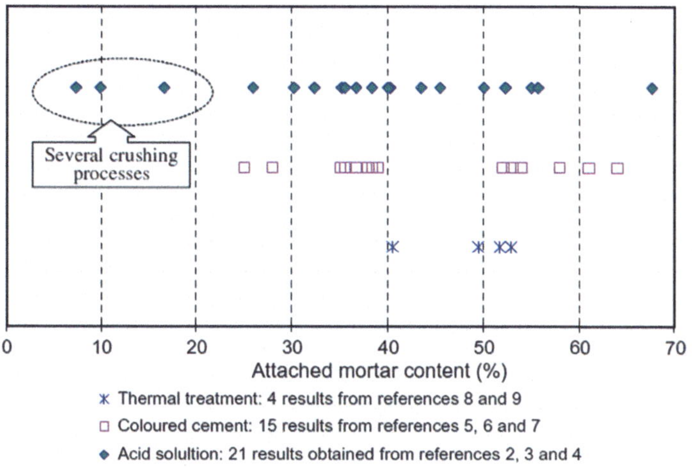

Fig. 3 Attached mortar content for different test methods

of smaller-sized RA, the presence of AM is more due to the higher surface area available for an equal volume of aggregate (Padmini et al. 2009; Silva et al. 2014).

It needs to be noted that the acid treatment cannot be used with limestone because acid also attacks the aggregate (Juan and Gutiérrez 2009). During thermal treatment, when the temperature is higher than 500 °C, RA properties may be degraded (Shi et al. 2016). RA suffers from decarbonization which causes severe microcracking of cement matrix between 600~800 °C (Al-Bayati et al. 2016).

2.2.3 Absorption

Absorption (or water absorption) of RA is higher than that of NA due to porosity of mortar attached to RA (Hansen and Narud 1983; Katz 2003; Nagataki et al. 2004; Tam and Tam 2008; Juan and Gutiérrez 2009; Padmini et al. 2009; Zega et al. 2010; Ismail and Ramli 2013; Silva et al. 2014; Al-Bayati et al. 2016; Shi et al. 2016; Koper et al. 2017; Salesa et al. 2017). The water absorption of RA ranges from 3–12% compared with 1–5% for NA (Shi et al. 2016). Figure 4 shows the relation between AM content and water absorption of RA: when mortar content increases, water absorption increases too (Juan and Gutiérrez 2009).

Concrete with 28-day compressive strength of 28 MPa was crushed using a mini jaw crusher at ages 1, 3, and 28 days to serve as a source of aggregate for new concretes, simulating the situation in pre-cast concrete plants (Katz 2003). The crushed concrete was divided into fractions of the following sizes: coarse (larger than 9.5 mm), medium (smaller than 9.5 mm but more significant than 2.36 mm) and fine (smaller than 2.36 mm). The porosity of the RA expressed by absorption values ranged from 3.2 to 12% for the coarse and the fine fraction, respectively, while the porosity of the NAs used in the production of the old concrete was low (0.5~1.5%).

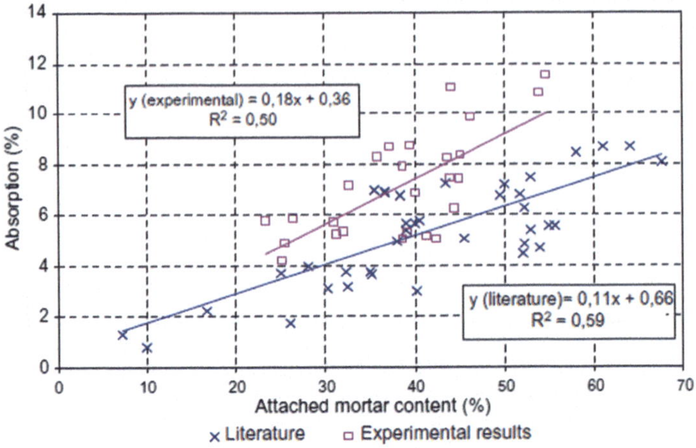

Fig. 4 Relationship between mortar content and absorption (Juan and Gutiérrez 2009)

The strength of the original concrete may cause a reduction in the water absorption of the resulting aggregate. The amount of AM is more in RA derived from higher strength old concrete than lower strength old concrete. Padmini et al. (2009) used three different maximum size fresh crushed granite aggregates (10, 20, and 40 mm). Three different water-cement ratios (0.58, 0.43, and 0.34) were used to produce concrete using three different coarse aggregates, respectively, resulting in nine different original concretes. The compressive strengths ranged between 35 and 56 MPa for concretes using 10-mm NA, 37 MPa and 58 MPa for 20-mm NA, and 31 MPa and 52 MPa for 40-mm NA. RA of maximum size 10, 20, and 40 mm was then produced from original concretes. During the crushing of lower strength parent concrete, as the bond between mortar and aggregate was weaker, most of the mortar phase separated from the aggregate surface. As the strength of parent concrete increased, the quantity of mortar adhering to the aggregates increased due to the comparatively larger bond between the aggregate and the mortar phase. The percentage of water absorption increased with the increased strength of parent concrete from which the RA has been derived due to the higher quantity of AM in RA obtained from higher strength parent concrete. (Juan and Gutiérrez 2009).

Data were collected from 236 publications published over 38 years (1977–2014), and data were subjected to statistical analysis by Silva et al. (2014). Parameters affecting water absorption were identified as (1) recycling procedure, (2) quality of the original material, and (3) size. Concrete subjected to a secondary (or tertiary) crushing procedure typically resulted in RA with less AM (less water absorption). While RA's water absorption decreased as the strength of the original material increased in general, it remained unaffected as the strength of the original material increased. Therefore Silva et al. suggested that the strength of the original material may cause a reduction in the water absorption of the resulting aggregate. Coarse RA had lower water absorption than the corresponding fine fraction. Figure 5 and

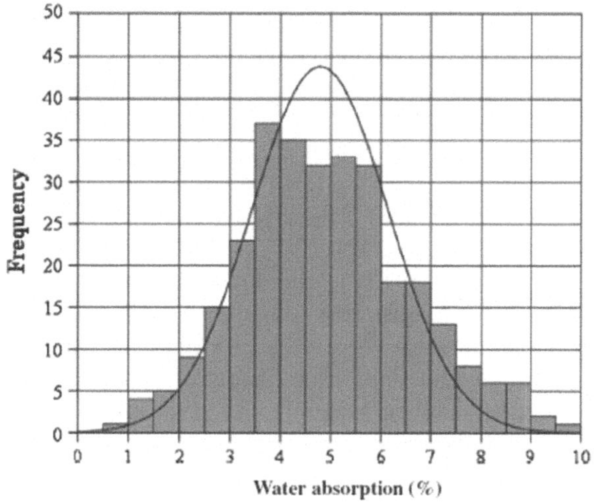

Fig. 5 Normal distribution of water absorption of coarse RA (Silva et al. 2014)

Table 5 Statistical data of the normal distribution of RA's water absorption (Silva et al. 2014)

Aggregate type	RA	
Aggregate size	Fine	Coarse
Sample size	43	298
Mean	9.5	4.9
95% Confidence interval for mean		
Lower bound	8.7	4.7
Upper bound	10.3	5.1
Standard deviation	2.6	1.7
Coefficient of variation (%)	27.6	34.7

Table 5 shows the normal distribution of water absorption of coarse RA and related statistical data, respectively, where the mean is 4.9% with a standard deviation of 1.7% (sample size: 298). The mean is 9.5% for the fine RA with a standard deviation of 2.6% in Table 5 (sample size: 43). It needs to be noted that some researchers claimed that the higher the compressive strength of the original concrete, the lower the water absorption of the aggregate obtained from the concrete (Kou and Poon 2015; Koper et al. 2017). This may occur in RA made from high-strength concrete with a compressive strength of at least 60 MPa, although more test data is needed to confirm this claim.

2.2.4 Unit Weight, Bulk Specific Gravity, and Density

The unit weight, bulk specific gravity, and density of RA are lower than those of NA due to AM to RA as mortar is less dense than NA (Hansen and Narud 1983; Katz

2003; Nagataki et al. 2004; Tam et al. 2008; Juan and Gutiérrez 2009; Padmini et al. 2009; Zega et al. 2010; Ismail and Ramli 2013; Silva et al. 2014; Al-Bayati et al. 2016; Shi et al. 2016; Koper et al. 2017; Salesa et al. 2017). Katz (2003) reported that the bulk density of RA was relatively low, approximately 1,300 kg/m^3. The bulk specific gravity of RA was 2.23, 2.32, and 2.55 for the fine fraction (smaller than 2.36 mm), a medium fraction (between 2.36 mm and 9 mm), and the coarse fraction (larger than 9.5 mm), respectively. The bulk specific gravity of the corresponding NA was 2.7.

Zega et al. (2010) used four different coarse NAs (three crushed stones; granite—G, basaltic—B, quartzitic—Q, and a rounded aggregate, siliceous river gravel—S) with a nominal size of 6–20 mm. Concretes with water-cement ratios of 0.45 and 0.65 were made using each of four NAs, cured for 28 days, and crushed using a jaw crusher and subsequently sieved to obtain RA 6–20 mm in nominal size. The physical properties of natural coarse aggregate (NCAs) and RA are summarized in Tables 6 and 7, respectively. In Table 7, the RA ranges between 1.100 and 1,290 kg/m^{3,} while the unit weight of corresponding NAs ranges between 1,310 and 1,580 kg/m^3 in Table 6. In the same table, RA's bulk specific gravity (SSD) is 2.35–2.66, while it is 2.48–3.03 for NAs. Agrela et al. (2011) investigated a total of 35 coarse RA in terms of composition, water absorption, density, sulfate and sulfur contents, taken from 13 CDW recycling plants in Spain. RA were grouped into three types in this study: concrete RA (concrete content > 90%), mixed RA (10% < ceramic content < 30%), and ceramic RA (ceramic content > 30%). SSD density of RA ranged from 2,060 kg/m^3 to 2,580 kg/m^3: from 2,340 kg/m^3 to 2,470 kg/m^3 for concrete RA, from 2,220 kg/m^3 to 2,580 kg/m^3 for mixed RA, and from 2,060 kg/m^3 to 2,390 kg/m^3 for ceramic RA. A relatively good correlation ($R^2 = 0.83$) was determined by water absorption and density, as shown in Fig. 6, which clearly shows that the absorption decreases with decreasing density of RA.

Silva et al. (2014), using data from the literature, reported that coarse RA's mean oven-dry density (ODD) is 2,327 kg/m^3 with a standard deviation of 117 kg/m^3

Table 6 Properties of natural coarse aggregates (Zega et al. 2010)

Properties	G	Q	B	S
Specific gravity (kg/dm^3)	2.72	2.48	3.03	2.60
Absorption (%)	0.3	2.0	0.8	0.5
Los-Angeles abrasion (%)	25.0	59.8	9.1	18.8
Material finer than 75 μm (%)	0.6	0.9	0.3	0.2
Unit weight (kg/m^3)	1410	1310	1530	1580
Voids (%)	48.0	46.2	49.3	38.4
Flakiness index (%)	19.2	25.3	26.7	9.9
Weight loss in ethylene glycol (%)	–	–	2.5	–

Note G—crushed granite, Q—crushed quartzitic stone, B—crushed basaltic stone, S—siliceous river gravel

Table 7 Properties of recycled coarse aggregates (Zega et al. 2010)

Properties	RG45	RG65	RQ45	RQ65	RB45	RB65	RS45	RS65
Specific gravity (kg/dm^3)	2.52	2.51	2.37	2.35	2.66	2.65	2.45	2.44
Absorption (%)	4.0	4.1	5.9	6.0	3.9	4.5	3.9	4.4
Los-Angeles abrasion (%)	34.8	37.4	52.2	55.4	25.3	30.3	31.6	37.0
Material finer than 75 μm (%)	0.6	0.1	0.2	0.2	0.2	0.4	0.1	0.1
Unit weight (kg/m^3)	1220	1190	1100	1140	1260	1290	1190	1210
Voids (%)	49.4	50.9	50.9	48.5	50.9	49.2	49.5	48.1
Flakiness index (%)	12.9	10.1	14.9	12.7	11.3	9.9	11.2	10.4
Mortar content (%)	45.6	41.8	64.4	57.6	48.4	42.0	49.3	44.2

Note RG45—recycled granite coarse aggregate made from original concrete of w/c = 0.45

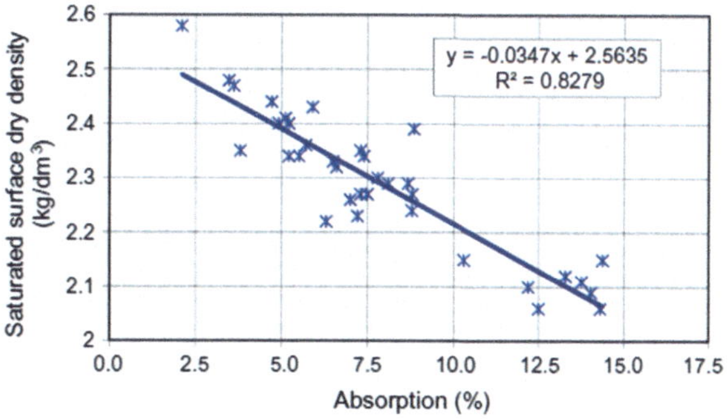

Fig. 6 Correlation between water absorption and SSD density of RA (Agrela et al. 2011)

(sample size: 292) and mean SSD density is 2,442 kg/m^3 with a standard deviation of 84 kg/m^3 (sample size: 288) as shown in Fig. 7, Tables 8, and 9. For the fine RA, mean ODD is 2,065 kg/m^3 with a standard deviation of 158 kg/m^3 (sample size: 46) and mean SSD density is 2,300 kg/m^3 with a standard deviation of 101 kg/m^3 (sample size: 45) in Tables 8, and 9, respectively. It can be concluded that the unit weight, bulk specific gravity (and density) of RA are all lower than those of NAs. The unit weight, bulk specific gravity (and density) of fine RA are lower than coarse RA due to the increased amount of AM and a corresponding increase of porosity.

2.2.5 Strength and Abrasion Resistance

Compared to NA, RA has a higher Los-Angeles abrasion loss percentage and higher aggregate crushing value (Hansen and Narud 1983). In a study by Juan and Gutiérrez (2009), 15 samples of different RA were obtained. Los-Angeles abrasion test was performed on fifteen different RA of 4–8 mm fractions. Test results showed that the Los-Angeles abrasion loss ranged between 35 and 42%.

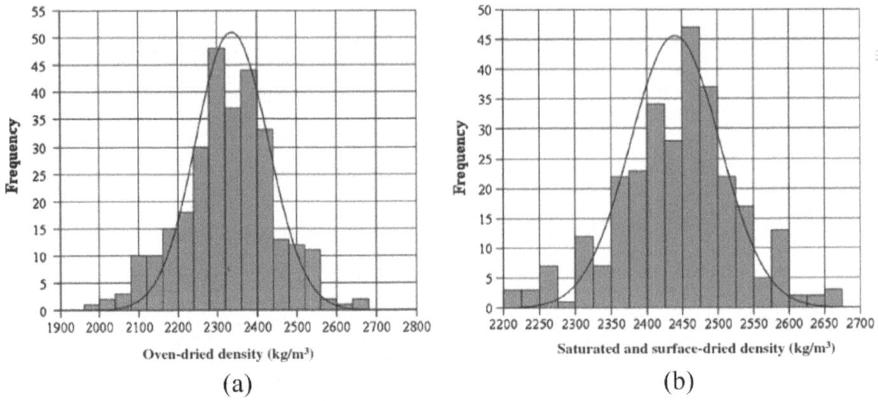

Fig. 7 Normal distribution of ODD and SSD density (Silva et al. 2014)

Table 8 Statistical data of the normal distribution of RA's ODD (Silva et al. 2014)

Aggregate type	RA	
Aggregate size	Fine	Coarse
Sample size	46	292
Mean	2065	2327
95% Confidence interval for mean		
Lower bound	2018	2314
Upper bound	2112	2341
Standard deviation	158	117
Coefficient of variation (%)	7.7	5.0

Table 9 Statistical data of the normal distribution of RA's SSD (Silva et al. 2014)

Aggregate size	Fine	Coarse
Sample size	45	288
Mean	2300	2442
95% Confidence interval for mean		
Lower bound	2270	2432
Upper bound	2331	2452
Standard deviation	101	84
Coefficient of variation (%)	4.4	3.4

Padmini et al. (2009) used three different maximum size fresh crushed granite aggregates (10, 20, and 40 mm). Three different water-cement ratios (0.58, 0.43 and 0.34) were used to produce concretes using three different coarse aggregates, respectively, resulting in nine different originals concretes. RA of maximum size 10, 20, and 40 mm was produced from original concretes. For 10-mm maximum size RA derived from original concretes with the strength of 35, 49, 56 MPa, the LA abrasion loss ranges between 46 and 48% as shown in Table 10, while 29% for corresponding 10-mm NA. For 20-mm maximum size RA derived from original concretes with the strength of 37, 50, 58 MPa, the LA abrasion loss ranges between 33 and 38%. For 40-mm maximum size RA derived from original concretes with the strength of 31, 45, 52 MPa, the LA abrasion loss ranges between 29 and 33%. The LA abrasion loss for corresponding 20-mm NA is 26%, while it is 26% for 40-mm NA.

Zega et al. (2010) used four different coarse NAs (three crushed stones: granite—G, basaltic—B, quartzitic—Q, and rounded siliceous river gravel - S) with a nominal size of 6–20 mm with properties as shown in Table 6. Concretes with water-cement ratios of 0.45 and 0.65 were made using each of four NAs to obtain eight RA 6–20 mm in nominal size with properties shown in Table 8. Tables 6 and 7 indicate that LA abrasion loss of granite RA ranges between 34.8 and 37.4%, while that of corresponding NA is 25.0%. LA abrasion loss of quartzitic RA ranges between 52.2 and 55.4%, while that of corresponding quartzitic NA is 59.8%. The values for basaltic RA range between 25.3 and 30.3%, while that of corresponding basaltic NA is only 9.1%. Finally, the LA abrasion loss for siliceous river gravel RA ranges between 31.6 and 37.0%, while corresponding NA is 18.8%. Zega et al. concluded that RA obtained from concrete with a water-cement ratio of 0.45 might exhibit a higher abrasion value than from 0.65, which shows that the water-cement ratios of original concrete are relatively less important than the abrasion loss value of NAs it contains. From data collected from existing publications (78 samples), Silva, De Brito et al. (2014), reported that the mean LA abrasion loss is 32.5% with a standard deviation of 8.5%, as shown in Tables 8 and 9. It can be concluded that the LA abrasion loss of RA is more significant than that of NA in general. ASTM (2016) sets a limit on the LA abrasion loss of 50% for concrete, and, based on statistical data in Fig. 8 and Table 9, most RA satisfies the limit on abrasion resistance.

In a study by Koper et al. (2017), concretes with five different water-cement ratios (0.35, 0.45, 0.50, 0.60, 0.7) were made in the laboratory. The 28-day compressive strengths of the original concretes (or primary concretes) ranged between 60.2 MPa (w/c = 0.35) and 34.8 MPa (w/c = 0.7) in Table 12. The primary concretes were subjected to double crushing, which allowed the gravel fractions of 2–16 mm from each concrete. The physical properties of five different RA were investigated. In Table 12, the crushing index ranges between 16.5 and 21.9%. The crushing index increases as the water-cement ratio of primary concrete increases (i.e., concrete strength decreases). The crushing values were also investigated by Padmini et al. (2009) as shown in Table 10. RA of 10-mm maximum size derived from 35-MPa, 49-MPa, 56-MPa concrete has a crushing value of 32%, 30%, 30%, respectively, while the crushing value is 25% of the corresponding 10-mm NA. RA of 20-mm maximum size derived from 37-MPa, 50-MPa, 58-MPa concrete has a crushing value of 26%, 25%, 23%, respectively, while 22% in the corresponding 20-mm NA. It can be concluded that the RA crushing value is higher than that of the corresponding NA.

Table 10 Properties of natural and recycled coarse aggregate (Padmini et al. 2009)

Property		Freshly crushed granite aggregate			RCA maximum size								
					10 mm Derived from parent concrete of compressive strength			20 mm Derived from parent concrete of compressive strength			40 mm Derived from parent concrete of compressive strength		
		10	20	40	35	49	56	37	50	58	31	45	52
		(mm)			(MPa)								
Physical properties													
Specific gravity		2.8	2.8	2.8	2.46	2.4	2.38	2.52	2.51	2.48	2.56	2.53	2.52
Water absorption		0.3	0.3	0.3	4.60	4.8	5.0	3.65	4.1	4.86	2.2	2.5	2.8
Bulk density kg/m^3	Loose	1408	1462	1406	1338	1327	1324	1432	1421	1394	1341	1334	1329
	Rodded	1561	1625	1590	1468	1438	1427	1568	1536	1498	1480	1474	1470
Percentage	Loose	50	48	49	46	45	44	43	43	44	48	47	47
	rodded	44	42	43	40	40	40	38	39	40	42	42	42
Mechanical properties													
Crushing value (%)		25	22	–	32	30	30	26	25	23	–	–	–
Impact value (%)		18	17	–	38	32	31	25	24	21	–	–	–
Abrasion value (%)		29	26	26	48	46	46	38	35	33	30	29	29

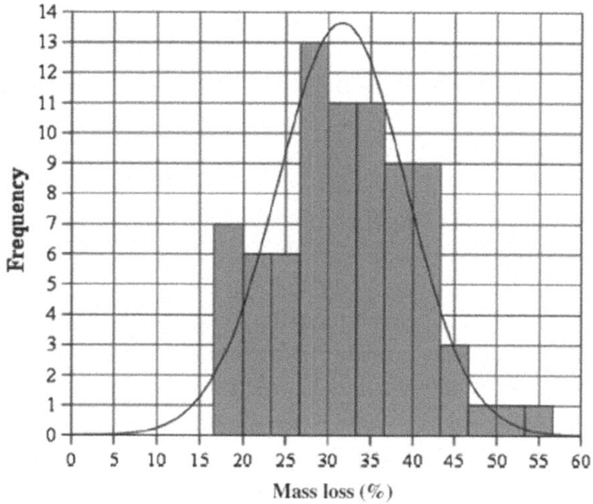

Fig. 8 Normal distribution of the LA abrasion loss of coarse RA (Silva et al. 2014)

2.2.6 Gradation

As discussed previously, Zega et al. (2010) used four different coarse NAs (three crushed stones: granite—G, basaltic—B, quartzitic—Q, and rounded siliceous river gravel—S) with a nominal size of 6–20 mm with properties as shown in Table 6. Concretes with water-cement ratios of 0.45 and 0.65 were made using each one of four NAs to obtain eight RA 6–20 mm in nominal size with properties shown in Table 7. Figure 9 a shows the grading of the four NAs together with limits set by the regulations in Argentina (CIRSOC).

The grading determined on each of the eight RA is shown in Fig. 9 b. It can be noted that the eight RA have similar particle size distribution, regardless of the type of NA and the water-cement ratio of the source concrete from which they were derived. The grading meets the limits prescribed for the NAs. Therefore the grading

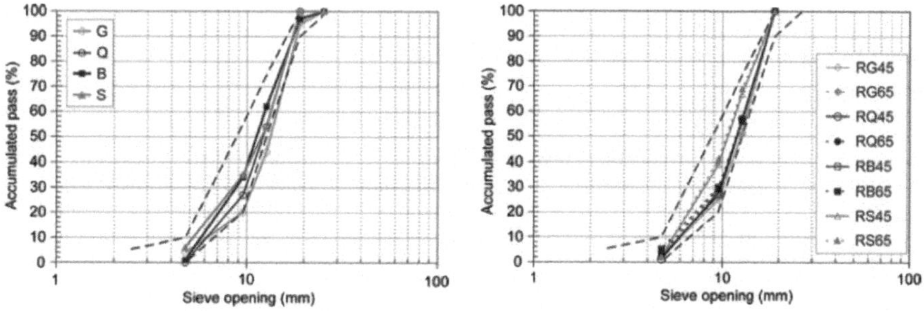

Natural coarse aggregate — Recycled coarse aggregate

Fig. 9 Grading of natural coarse aggregates and recycled coarse aggregates derived from the NA of RA is not influenced by the different characteristics of NAs in the other source concrete (Table 11).

In an investigation by Huda and Alam (2014), 100% RA repeatedly used in the first, second, and third-generation was considered in Canada. The first Recycled coarse aggregates were made from original concrete with 43 MPa compressive strength with natural coarse and fine aggregates, cured for 56 days and then crushed to produce the first generation recycled coarse aggregate (RA1). The crushing procedure consisted of a jaw crusher followed by a sieving and an impact crusher. RA2 and RA3 were generated using the same procedure: i.e., concretes were made using coarse aggregate generated from the previous generation concrete, cured for 56 days, and crushed. The gradation of natural fine and coarse aggregates are shown in Fig. 10 a, b, respectively. Figure 11 a–c shows the sieve analysis results of three different generations of repeated RAC aggregate. It can be observed that all the different generations of repeated recycled coarse aggregates fall within the acceptable range.

Hansen and Narud (1983) have suggested that the crushing characteristics of hardened concrete are similar to those of many natural rocks and that a single passing can produce reasonably well-graded concrete coarse aggregate of good shape through a jaw crusher or granulator. More recent graduation studies in Fig. 9 through Fig. 11 show the same results (Zega et al. 2010; Huda and Alam 2014).

Table 11 Statistical data of the normal distribution of RA's LA abrasion loss (Silva et al. 2014)

Aggregate type	RA
Sample size	78
Mean	32.5
95% Confidence interval for mean	
Lower bound	30.6
Upper bound	34.4
Standard deviation	8.5
Coefficient of variation (%)	26.1

Table 12 Test results of RA properties by (Koper et al. 2017)

RCA labeling	Primary concrete	RCA properties			
	w/c ratio	Bulk density/ kg/dm^3	Water absorption/%	Mineral dust content/%	Crushing index w_{rm}/%
RCA I	0.35	2.460	5.6	0.30	16.5
RCA II	0.45	2.424	6.6	0.52	19.0
RCA III	0.50	2.434	7.0	0.49	19.9
RCA IV	0.60	2.439	7.5	0.72	21.5
RCA V	0.70	2.423	7.9	0.97	21.9

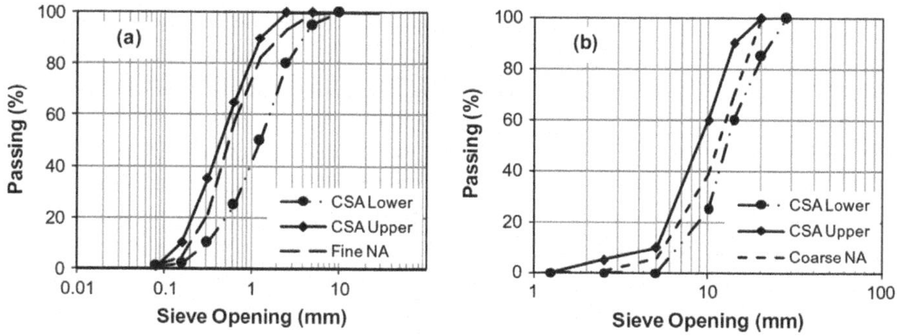

Fig. 10 Gradation curves of **a** natural fine aggregate and **b** natural coarse aggregate (Huda and Alam 2014)

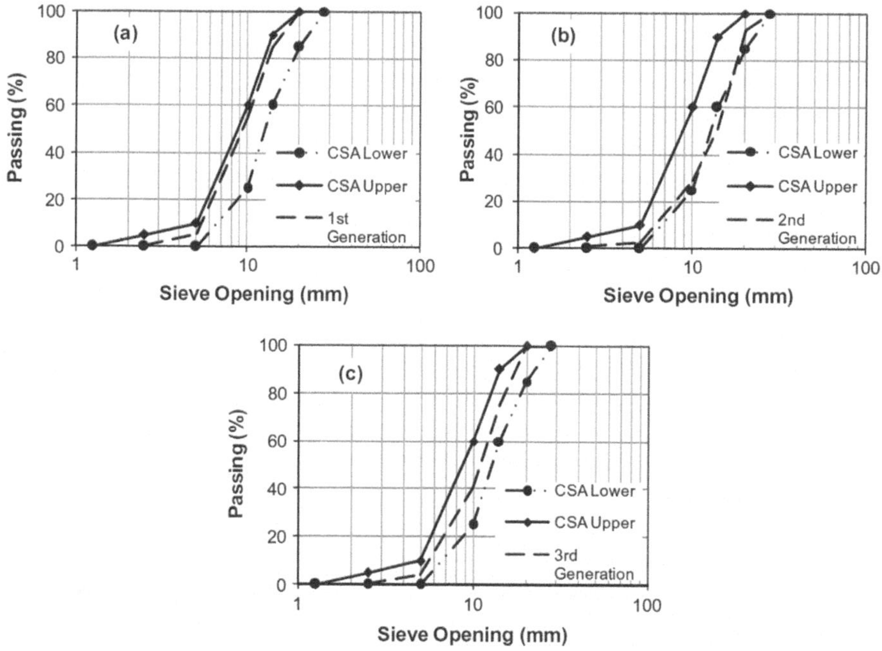

Fig. 11 Sieve analyses of different generations of repeated recycled coarse aggregates: **a** 1st generation RA, **b** 2nd generation RA, **c** 3rd generation RA (Huda and Alam 2014)

2.2.7 Hazardous Substances and Impurities Content

Recycled aggregates may have higher sulfate content than the NAs, while water-soluble sulfates in RA are potentially reactive and give rise to expansive reactions (Juan and Gutiérrez 2009; Agrela et al. 2011; Silva et al. 2014). Source of sulfate includes cement of AM, gypsum plaster from interior faces of the building, and crushed clay brick. In a study by Juan and Gutiérrez 2009, the sulfate content SO_3-increased with the increasing amount of AM, but the highest sulfate amount was smaller than 0.8%. In a study by Agrela et al. 2011, nine out of 35 RA investigated in Spain presented more than 2% gypsum. Many RA standards and specifications of RAC limit the water-soluble (or acid-soluble) sulfate content calculated as SO_3-to 0.8% (Germany), 1.0% (RILEM, Belgium, Netherlands, UK, Brazil, Hong Kong) by wt. Chlorides can also contaminate RA. The chloride ions can penetrate to original concrete from outside in marine structures or highway bridges and pavements exposed to deicing salts. Chlorides can also be original concrete in accelerating admixtures or poorly washed sea fine aggregates (Hansen and Narud 1983). In many RA standards and specifications, the chloride limit is set to 0.04% (Germany)-0.06% (Belgium, China) or 0.1% for coarse RA and 0.05% for fine RA (Netherlands) by weight.

Fine soil and clay particles can remain attached to the RA surface when RA is not adequately washed before use. If not removed, their presence may adversely affect the bond between aggregate and cement paste by forming a thin film on the surface of RA. Many standards limit the maximum content of fines (<65–80 μm) in coarse RA to 1–3% (RILEM, Belgium, Brazil, China, South Korea, UK) by wt. Asphalt can also decrease the compressive strength of concrete, and a limit of 1% by wt. can be set. Lightweight and organic contaminants such as paper, cloth, wood, and plastics can be included in RA by a small quantity, so the amount is limited by 0.1–0.5% by wt. in many standards and specifications. The density of glass is similar to that of concrete, and separation can be difficult. Glass can be potentially dangerous because it can take part in alkali-aggregate reactions (Hansen and Narud 1983). Glass is usually treated as a lightweight material, so the glass is also included in this category.

Park and Noguchi (2013) studied minor metal content in RA, steel, and aluminum, on RAC. The presence of a small amount of steel in RAC did not affect the mechanical properties of RAC in terms of compressive strength and elastic modulus, but rust stains were observed in concretes with not less than 2% steel impurity by wt. It was suggested that an upper limit for steel impurity be set for 1%. Unlike steel, aluminum cannot be sorted by a magnetic separator. The reaction of aluminum with calcium hydroxide in an aqueous solution produces hydrogen, and therefore the chemical reaction between aluminum in RA and alkaline concrete can generate hydrogen gas. As the hydrogen generation reactions consume calcium hydroxide, the pH of the aqueous solution also decreases, increasing the potential for carbonation. After hardening, the hydrogen gas generated by aluminum impurity was observed as cracks, rock pockets around aggregates, and concrete surfaces. The compressive strength of concrete decreased, especially for concretes containing smaller sizes and a more significant aluminum impurity. A new test method was suggested by Park and Noguchi (2013) by which the aluminum impurity can be inspected.

Radioactive substances in RA should not contaminate RAC while the source can be a nuclear power plant, hospitals, etc.) Original concrete with the possibility of radioactive substances should be considered suspect until proved innocuous (Hansen and Narud 1983).

2.3 Repeated Recycling

The effect of multiple recycling on the properties of coarse RA was investigated in recent studies (Marie and Quiasrawi 2012; Huda and Alam 2014; Salesa et al. 2017). In an investigation (Huda and Alam 2014), as introduced in the previous section, the use of 100% RA to be used repeatedly in first (RA1), second (RA2), and third (RA3) generation was considered. Bulk density, specific gravity, and absorption capacity were compared. The bulk density of RA1, RA2, and RA3 was 1,396 kg/m^3, 1,251 kg/m^3, and 1,196 kg/m^3, respectively, while that of original coarse NA was 1,622 kg/m^3. The bulk specific gravity of RA1, RA2, and RA3 was 2.55, 2.33, and 2.23, respectively, while the bulk specific gravity of the original coarse NA was 2.69. The absorption capacities of RA1, RA2, and RA3 were 5.2%, 7.1%, and 9.4%, respectively, while the absorption capacity of the original coarse NA was 1.2%. The quality of coarse RA degraded with several repetitions. Another research considered RA to be used repeatedly (2 cycles of recycling) in pre-cast concrete production (Salesa et al. 2017). Some of the concrete elements produced were rejected in the pre-cast industry due to quality test failures. The RA replacement was 100% of coarse NA. CEM II cement, crushed limestone sand, and grit and gravel coarse aggregates were used. The RA1 was obtained by crushing non-quality pre-cast concrete pieces. RA2 was obtained from RC1 prisms (i.e., concrete made of RA1), which, once tested, were crushed. The properties of the natural and recycled coarse aggregates were compared using density, water absorption, and Los-Angeles abrasion coefficient as shown in Table 13, which shows that the true density of RA is less than that of NA and decreases with repeated recycling: the water absorption of RA is larger than that of NA and increases with repeated recycling. In this particular data, the LA abrasion coefficient of RA is smaller than that of NA, and it is similar between RA1 (RA) and RA2 (re-RA).

Table 13 Properties of coarse aggregates (Salesa et al. 2017)

Coarse aggregate		Test		
Type	Size	True density/g/cm^3	Water absorption/%	Los-Angeles abrasion/%
NCA	Gravel	2.60	1.1	32
	Grit	2.54	2.2	31
RCA1	Gravel	2.33	5.2	24
	Grit	2.38	4.4	21
RCA2	Gravel	2.22	6.2	26
	Grit	2.28	5.7	23

3 Quality Control Parameter for Recycled Concrete Aggregates

One of the main issues related to the promotion and the diffusion of recycling processes for CDW is the lack of confidence of consumers and builders regarding the quality of the generated aggregates due to the heterogeneity of the recycled waste. In general, CDW is a mixed residue, but the degree of heterogeneity depends mainly on its origin (Al-Bayati et al. 2016). Nevertheless, despite the very heterogeneous composition of the CDW, there is a high potential for its recycling in the form of aggregate since around 40–85% of these materials are composed of concrete, mortar, rocks and ceramic materials. As a matter of principle, RCA particles can be obtained from different types of original concretes (e.g., demolition of existing concrete frames and elements or residues of fresh concrete from ready mix companies), which are commonly more porous, rougher and more irregular than the corresponding NA due to the presence of the attached mortar within the particles' composition (Pepe et al. 2016). Thus, the RCAs are characterized by considerably higher water absorption and lower specific mass than NA (Koenders et al. 2014). Consequently, the properties of the RCAs are directly related to the characteristics and quantity of this mortar adhered to the grains. In view of these peculiarities, a protocol processing procedure to identify critical quality control parameters for the RCAs' characterization should be defined to promote high performance products worldwide, including those secondary raw materials.

In this context, a case study was carried out where approximately 20 tons of concrete waste from three different concrete waste sources were processed, simulating a recycling process in an industrial pilot plant. The key innovative aspect of the present study is that it aims at identifying and defining quality management of the concrete waste derived from different sources in order to promote their recycling for "high-end" application and, at the same time, demonstrate a pilot-scale application for their usage on a large scale in the market (Rangel et al. 2019).

3.1 Processing Procedures for RCAs

The first concrete debris was produced in the laboratory (*L-waste*) to simulate concrete construction waste: with an early age, with dosage and controlled properties and without contamination. The second consists of a fresh concrete residue (*C-waste*) from the washing of remaining material adhered to the wall of concrete-mixer trucks from a concrete production plant: with unknown dosage and properties, recent age and high addition of water to the mixture (high w/c ratio). The third consists of waste obtained from a concrete recycling plant. It is demolition waste with a mixture of different types of concrete (*D-waste*): with unknown dosage and properties and advanced age.

For each residue, RCAs of four granulometric size fractions were produced: coarse 1 (19 mm > d > 9.5 mm), coarse 0 (9.5 mm > d > 4.75 mm), sand (4.75 mm > d > 150 μm) and powder (d < 150 μm). The quantities produced from each fraction were measured for all three concrete waste sources to quantify the proportion of each aggregate and the losses involved in the process. The RCA's fractions were then thoroughly characterized to establish quality control parameters that could classify different types of RA generated.

L-, C-, and D- waste debris was processed in the following three stages: crushing, grain size classification by mechanical sieving and homogenization (Fig. 12). First of all, the debris of concrete waste was fragmented using a jaw crusher. Its mandibles opening was set to produce coarse aggregates of a maximum diameter of 19 mm (see d). As a result of the crushing process, besides the RCAs coarse fraction (see e), fine particles (d < 4.75 mm) were also obtained (see f). The two granular classes of aggregates were then stored in bags and conducted to the sieving stage.

Then, the grain size classification was carried out in two stages: (i) industrial mechanical sieving, dividing the coarse particles into two granulometric sizes ("C1" with 19 mm > d > 9.5 mm and "C0" with 9.5 mm > d > 4.75 mm) and (ii) laboratory mechanical sieving, employed for the classification of the fine particle (sand "S" with 4.75 mm > d > 150 μm and powder "P" with d < 150 μm). Figure 12g–i illustrate the industrial mechanical sieving phase in which, first of all, the stored bags (containing the coarse particles resulting from the previous crushing stage) fed the sieve belt (g) passing through sieving screens (h) until the classified RCAs (i) obtained. It was observed that, although the jaw crusher was set to obtain a maximum nominal diameter equal to 19 mm, a certain amount of particles was retained in the 19 mm mesh. A sieve shaker was used for the laboratory mechanical sieving for the RCA's fine particles classification., In order to prevent the dust from being lost during sieving, a proof cover was employed. Further, it was observed that the powder loss was relatively low for the coarse fraction (i.e., subjected to the industrial mechanical sieving).

After the sieving stage, the classified RCAs were homogenized using the "cells process" (see j–l). With this process, the RCAs were spread on a plastic sheet in different layers throughout the whole length of the sheet. Thus, the aggregates are homogeneously distributed over the length of the sheet, and, starting from the middle section, layers of raw materials are subdivided into several sections called "cells" (Pepe et al. 2014).

One of the objectives of the present study is to analyze the influence of the concrete waste source on the resulting amount generated for each RCA class by maintaining the same processing stages. For this reason, at the end of the RCAs production process, each produced class was weighed to determine the obtained amount of coarse RAs, sand and powder.

The percentage obtained for each class of RCAs produced in this study is illustrated in Fig. 13 (Rangel et al. 2019). From the presented results, the overall quantity of the coarse fraction was higher for the debris of D-waste (i.e., 68.2 %) followed by the L-waste (i.e., 46.6 %) and then the C-waste with a coarse percentage equal to 37.2%.

Fig. 12 The processing procedure: **a–c** visual aspect of the used concrete waste debris; **d–f** crushing stage; **g–i** industrial mechanical sieving stage; **j–l** homogenization stage (Rangel et al. 2019)

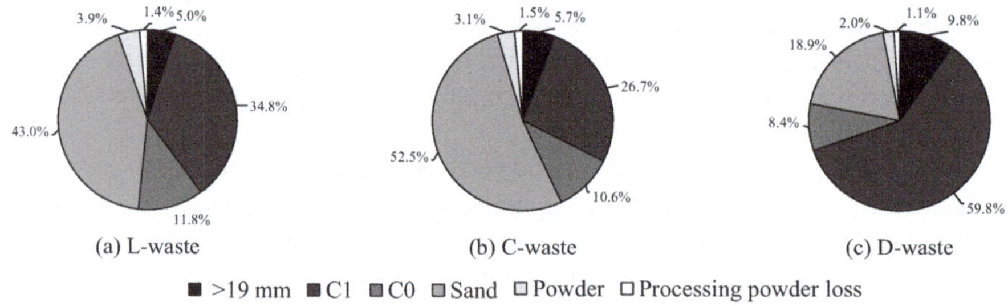

Fig. 13 Quantity of each granulometric fraction produced from each concrete waste (Rangel et al. 2019)

Then, focusing on the results obtained within the coarse fraction, it was observed that, since the crusher mandibles opening was set for obtaining a maximum nominal diameter equal to 19 mm, the relative percentage of C1 class aggregates was higher for the studied wastes (85%, 75% and 70% for sources D, L and C, respectively). This result is confirmed by observing the feed materials presented in c where debris of large dimension is predominant for D-waste. The difference in the percentages is due to the different sources of recycled concrete employed in this study. Depending on the original concrete composition/quality, the amount of coarse and fine fractions changes when keeping constant the processing procedure. The graphs presented herein clearly emphasize this behavior. Another finding that stands out is the higher amount of the RCAs sand class produced from C-waste (i.e., 52.5%), L-waste (i.e., 43.0%) D-waste source, where only 18.9% of fine aggregates were generated. Also, in this case, this evidence is supported by the pictures shown in b, where a predominant presence of small debris in the C-waste source emerges. The higher amount of acceptable debris in C-waste is mainly due to its composition as it is obtained from the remains of mixer trucks.

The percentage of recycled powder ($d < 150\ \mu m$) present in the overall RCAs reached 3.9%, 3.1% and 2.0% for L-, C- and D-waste, respectively. If only the fine fraction ($d < 4.75$ mm) is taken into account, the relative percentages of powder are of the following values: 8.3%, 5.6% and 9.5% for L-, C- and D-waste, respectively.

The total amount of powder lost during the processing procedure was, in all cases, lower than 2% and, with the highest losses being observed for C-waste source. It must empathize that the powder lost during the crushing stage was not recorded.

It is worth noting that, considering the three origin sources, the D-waste, presenting the higher percentage of coarse particles, is also the most representative residue of the construction industry. Finally, it can be concluded that by keeping constant the processing procedure for transforming concrete debris into RAs, the amount produced for each class depends mainly on the source and debris size. This evidence emphasizes the importance of a fundamental characterization for each obtained class to promote integral and adequate re-use of these secondary raw materials.

3.2 Recycled Concrete Aggregates Properties and Quality Parameter Control

The main results summarizing the properties of the produced RCAs are presented in Fig. 14. The experimental results make defining a clear trend between the oven-dried particle density values within each class or the source is impossible. For example, as the coarse particles of class C1 present higher particle density values than class C0, for CDW aggregates, an opposite trend is observed in the case of RCAs generated from *L-waste* sources. Similar conclusions are achieved based on the results for the water absorption capacity at 24 hours, Los-Angeles abrasion and shape index.

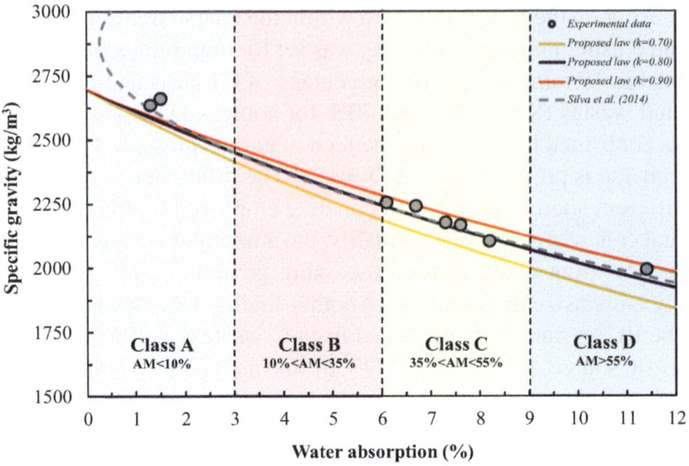

Fig. 14 Proposed classification for coarse RCAs (Rangel et al. 2019)

These results confirm that, as reported in the literature (Pepe 2015), the RCA's characterization should complement the attached mortar's evaluation content for a fundamental understanding of the aggregate mentioned above properties.

Specifically, there is a clear correlation between the AM content and the water absorption capacity and the oven-dried particle density (Pepe et al. 2016).

$$A = A_{NAT} \cdot (1 - AM) + A_{AM} \cdot AM \quad (1)$$

A_{NAT} (equal to 1%) and A_{AM} (equal to 15%) represent the porosity of the two phases present in the RCAs: NA and attached mortar, respectively.

It is worth highlighting that the above correlations demonstrate that the evaluation of the AM content allowed a fundamental interpretation of key aggregate properties independently of the RCAs source and their granulometric class.

Since the AM content defines the values of both the particle density and the water absorption capacity of coarse RCAs, as presented in Eq. (1), a clear existing relationship can be identified between these two parameters.

Silva et al. (2014) has proposed an empirical relationship for describing this trend in the literature. However, as these two properties are physically correlated, a physically-based approach is proposed to identify them.

In fact, by assuming that the oven-dried density (γ_{DRY}) is a function of the total porosity of the particles (p) and the specific density of the constituent material (γ_m), it results in:

$$\gamma_{DRY} = (1 - p) \cdot \gamma_m \Rightarrow p = 1 - \frac{\gamma_{DRY}}{\gamma_m} \quad (2)$$

In addition, assuming that the open porosity accessible by the water in 24 hours (p_{open}) is equal to “k” times the total porosity (p), the definition of the water absorption capacity is equal to

$$A = \frac{p_{open} \cdot \gamma_{water}}{\gamma_{DRY}} = \frac{k \cdot p \cdot \gamma_{water}}{\gamma_{DRY}} \tag{3}$$

Then, putting equation (2) in equation (3), the following expression can be obtained:

$$\gamma_{DRY} = \gamma_m \cdot \frac{1}{1 + \frac{A}{k} \cdot \frac{\gamma_m}{\gamma_{water}}} \tag{4}$$

where the “k” parameter represents the amount of open porosity (i.e., accessible by the water during the water absorption capacity test) of the coarse particles and, this value has been calibrated to be between 0.9 and 0.70. Based on the use of Eq. (4), in combination with the relationship proposed by Pepe et al. (2016), a generalized coarse RCAs classification can be proposed:

- *Class A*: RCAs coarse particles characterized by A values ranging between 0 and 3% and an AM content lower than 10%. This class also includes NA;
- *Class B*: RCAs coarse particles characterized by A values ranging between 3 and 6% and an AM content between 10 and 35%;
- *Class C*: RCAs coarse particles characterized by A values ranging between 6 and 9% and an AM content between 35 and 55%;
- *Class D*: RCAs coarse particles A values above 9% and an AM content above 55%.

The introduction of the above-presented classification can be identified as a “quality control” graph allowing to determine, with conventional parameters, the amount of AM independently from their source. It is believed that such a classification could promote the integral and rational re-use of these secondary raw materials in different cement-based products of the construction sector.

4 Variability in the Properties of RA in Different Countries and its Effect on the Properties of the Concrete

Recycled aggregate uses materials that would otherwise be considered construction waste as a replacement for natural aggregates (NAs). This provides a two-fold benefit of waste management and natural resource conservation. Considering the long-term environmental benefits, RA is recognized as an efficient strategy for the sustainable growth of the construction industry (Fig. 15).

Global awareness of the sustainable aspect of RA is the collective result of several researchers investigating the potential of recycling CDW waste as RA. However,

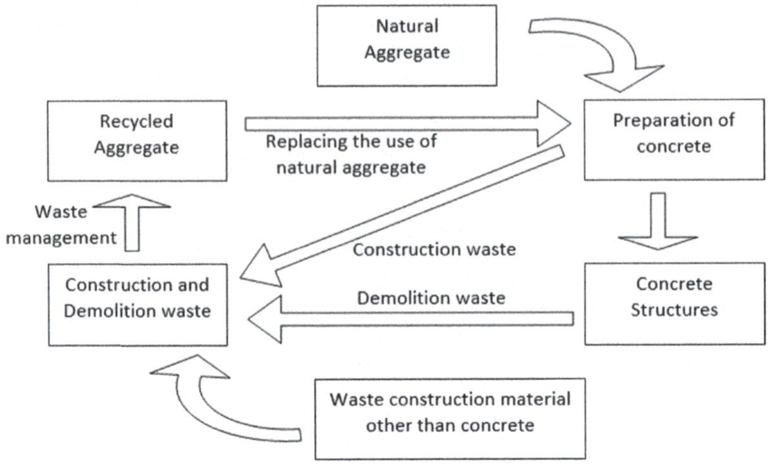

Fig. 15 Recycling of CDW as aggregates

appreciation for the benefits of RA does not guarantee its success in the construction industry. It can be seen that China, despite being the largest consumer of NA, has a limited recycling rate of 5% for CDW, as opposed to countries like Germany and South Korea, which have achieved a recycling rate of over 70% (Huang et al. 2018). Figure 16, presents the CDW generation and recycling scenario for the countries of the European Union. Despite having similar regulations, there is a significant variation in the recycling rates of CDW, with Holland having a recycling rate as high as 98% and Italy having a 0% recycling rate. The limitations in industrial applications of RA have been attributed to variation in RA properties, resulting in a lack of industrial confidence (Rangel et al. 2019).

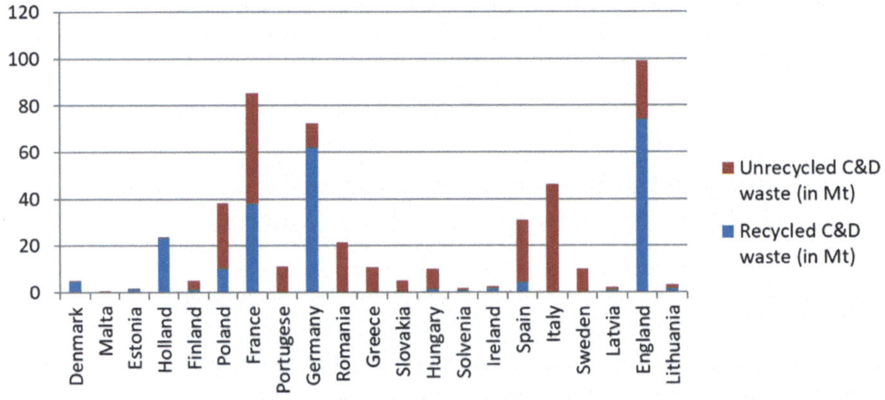

Fig. 16 CDW generation and recycling (Özalp et al. 2016)

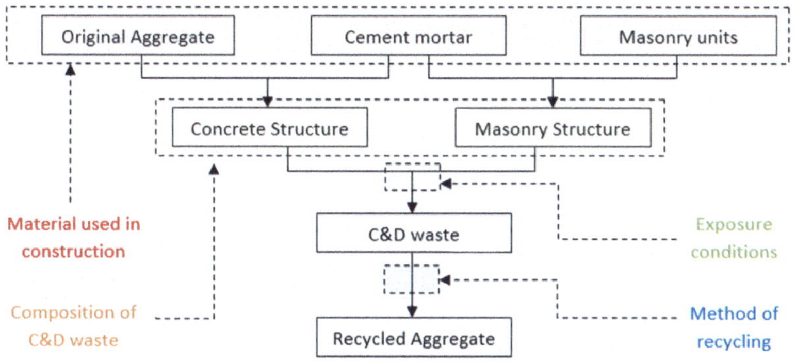

Fig. 17 Factors affecting properties of RA

Properties of RA and its effect in concrete vary significantly on the source of RA and the recycling process of CDW (Alexandridou et al. 2018; Rangel et al. 2019) (Fig. 17). Source characteristics affecting RA properties are significantly affected by regional construction practices and associated characteristics of CDW. This section covers the variability in RA properties in different countries and their effect on the properties of concrete. This will assist stakeholders in understanding the variation in RA properties and ensure reliability in its use in the industry.

4.1 *Effect of Regional Construction Practices*

Construction practices vary among countries because of resource availability, economic status and geographical conditions. Different construction practices result in different types of structures in terms of construction, quality of masonry blocks and grade of concrete. This section presents an overview of the differences in construction practices and their influence on RA properties.

4.1.1 Construction and Demolition Waste Composition

Construction practices have gradually transitioned from brick-based to concrete-based or composite construction. Comparing Nepal, India, Turkey, Indonesia, Egypt, and Pakistan shows that masonry construction varies significantly in wall thickness, mortar to brick thickness, and unconfined masonry (Okazaki and Pribadi 2016). The type of construction influences the characteristics of CDW generated in a country.

Table 14 compares the composition of CDW in different countries and highlights different dependencies on brick and concrete structures. An estimate on waste composition in India conducted before 2019 estimated 23% concrete and 31% masonry in CDW and was similar to a field investigation published in 2020 (Table 14). A case

Table 14 Composition of CDW in different countries

S. No.	Country/region	Sub-region	Percent of concrete waste/%	Percent of masonry waste/%	Ref.
1	Iran	Yazad	38	20	Ansari and Ehrampoush (2018)
2	Spain	Alhendin, Granada	87	2.50	Cuenca-Moyano, Martín-Pascual et al. (2020)
3	Greece	–	40	30	Oikonomou (2005)
4	India	Chennai	23	30	Ram, Kishore et al. (2020)

study of India highlights a regional consistency that can estimate the waste generation characteristics for different regions.

CDW, when recycled, results in different aggregate types, which broadly consist of concrete and masonry (Wu et al. 2019). RAC properties differ from recycled masonry, resulting in RA properties based on the CDW composition in different countries. Several classifications have been developed to compensate for CDW composition to group RA with similar properties. Specifications covering the classification of RA are discussed separately in Part III of this report.

4.1.2 Materials Used in Construction

Countries follow different standards governing the material used in constructions, creating regional patterns of materials being recycled as aggregate. Even concrete structures, not designed as per their respective standards in Nepal, India, Turkey, Indonesia, Egypt and Pakistan, have shown a consistent regional pattern in terms of strength of bricks and proportion of concrete (Okazaki and Pribadi 2016). Although material use is not uniform in a country, similar local site specifications will result in similar CDW. This can be used for managing recycling facilities to have consistency in the properties of RA.

A. Effect of RAC

RA produced from concrete recycling typically consists of aggregate, AM and interfacial transition zone (ITZ) (Evangelista and Guedes 2019), as depicted in Fig. 18. AM refers to the mortar attached to the surface of RA. The quantity of AM depends on the properties of source concrete (Ouyang et al. 2020). In practice, weaker concrete is associated with a more porous cement matrix, and thus the quality of AM can also be associated with the grade of RAC. As already discussed, regional design patterns have shown concrete design consistency, and AM similarities can be expected in RCA. AM shows poor performance compared to the aggregates and is considered an undesirable aspect in RA (Ouyang et al. 2020).

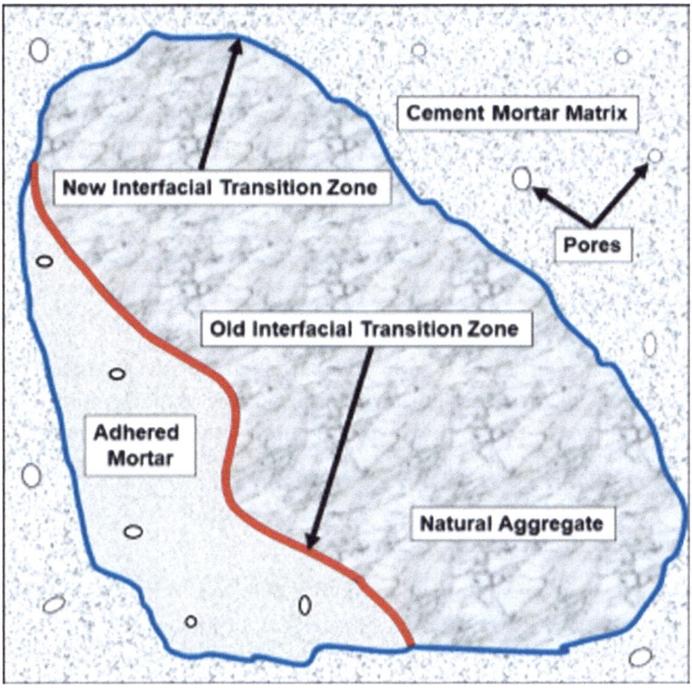

Fig. 18 Recycled coarse aggregate (Jayasuriya et al. 2019)

B. Strength of masonry being used

Like concrete, porosity and strength are also inversely correlated in masonry units containing bricks, blocks, ceramic tiles, etc. Therefore poor grade of masonry units will result in higher porosity and weaker mechanical characteristics in the recycled masonry particles.

C. Aggregates used in construction

Aggregate used in original construction is subjected to mechanical and chemical actions during recycling. It has been observed that rounder aggregates produce a lesser quantity of AM than angular aggregates. This shows that the quantity of AM can be correlated with the shape of aggregates used. As aggregates used in construction are based on regional availability, the aggregate used can be easily factored in for RCA properties.

4.1.3 Method of Recycling

Recycling primarily involves the mechanical crushing of CDW to obtain RA of desired grading size. Different countries utilize different recycling processes to obtain

RA. The United States and Russia use impact crushers, while Germany, China and Japan use jaw crushers (Jayasuriya et al. 2019). In jaw-type crushing, there is a high possibility of micro-cracks in the resulting aggregates, lowering the performance of resulting RA (Jayasuriya et al. 2019). Moreover, recycling also involves different waste separation and cleaning practices, which may leave different degrees of impurities in RA (Jayasuriya et al. 2019). Different recycling methods in different countries can result in significant variation in RA properties.

4.1.4 Other Factors

Variability in RA properties is significantly affected by several other factors like exposure conditions of structures before demolition, age of the demolished structure, and the type of failure in the demolished structure. Although current knowledge on RA lacks specific investigations for the same, existing literature indicates the effect of these factors.

A. Exposure conditions

Material properties have been known to modify over time when exposed to extreme environments. Wu et al. (2019) observed that RA produced from industrial CDW contains both organic and heavy metal pollutants. Butera et al. (2014) observed that RA with prolonged exposure to urban environments had a higher concentration of certain organic pollutants. Specific exposure conditions have been shown to increase specific pollution concentrations, which will impart harmful properties to RA in significant concentrations. Understanding the effect of regional exposure conditions on RA will help identify critical pollutants that need to be monitored for the use of RA in a particular country.

B. Age of demolished structure

The age of demolished structures becomes significant in countries where systematic demolition of a structure is practiced. Construction practices of a region change with time, and characteristics of demolition waste vary based on the construction period of the structure (Cochran and Townsend 2010; Mastrucci et al. 2017). However, this has less practical significance as demolition of structures in most countries is not according to their age.

C. Socio-economic factors

Socio-economic do not directly influence the properties of RA but significantly influence other parameters. A comparative study among multiple countries, a case study of the UK highlights that waste processing facilities preferred economically fair use CDW as landfill over its high-value use as RA. Therefore, favorable socio-economic parameters in different countries can encourage stakeholders toward better practices for RA.

4.2 *Effect of Regional Variations on the Properties of RA*

Regional factors affect RA in terms of composition, AM, gradation and concentration of pollutants. The effect of regional variations subsequently affects the properties of RA.

4.2.1 Effect of Construction and Demolition Waste Composition on the Properties of RA

Masonry particles have lower strength and higher porosity than concrete particles (due to aggregate). An increase in masonry content increases water absorption and decreases the density of RA (Agrela et al. 2011; Abdulla 2015).

In terms of concrete design, higher water absorption by masonry particles in RA increases the requirement of water content at the time of design (Agrela et al. 2011, 2013). An increase in masonry content also lowers the density of RA, reducing the overall weight of concrete (Agrela et al. 2011, 2013).

Mechanical properties of concrete are more affected by recycled masonry particles than RAC particles. Concretes made with RA having low masonry content (< 5%) have shown better strength characteristics than RA having high masonry content (5–25%) (Agrela et al. 2013). Further cement is recommended for concrete made using RA with higher masonry content (Agrela et al. 2013). The poor mechanical quality of masonry particles lowers aggregates' abrasion resistance, as observed in the Los-Angeles abrasion test (Barbudo et al. 2012). The fracture occurred mainly around masonry particles in flexural testing, indicating masonry content as the limiting factor for flexural strength (Hoffmann et al. 2012).

A higher concentration of masonry particles in RA increases water-based transportation due to their porous nature and lowers durability (Hoffmann et al. 2012). RA containing masonry wastes has shown a higher concentration of certain inorganic pollutants (Barbudo et al. 2012; Butera et al. 2014). Inorganic pollutants, like sulfates, lower the mechanical performance of concrete by forming additional chemical products, like ettringite, limiting RA application (Agrela et al. 2014).

4.2.2 Effect of AM on the Properties of RCA

Adhered mortar having higher porosity and lower density than the aggregate increases the porosity and lowers the density of RCA. Quantity of AM and its effect increases for finer RA AM increases the water absorption in aggregate and increases the water content requirement of concrete. Higher water content increases the thickness of the new ITZ and lowers the strength of concrete (Evangelista and Guedes 2019). High water concentration also affects drying shrinkage, which can go 50% higher for RCA than NA. High porosity in AM increases the overall porosity of concrete and influences the durability of concrete (Jayasuriya et al. 2019).

Higher AM porosity can also result in stress concentrations and a reduction in compressive strength. AM also provides a rough geometry, which imparts better flexural performance than flexural performance. Under abrasive action, AM easily breaks off at ITZ in the Los-Angeles test, giving a mass loss as high as 63% (Jayasuriya et al. 2019).

Multiple recycling of RCA increases the content of AM on the aggregate, which can act as the limiting factor on long-term recycling (Thomas et al. 2018).

4.2.3 Accounting Effect of Regional Variations in the Properties of RA

To account for the effect of regional variations, most countries have adopted their standards for RA. These standards broadly involve the classification of RA based on its composition and permissible limits for different uses. Comparing these standards shows a significant variation in maximum permissible replacement, the type of concrete used, and the limits of masonry and concrete content (Agrela et al. 2013; Martín-Morales et al. 2013).

Table 15 presents an overview of classification based on standards and highlights the variation in classification. Table 15 also reflects that many countries have not adopted separate classification for recycled masonry aggregate, indicating a lack of masonry based aggregate in their practice. In terms of pollutants, all countries have specifications for chloride, and there are regional variations for organic matter and sulfates.

A closer look at the existing standards reveals that most standards are limited to aggregating and defining specifications for a limited number of pollutants. Authors have suggested classification for RA based on AM content (Rangel et al. 2019). However, standards in their current form have not extended the discussion to AM.

In current practice, AM is accounted for using treatment methods, like acid treatment (Juan and Gutiérrez 2009). Since AM depends on regional construction practices, a systematic investigation is needed for the suitability treatment methods in different countries. Variation in the properties of RA resulting from regional variations has a higher significance for the countries transitioning toward the use of RA. For example, countries recycling waste with higher concrete content, like Spain, will have better quality RA than those with higher masonry content, like India. Understanding the regional variation in RA will assist in the development of suitable standards and increase reliability among stakeholders.

5 Summary

Regional variations in construction practice and recycling method results can be observed among different countries, resulting in variations in RA properties produced. Variation in RA properties lowers the confidence of stakeholders and limits the use of RA in construction. It is imperative to understand regional variations in

Table 15 Review of standards on RA

Standard	Region covered	No of classification for RA	Separate class for masonry aggregate	Specification for sulfates	Specification for chloride	Specifications for organic matter
RILEM	–	3	Y	Y	Y	Y
CSIRO	Australia	2	N	N	Y	Y
PTV 406	Brazil	3	Y	Y	Y	Y
DG/TJ07/008	China	2	N	Y	Y	Y
DS 2426	Denmark	2	N	N	Y	N
DIN 4226-100	Germany	4	Y	Y	Y	N
WBTC 12	Hong Kong	1	N	Y	Y	N
JIS A 5021	Japan	1	N	N	Y	Y
CUR NEN 5905	Netherlands	2	N	Y	Y	Y
NB 26	Norway	2	N	N	Y	Y
LNEC E 471 3	Portugal	3	N	Y	Y	Y
EHE 08	Spain	1	N	Y	Y	N
SIA 2030	Switzerland	2	N	Y	Y	N
BRE Digest 433	United Kingdom	3	N	Y	Y	N

Source (Martín-Morales et al. 2013)

terms of factors, like CDW composition, AM content, exposure conditions and recycling methods, and their effect on RA properties. Through proper understanding, suitable standards and regional practices can be developed to ensure consistency in RA properties. This will simplify the use of RA in construction and help in the sustainable development of construction practices.

References

Abdulla, N.A.: Effect of recycled coarse aggregate type on concrete. J. Mater. Civ. Eng. **27**(10), 04014273 (2015)

Afshar, T., Disfani, M.M., Arulrajah, A., Narsilio, G.A., Emam, S.: Impact of particle shape on breakage of recycled construction and demolition aggregates. Powder Technol. **308**, 1–12 (2017)

Agrela, F., Sánchez de Juan, M., Ayuso, J., Geraldes, V.L., Jiménez, J.R.: Limiting properties in the characterization of mixed recycled aggregates for use in the manufacture of concrete. Constr. Build. Mater. **25**(10), 3950–3955 (2011)

Agrela, F., De Juan, M.S., Ayuso, J., Geraldes, V., Jiménez, J.: Limiting properties in the characterization of mixed recycled aggregates for use in the manufacture of concrete. Constr. Build. Mater. **25**(10), 3950–3955 (2011)

Agrela, F., Alaejos, P., De Juan M.S.: Properties of concrete with recycled aggregates. Handb. Recycl. Concr. Demolition Waste 304–329 (2013)

Agrela, F., Cabrera, M., Galvín, A., Barbudo, A., Ramirez, A.: Influence of the sulphate content of recycled aggregates on the properties of cement-treated granular materials using sulphate-resistant Portland cement. Constr. Build. Mater. **68**, 127–134 (2014)

Alexandridou, C., Angelopoulos, G.N., Coutelieris, F.A.: Mechanical and durability performance of concrete produced with recycled aggregates from Greek construction and demolition waste plants. J. Clean. Prod. **176**, 745–757 (2018)

Al-Bayati, H.K.A., Das, P.K., Tighe, S.L., Baaj, H.: Evaluation of various treatment methods for enhancing the physical and morphological properties of coarse recycled concrete aggregate. Constr. Build. Mater. **112**, 284–298 (2016)

Ansari, M., Ehrampoush, M.H.: Quantitative and qualitative analysis of construction and demolition waste in Yazd city, Iran. Data Brief **21**, 2622–2626 (2018)

ASTM, A.: C33/C33M-16e1: standard specification for concrete aggregates. In: ASTM international. West Conshohocken, Pennsylvania, United States (2016)

Barbudo, A., Agrela, F., Ayuso, J., Jiménez, J.R., Poon, C.S.: Statistical analysis of recycled aggregates derived from different sources for sub-base applications. Constr. Build. Mater. **28**(1), 129–138 (2012)

Butera, S., Christensen, T.H., Astrup, T.F.: Composition and leaching of construction and demolition waste: inorganic elements and organic compounds. J. Hazard. Mater. **276**, 302–311 (2014)

Cochran, K., Townsend, T.G.: Estimating construction and demolition debris generation using a materials flow analysis approach. Waste Manag. **30**(11), 2247–2254 (2010)

Courard, L., Rondeux, M., Zhao, Z., Michel, F.: Use of recycled fine aggregates from C&DW for unbound road sub-base. Materials **13**(13), 2994 (2020)

Cuenca-Moyano, G.M., Martín-Pascual, J., Martín-Morales, M., Valverde-Palacios, I., Zamorano, M.: Effects of water to cement ratio, recycled fine aggregate and air entraining/plasticizer admixture on masonry mortar properties. Constr. Build. Mater. **230**, 116929 (2020)

da Silva, W.R.L., Lucena, D.S., Štemberk, P., Prudêncio, L.R.: Evaluation of the effect of concrete compositional changes and the use of ethyl-alcohol and biodegradable-oil-based release agents on the final surface appearance of self-compacting concrete pre-cast elements. Constr. Build. Mater. **52**, 202–208 (2014)

Evangelista, L., Guedes, M.: Microstructural studies on recycled aggregate concrete. New Trends Eco Effic. Recycl. Concr. Elsevier 425–451 (2019)

Hansen, T.C., Narud, H.: Strength of recycled concrete made from crushed concrete coarse aggregate. Concr. Int. **5**(1), 79–83 (1983)

Hansen, T.C.: Recycled aggregates and recycled aggregate concrete second state-of-the-art report developments 1945–1985. Mater. Struct. **19**(3), 201–246 (1986)

Hoffmann, C., Schubert, S., Leemann, A., Motavalli, M.: Recycled concrete and mixed rubble as aggregates: Influence of variations in composition on the concrete properties and their use as structural material. Constr. Build. Mater. **35**, 701–709 (2012)

Huang, Y., He, X., Wang, Q., Sun, Y.: Mechanical properties of sea sand recycled aggregate concrete under axial compression. Constr. Build. Mater. **175**, 55–63 (2018)

Huda, S.B., Alam, M.S.: Mechanical behavior of three generations of 100% repeated recycled coarse aggregate concrete. Constr. Build. Mater. **65**, 574–582 (2014)

Ismail, S., Ramli, M.: Engineering properties of treated recycled concrete aggregate (RCA) for structural applications. Constr. Build. Mater. **44**, 464–476 (2013)

Jamkar, S., Rao, C.: Index of aggregate particle shape and texture of coarse aggregate as a parameter for concrete mix proportioning. Cem. Concr. Res. **34**(11), 2021–2027 (2004)

Jane, N., Grayson, D.: World business council for sustainable development (WBCSD). In Corporate responsibility coalitions, 300–317. Routledge. (2017)

Jayasuriya, A., Adams, M., Bandelt, M.: Computational generations for recycled concrete aggregate systems. Code Ocean (2019)

De Juan, M.S., Gutiérrez, P.A.: Study on the influence of attached mortar content on the properties of recycled concrete aggregate. Constr. Build. Mater. **23**(2), 872–877 (2009)

Katz, A.: Properties of concrete made with recycled aggregate from partially hydrated old concrete. Cem. Concr. Res. **33**(5), 703–711 (2003)

Kim, H.-S., Kim, B., Kim, K.-S., Kim, J.-M.: Quality improvement of recycled aggregates using the acid treatment method and the strength characteristics of the resulting mortar. J. Mater. Cycles Waste Manag. **19**(2), 968–976 (2017)

Koenders, E.A.B., Pepe, M., Martinelli, E.: Compressive strength and hydration processes of concrete with recycled aggregates. Cem. Concr. Res. **56**, 203–212 (2014)

Koper, A., Koper, W., Koper, M.: Influence of raw concrete material quality on selected properties of recycled concrete aggregates. Procedia Eng. **172**, 536–543 (2017)

Kou, S.-C., Poon, C.-S.: Effect of the quality of parent concrete on the properties of high performance recycled aggregate concrete. Constr. Build. Mater. **77**, 501–508 (2015)

Li, L., Xiao, J., Poon, C.S.: Dynamic compressive behavior of recycled aggregate concrete. Mater. Struct. **49**(11), 4451–4462 (2016)

Mamlouk, M.S., Zaniewski, J.P.: Materials for civil and construction engineers. Pearson Education Limited London (2018)

Martín-Morales, M., Zamorano, M., Valverde-Palacios, I., Cuenca-Moyano, G., Sánchez-Roldán, Z.: Quality control of recycled aggregates (RAs) from construction and demolition waste (CDW), pp. 270–303. Elsevier, Handbook of recycled concrete and demolition waste (2013)

Marie, I., Quiasrawi, H.: Closed-loop recycling of recycled concrete aggregates. J. Clean. Prod. **37**, 243–248 (2012)

Mastrucci, A., Marvuglia, A., Popovici, E., Leopold, U., Benetto, E.: Geospatial characterization of building material stocks for the life cycle assessment of end-of-life scenarios at the urban scale. Resour. Conserv. Recycl. **123**, 54–66 (2017)

Nagataki, S., Gokce, A., Saeki, T., Hisada, M.: Assessment of recycling process induced damage sensitivity of recycled concrete aggregates. Cem. Concr. Res. **34**(6), 965–971 (2004)

Oikonomou, N.D.: Recycled aggregate concretes. Cem. Concr. Compos. **27**(2), 315–318 (2005)

Okazaki, K., Pribadi, K.S.: Resilient homes make cities resilient. Urban Disasters Resil. Asia Elsevier 49–68 (2016)

Ouyang, K., Shi, C., Chu, H., Guo, H., Song, B., Ding, Y., Guan, X., Zhu, J., Zhang, H., Wang, Y.: An overview on the efficiency of different pretreatment techniques for recycled concrete aggregate. J. Clean. Prod. 121264 (2020)

Özalp, F., Yılmaz, H.D., Kara, M., Kaya, Ö., Şahin, A.: Effects of recycled aggregates from construction and demolition wastes on mechanical and permeability properties of paving stone, kerb and concrete pipes. Constr. Build. Mater. **110**, 17–23 (2016)

Padmini, A.K., Ramamurthy, K., Mathews, M.S.: Influence of parent concrete on the properties of recycled aggregate concrete. Constr. Build. Mater. **23**(2), 829–836 (2009)

Park, W., Noguchi, T.: Influence of metal impurity on recycled aggregate concrete and inspection method for aluminum impurity. Constr. Build. Mater. **40**, 1174–1183 (2013)

Pepe, M.: A conceptual model for designing recycled aggregate concrete for structural applications. Springer. (2015)

Pepe, M., Toledo Filho, R.D., Koenders, E.A., Martinelli, E.: Alternative processing procedures for recycled aggregates in structural concrete. Constr. Build. Mater. **69**, 124–132 (2014)

Pepe, M., Toledo Filho, R.D., Koenders, E.A.B., Martinelli, E.: A novel mix design methodology for recycled aggregate concrete. Constr. Build. Mater. **122**, 362–372 (2016)

Poon, C., Ann, T., Ng, L.: On-site sorting of construction and demolition waste in Hong Kong. Resour. Conserv. Recycl. **32**(2), 157–172 (2001)
Ram, V., Kishore, K.C., Kalidindi, S.N.: Environmental benefits of construction and demolition debris recycling: evidence from an Indian case study using life cycle assessment. J. Clean. Prod. **255**, 120258 (2020)
Rangel, C.S., Toledo Filho, R.D., Amario, M., Pepe, M., de Castro Polisseni, G., de Andrade, G.P.: Generalized quality control parameter for heterogenous recycled concrete aggregates: a pilot scale case study. J. Clean. Prod. **208**, 589–601 (2019)
Salesa, Á., Pérez-Benedicto, J.A., Colorado-Aranguren, D., López-Julián, P.L., Esteban, L.M., Sanz-Baldúz, L.J., Sáez-Hostaled, J.L., Ramis, J., Olivares, D.: Physico—mechanical properties of multi—recycled concrete from pre-cast concrete industry. J. Clean. Prod. **141**, 248–255 (2017)
Shi, C., Li, Y., Zhang, J., Li, W., Chong, L., Xie, Z.: Performance enhancement of recycled concrete aggregate—a review. J. Clean. Prod. **112**, 466–472 (2016)
Silva, R.V., De Brito, J., Dhir, R.: Properties and composition of recycled aggregates from construction and demolition waste suitable for concrete production. Constr. Build. Mater. **65**, 201–217 (2014)
Sim, J., Park, C.: Compressive strength and resistance to chloride-ion penetration and carbonation of recycled aggregate concrete with varying amount of fly ash and fine recycled aggregate. Waste Manag. **31**(11), 2352–2360 (2011)
Sui, Y., Mueller, A.: Development of thermo-mechanical treatment for recycling of used concrete. Mater. Struct. **45**(10), 1487–1495 (2012)
Tam, V.W.Y., Tam, C.M.: Diversifying two-stage mixing approach (TSMA) for recycled aggregate concrete: TSMAs and TSMAsc. Constr. Build. Mater. **22**(10), 2068–2077 (2008)
Tam, V.W., Gao, X., Tam, C.M., Chan, C.: New approach in measuring water absorption of recycled aggregates. Constr. Build. Mater. **22**(3), 364–369 (2008)
Thomas, C., De Brito, J., Gil, V., Sainz-Aja, J., Cimentada, A.: Multiple recycled aggregate properties analyzed by X-ray micro-tomography. Constr. Build. Mater. **166**, 171–180 (2018)
Townsend, T., Wilson, C., Beck, B.: The benefits of construction and demolition materials recycling in the United States. University of Florida, Gainesville, FL, USA (2014)
Wu, H., Zuo, J., Zillante, G., Wang, J., Yuan, H.: Status quo and future directions of construction and demolition waste research: a critical review. J. Clean. Prod. **240**, 118163 (2019)
Zega, C., Villagran-Zaccardi, Y., Di Maio, A.: Effect of natural coarse aggregate type on the physical and mechanical properties of recycled coarse aggregates. Mater. Struct. **43**(1–2), 195–202 (2010)
Zhao, Z., Remond, S., Damidot, D., Xu, W.: Influence of fine recycled concrete aggregates on the properties of mortars. Constr. Build. Mater. **81**, 179–186 (2015)
Zhao, Z., Courard, L., Michel, F., Remond, S., Damidot, D.: Influence of granular fraction and origin of recycled concrete aggregates on their properties. Eur. J. Environ. Civ. Eng. **22**(12), 1457–1467 (2018)
Zhao, Z., Courard, L., Groslambert, G., Jehin, T., Léonard, A., Xiao, J.: Use of recycled concrete aggregates from block wastes for the production of new concrete blocks: an industrial-scale study. Resour. Conserv. Recycl. **157**, 104786 (2020a)
Zhao, Z., Courard, L., Groslambert, G., Jehin, T., Léonard, A., Xiao, J.: Use of recycled concrete aggregates from pre-cast block for the production of new building blocks: an industrial-scale study. Conserv. Recycl. **157**, 104786 (2020b)

New Technology to Improve the Properties of Recycled Aggregate Concrete

Amardeep Singh, Jianzhuang Xiao, Romildo Dias Toledo Filho, Surender P. Shah, and Wengui Li

1 Introduction

The re-use of RCA still has shortcomings, and their applications are often limited to low-level civil works like road sub-bases and non-structural concrete products. This is mainly because compared with NA, RCA has a lower density, higher water absorption, higher porosity, and lower mechanical properties, which have adverse impacts on recycled aggregate concrete (RAC) quality, resulting in lower strength and elasticity; higher shrinkage, and creep; and more inferior durability.

To efficiently facilitate the effective re-use of RCA in new concrete, ways to improve the quality of RCA and the performance of RAC have attracted much research attention in the past decades. This chapter summarizes some new technologies to improve RCA's qualities and enhance RAC, mainly focusing on CO_2 curing for enhancement of RCA, mechanical processes for high-quality RCA production, and RAC's enhancement by modifying with nano-silica particles. This would provide

A. Singh
School of Civil Engineering and Architecture, Changzhou Institute of Technology, Changzhou, China

J. Xiao
Department of Structural Engineering, College of Civil Engineering, Tongji University, Shanghai, China

R. D. Toledo Filho
Civil Engineering Program, COPPE, Federal University of Rio de Janeiro, Rio de Janeiro, Brazil

S. P. Shah (✉)
Department of Civil Engineering, Northwestern University, Evanston, USA
e-mail: s-shah@northwestern.edu

W. Li
School of Civil and Environmental Engineering, The University of New South Wales, Kensington, NSW, Australia

J. Xiao et al. (eds.), *Structural Behaviour and Innovation of Recycled Aggregate Concrete*, RILEM State-of-the-Art Reports 41,
https://doi.org/10.1007/978-3-031-84136-1_2

technical feasibility for engineers to re-use more RCAs as resources and to ensure the performance of RAC is comparable to conventional concrete.

2 CO_2 Curing for Enhancement of RCA

2.1 *Technical Background of CO_2 Curing*

Considering that the hardening process of cement-based materials can be accelerated by promoting the carbonation reaction between clinker phase and CO_2 in cement, CO_2 gas has been recommended as curing agent. During the CO_2 curing process, the cement-based materials are cured at different pressure levels (viz, static and pressurized, static and ambient, or flow through and ambient) in a container that is filled with CO_2. In the 1960s, this technology was developed firstly and several patents on the production of rapid-hardening cement-based material using CO_2 curing were granted (Bierlich 1969; Stuart 1970). In addition, due to the added value of CO_2 sequestration, people have made continuous efforts and development. Besides this, efforts and developments have been continuously made because of the added value of CO_2 sequestration.

Kauer and Freeman (1955) reported that the strength of concrete increased after adopting CO_2 curing in the new stage. Consider that excessive exposure to CO_2 caused the concrete to have "a soft, chalk-like, non-serviceable carbonated surface", authors did not recommend the CO_2 curing method. Since then, the University of Illinois has conducted several investigations on CO_2 curing of anhydrous cement phase. It was found that the calcium silicate phase (including C_2S and C_3S) in ordinary Portland cement (OPC) clinker reacted with CO_2 and water to form calcium carbonate and calcium silicate hydrate (C-S-H) (Berger et al. 1972). However, the morphological difference between C-S-H produced by carbonation and C-S-H produced by hydration could not be distinguished by the scanning electron microscope (SEM) photomicrographs.

The main reactions are:

$$3CaO \cdot SiO_2 + yH_2O + (3 - x)CO_2 \rightarrow xCaO \cdot SiO_2 \cdot yH_2O + (3 - x)CaCO_3 \tag{1}$$

$$2CaO \cdot SiO_2 + yH_2O + (2 - x)CO_2 \rightarrow xCaO \cdot SiO_2 \cdot yH_2O + (2 - x)CaCO_3 \tag{2}$$

The results supported the formation of C-S-H gel by accelerated carbonation (Goodbrake et al. 1979) in which a moist CO_2 gas was used to cure β-C_2S and C_3S powders. The heat released from the carbonation reactions was 347 kJ/mol for C_3S and 184 kJ/mol for β-C_2S, respectively. The results also showed that the compressive strength of the CO_2-cured mortars was significantly improved, and the

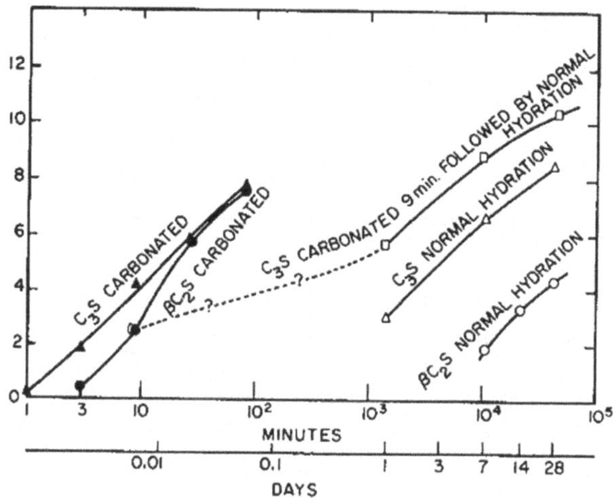

Fig. 1 Comparison of compressive strength development in water-cured and CO_2-cured mortar samples (Goodbrake et al. 1979)

strength of CO_2-cured mortars was further improved during the subsequent water curing process, as shown in Fig. 1. Other studies reported the similar results (Klemm and Berger 1972; Sorochkin et al. 1975). Under the optimum water-cement ratio, the compressive strength of cement cured with CO_2 for 5 min was higher than that cured with water for 1 day.

Some researchers have studied the CO_2 curing for non-hydraulic calcium silicates (Bukowski and Berger 1979). The reaction behavior of γ-Ca_2SiO_4 and $CaSiO_3$ with CO_2 was similar to that of β-C_2S and C_3S (Klemm and Berger 1972; Goodbrake et al. 1979), including development of the strength and products of the reaction. Although the development for the strength of β-C_2S was comparable to that of C_3S during the CO_2 curing process, the aluminate phases had almost no strength development (Klemm and Berger 1972).

Since 1990s, more attention was paid to the CO_2 curing technique in cement-based materials due to the impact of global warming. An investigation on producing pre-cast concrete by using the CO_2 curing method was reported by a work supported by the US Department of Energy (Wagh et al. 1994). The concrete achieved a shortened setting time after CO_2 curing for 30 min, and it was speculated by the author that about 10–24% of the CO_2 released in cement production might again be captured by pre-cast concrete. Cement mixtures and cement-bond particleboards were produced by using supercritical and near-critical CO_2 (Knopf et al. 1999; Hermawan et al. 2001). They attained excellent mechanical properties and densified structures.

CO_2 curing of mature concrete, also known as carbonation, is always expected to reduce cement pastes' total pore volume and pore connectivity (Rimmelé et al. 2008; Han et al. 2014). Thus, it was reported that the carbonation of cement pastes might cause a decrease in the water absorption of concrete compared to the un-carbonated concrete samples (Houst and Wittmann 1994). Furthermore, the compressive strength of mature cement paste was enhanced by carbonation due to microstructure densification (Johannesson and Utgenannt 2001; Purnell et al. 2001).

Recycled coarse aggregate derived from CDW contains much old cement mortar, making the RCA more porous than the original aggregate, due to the porous nature and the attached cement mortar's presence.

2.2 *Properties of CO_2-Cured RCA*

The purpose of CO_2 curing pretreatment is to densify the old mortar attached to the aggregate and promote the reaction between CO_2 and the cement hydration products. However, the conventional pretreatment method of RA is to remove the attached cement mortar.

Zhan et al. (2013) investigated the CO_2 curing process's influence on RCA properties derived from lab-prepared concrete with different strength grades. Four kinds of RCA were produced experimentally with particle sizes.

Figure 3 illustrates the varying carbonation percentage of RCA30 with particle sizes. With smaller particle sizes and a larger specific surface, RCA can take up more CO_2. The carbonation percentage could reach 56% for RCA with a particle size of 5–10 mm, and this indicates there is good potential to capture CO_2 with the smaller RCA.

Kou et al. (2014) studied coarse recycled mortar aggregates with particle sizes of 20 mm and 10 mm and cured under a 99% CO_2 concentration and 0.1 bar pressure for 6, 12, 24, 48 and 72 h. The test results showed a significant increase in density, a 10% good value of RMA, and decreased water absorption with increased ranging from 5 to 20 mm from crushing laboratory prepared parent concretes with strength grades of 30 MPa, 45 MPa, 60 MPa, and 80 MPa, respectively. The CO_2 curing process was carried out using the setup shown in Fig. 2.

After 24 h of carbonation, the apparent density, water absorption and porosity of the carbonated RCA were measured, and the results are shown in Table 1. The apparent density of RCA60 and RCA80 increased slightly, but no significant change was found for the RCA30 and the RCA55 aggregates. However, the water absorption and porosity of the four strength grades of RCA decreased.

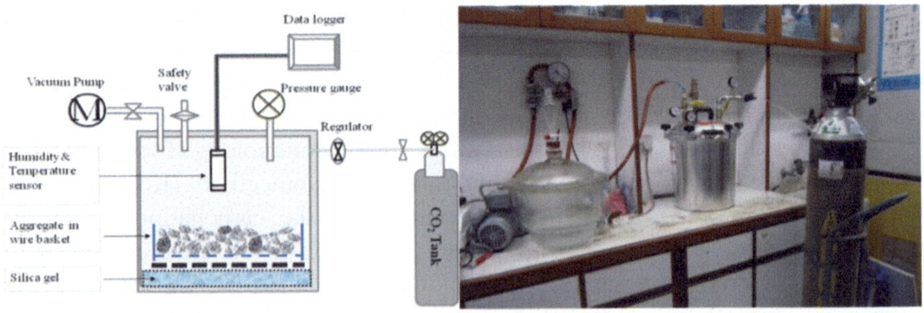

Fig. 2 Experimental setups of CO_2 curing

Fig. 3 Carbonation percentage of RCA with different size

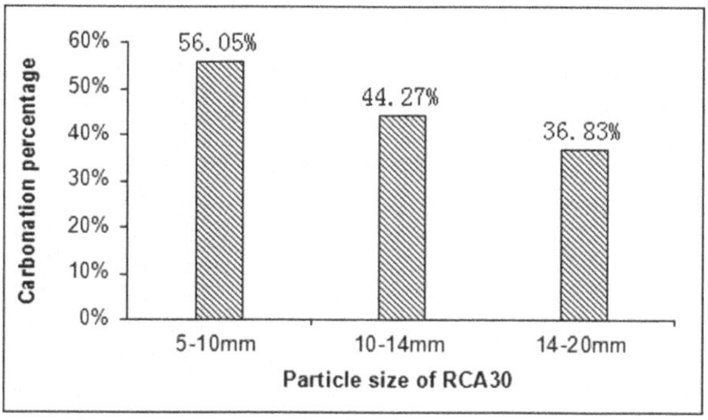

Table 1 Variation of physical properties of RCA before and after carbonation

	Apparent density, kg/m^3		Water absorption, %		Porosity, %	
	Before	After	Before	After	Before	After
RCA30	2639 ± 3.7	2619 ± 3.5	6.58 ± 0.6	5.25 ± 0.4	15.7 ± 1.3	12.68 ± 0.9
RCA45	2623 ± 7.1	2606 ± 7.7	5.98 ± 0.2	4.71 ± 0.2	14.3 ± 0.5	11.65 ± 0.7
RCA60	2620 ± 4.8	2670 ± 3.8	5.95 ± 0.2	4.61 ± 0.3	14.2 ± 0.4	11.18 ± 0.4
RCA80	2602 ± 13.7	2661 ± 5.7	5.81 ± 0.3	4.41 ± 0.1	13.8 ± 0.8	10.94 ± 0.3

CO_2 curing time. After 24 h of curing, there was only a marginal increase in the physical properties of CO_2-cured RMA with the further increase in curing time.

Zhao et al. (2018) studied the effect of accelerated carbonation on the properties of fine RCA. The aggregate was carbonized in pure CO_2 for 2 weeks, and the thermogravimetric analysis, the density, the porosity, and the water absorption of RCA were measured before and after accelerated carbonation (Fig. 4). The results clearly show that after accelerated carbonation, the density of RCA increases because of the transformation of portlandite into calcite. RCA's water absorption after carbonation decreased by 32.8, 30.9, 31.1, and 25.8% for the 2.5/5, 1.25/2·5, 0.63/1.25, and 0/0.63 mm fractions, respectively due to the decrease of porosity of the hardened cement paste.

Different carbonation pressures from 0.1 to 5.0 Bar were adopted in a study (Xuan et al. 2016), in which two types of RCAs (named new RCAs-NRCAs and old RCAs-ORCAs) with different particle sizes were used. The results indicated that the reduction of water absorption of the carbonated NRCAs was up to 16.7%. There was about a 4.0% increase of 10% good value and a 25.9% reduction of crushing value. At the same time, the enhancement of the properties of the ORCAs after 5-bar carbonation was limited because the natural carbonation process of ORCAs already happened during the demolition and storage stages of the ORCAs.

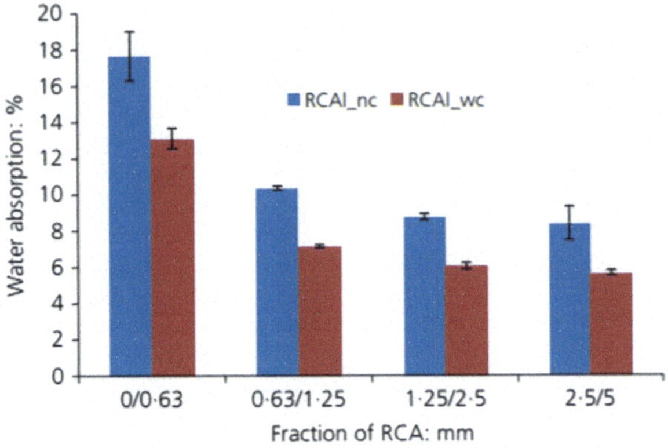

Fig. 4 Water absorption of RCAl_nc (before carbonation) and RCAl_wc (after carbonation) measured by the IFSTTAR method

2.3 Performance of RAC Prepared with Carbonated RCA

After the CO_2 curing process, the carbonated RMA (CIRMA) was subsequently used to prepare new concrete mixtures (Kou et al. 2014). It can be noticed that the compressive strength of concrete incorporated carbonated RMA is comparable to that of the concrete prepared with NAs, as shown in Fig. 5. Similar results can also be observed in Xuan et al. (2016), as shown in Fig. 6. When carbonated RCAs were used to replace all the NA in concrete, a 22.6% increase in compressive strength and 13.2% in the modulus of elasticity can be achieved (Fig. 7). Moreover, the cement paste's microhardness in the old mortar of NRCAs was enhanced after the carbonation treatment; the microhardness on the new ITZ around the carbonated NRCAs had higher values than that around the non-carbonated NRCAs. Due to the densification of microstructure and ITZ improvement, the resistance to chloride-ion penetration of concrete was also significantly improved (Fig. 8).

Zhang et al. (2015a, b, c) have also investigated the CO_2 curing of RCA on RCA and RCA mortar performance. An increase in apparent density decreases in

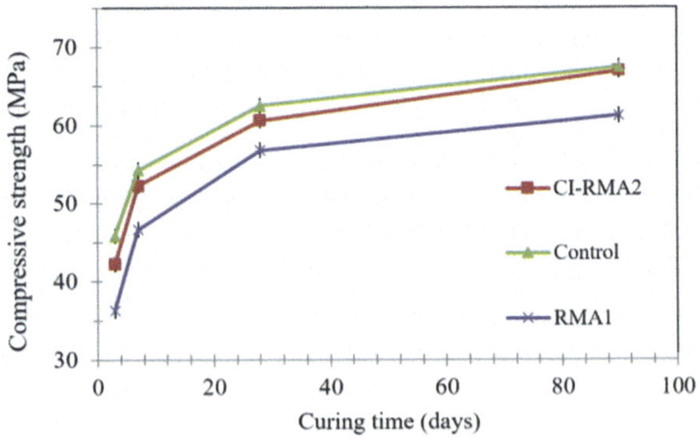

Fig. 5 Comparison of compressive strength of concrete prepared with NA (control), recycled mortar aggregate (RMA1), and carbonated recycled mortar aggregate (CI-RMA2) (Kou et al. 2014)

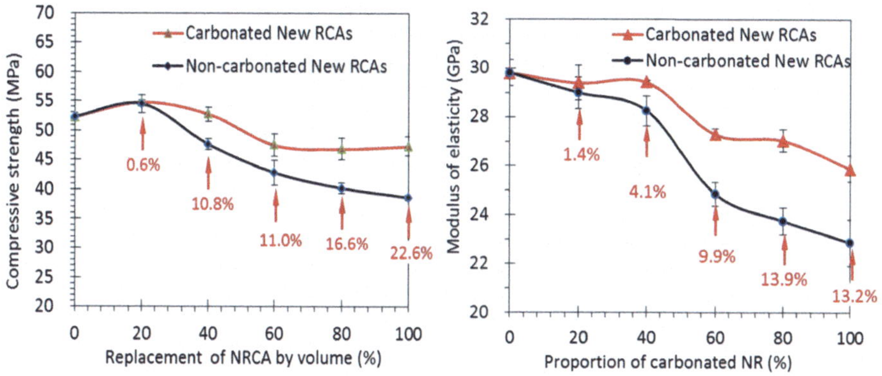

Fig. 6 Comparison of compressive strength (left) and modulus (right) of concrete prepared with non-carbonated and carbonated RCAs (Xuan et al. 2016)

Fig. 7 Vickers microhardness developments at new ITZs around carbonated and non-carbonated NRCAs (Xuan et al. 2016)

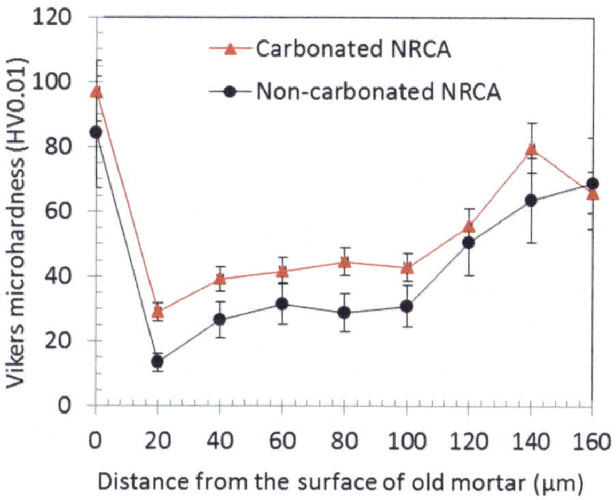

water absorption and crushing values can be found on recycled fine concrete aggregates. The cement mortars' performance incorporated with the carbonated RCAs was comparable to that of cement mortar prepared with natural sand. Besides, SEM examination on the ITZ in the RCA and RCA mortar revealed that CO_2 curing of RCA improved the original ITZ in the RCA and improved the newly formed ITZ in the RCA mortar (Fig. 9).

An accelerated carbonation technique to improve the quality of RCAs has been developed in recent years. When subjected to accelerated carbonation treatment, RCA's physical and mechanical properties can be improved. Incorporating the carbonated RCA in the new RAC enhances the compressive and flexural strengths of RAC and improves the durability of RAC.

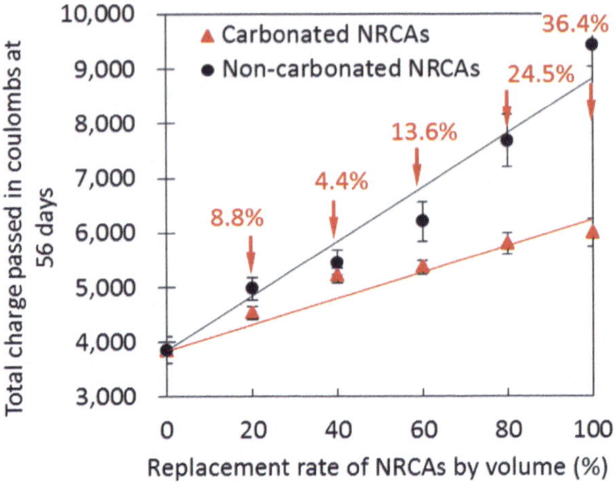

Fig. 8 Total charges passed of concrete with non-carbonated and carbonated RCAs at 56 days curing age (Xuan et al. 2016)

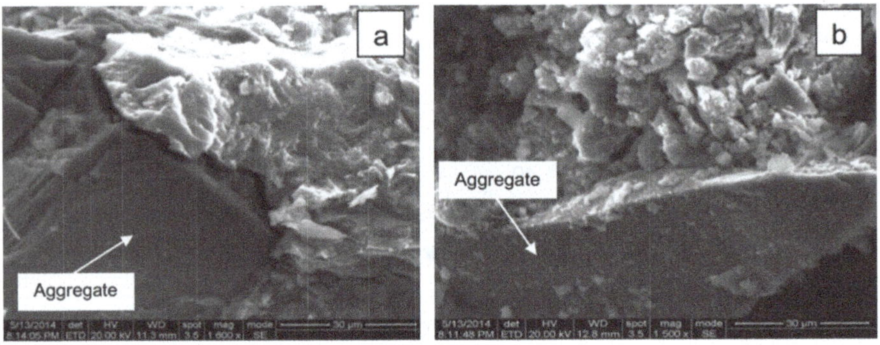

Fig. 9 Interfacial morphology of recycled gravel aggregate mortars at 28 days (**a** un-carbonated mortar; **b** carbonated mortar) (Zhang et al. 2015a, b, c)

3 Mechanical Processes for High-Quality RCA Production

Removal of RCA's residual mortars by improved mechanical or thermal–mechanical methods has been developed to reduce RCA's mortar content and improve mechanical properties and durability of RAC (Li et al. 2010; Quattrone et al. 2014).

3.1 Conventional Crushing Process

The normal crushing process for the production of RCA is similar to the production of crushed NAs. The system consists of grizzly feeders, a jaw crusher, and impact crusher, and a series of screens.

3.2 Improved Mechanical Process

Some improved mechanical processes have been developed to obtain high-quality RCA, as shown in Table 2. In the literature, five types of improved mechanical processes can be identified:

1. Use a combination of crushing processes with jaw and impact crushers to obtain RCA with lower adhered cement mortar contents (Nagataki et al. 2004).
2. A novel eccentric rotor crusher is used for producing high-quality RCA. The material is fed into the eccentric rotor crusher after the ordinary crushers have reduced the aggregate size. It is crunched, rolled, and tumbled by friction between the inner turning rotor and the drum's internal wall (Yonezawa et al. 2001).
3. A screw abrading crusher has been developed and used, where the comminution is done by friction. The coarse RA is fed into a circular tube that forces continuous contact between the aggregate particles and results in the enhanced removal of RCA's AM (Kasai and Yoshio 2006).
4. An innovative particle reshaping machine has been developed to remove the AM on RCA. The materials are fed in the main machine system from a feeding hole in

Table 2 Properties of RCA by Improved mechanical treatments

Technology for RCA production	Levels of processifng	Water absorption (%)	AM (%)	Recovery as coarse aggregates (%)
NA (Xuan et al. 2016)	–	0.80	–	–
Conventional crushing (Yonezawa et al. 2001)	0	5.61	–	–
Improved Jaw and Impact (Nagataki et al. 2004)	3	3.14	30.2	35
Novel eccentric crusher (Yonezawa et al. 2001)	3	1.00	–	27
Screw abrading crusher (Kasai and Yoshio 2006)	3	1.85	6.4	45.1
Particle reshaping machine (Li et al. 2010)	–	2.90	–	–
ADR separation system (Lotfi et al. 2015)	–	5.50	–	–

the spreading plate's upper end. The particles are then divided into two material flow parts: one part enters the spreading plate's inner cavity through the top of the vertical shaft type rotary impeller spreading plate, and the other part of materials falls from the periphery of the vertical shaft type rotary impeller spreading plate. Collision and abrasion between these particles produce low AM content in RCA (Li et al. 2010).

5. A novel dry classification technology called Advanced Dry Recovery (ADR) has been developed to obtain clean RCA products, which has been promoted in the C2CA project (Concrete to Cement and Aggregate) in Europe (Lotfi et al. 2015). The fine and light contaminants were removed by applying ADR. It can break the bonds formed by moisture and fine particles with the help of kinetic energy then materials can be classified into different sizes with different densities. After breaking up the material into a jet using a specifisalsied design rotation blade, the fine particles with a high concentration of contaminants such as wood, plastics, and foams are separated from the coarse particles. Figure 10 shows the ADR principle. ADR can help remove the contaminants and fines, but it would not help reduce the water absorption of RCA or the mortar content of RCA.

Compared to the conventional crushing process, improving the mechanical process may improve RCA's properties. When the eccentric crusher is adopted, RCA's water absorption is the lowest, close to NAs (Yonezawa et al. 2001). Figure 11 shows the strength of RAC prepared with RCA, which is produced by the eccentric crusher.

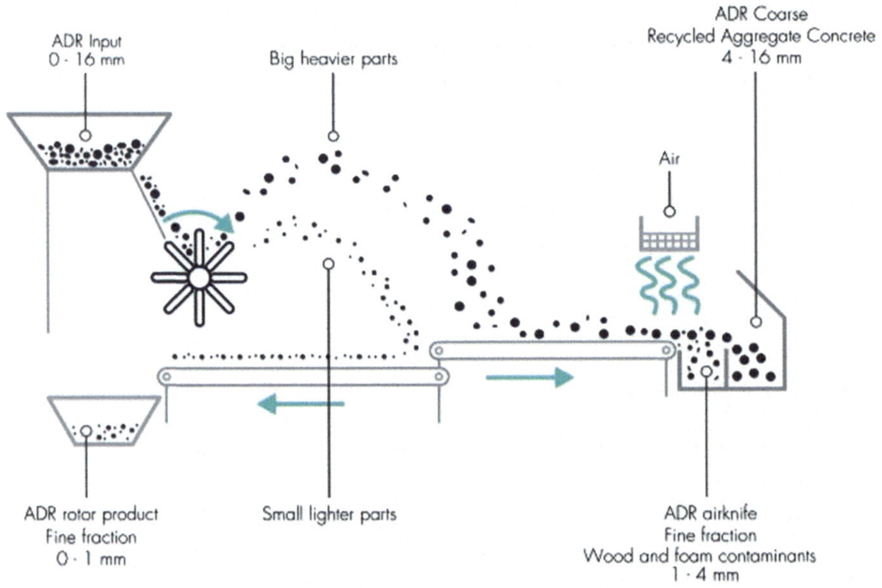

Fig. 10 Schematic of the ADR principle for the production of high-quality RCA (Lotfi et al. 2015)

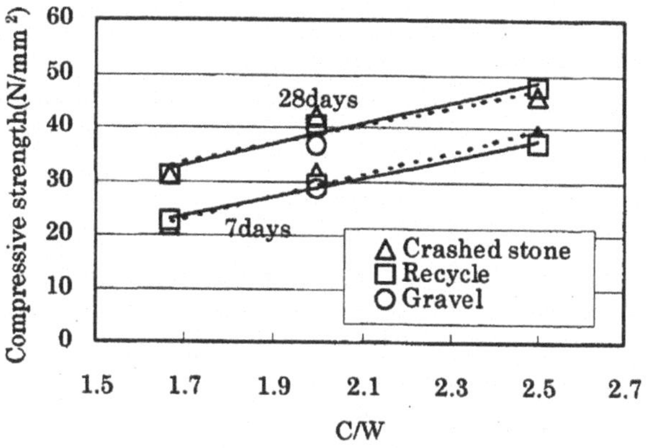

Fig. 11 Strength of RAC prepared with RCA produced by eccentric crusher (Yonezawa et al. 2001)

It can be observed that the strength of RAC prepared with RCA is comparable to the normal concrete prepared with natural crushed stone or gravel aggregates.

However, these techniques' demerits are the high energy input required and the corresponding increase in the generation of recycled fines from 40 to 70% (Li et al. 2010). The fine RCA with a smaller than 5 mm particle size has even more inferior properties with few application outlets in practice.

3.2.1 Thermo-mechanical Process

Some combined thermo-mechanical processes have been investigated to improve RCA quality (Li et al. 2010). The purpose of heating is to reduce the AM's strength and weaken the interface between the mortar and the NAs. When submitted to high temperatures, hydrated cement pastes in the old mortars decompose and lose their mechanical strength. C-S-H and AFt lose most water between 70 and 400 °C. Portlandite decomposes between 450 and 550 °C, and calcium carbonate decomposes above 600 °C. Due to thermal incompatibility and thermal-induced vapor pore stress, macro and micro-cracks in RCA are induced. After the heat treatment, mechanical crushing processes can be further applied to remove AM and obtain high-quality aggregates.

In the literature, four types of thermo-mechanical processes can be identified:

- Heating in combination with normal crushing and sorting. This combination of conventional mechanical and high-temperature thermal treatment generates RAC aggregates with a mortar's shallow content. The adhesive mortar is broken down after experiencing high-temperature treatment, and the constituents can be separated by simple sieving and air classifying (for powders) (Mulder et al. 2007).
- Heating in combination with rubbing process. The RCA is heated in a furnace for about 1 h, then crushed and rubbed in a tube mill. Due to the heating treatment process's low temperature (about 300 °C), rubbing is necessary to remove most

AM. The heating generates cracks in the old mortar, removing the remaining mortar in the RCA (Shima et al. 2005).

- Microwave heating in combination with rubbing process. The particles are heated for a few minutes in an industrial microwave oven, and the treated particles are subsequently rubbed by a mechanical machine (Akbarnezhad et al. 2011).
- A heating-Air classification system has been developed to separate the cementitious fines and the fine aggregates in the fine RCA particles (<4 mm) (Lotfi and Rem 2016). A combination of heating and air classification can properly separate finer fractions (0–0.250 mm) from the relatively coarser fractions (0.125–4 mm). The heated materials were further processed by ball milling to enhance the liberation of the cementitious fraction. The quality of the coarse recycled particles satisfies the required specifications.

Table 3 lists the properties of RCA by using different thermal–mechanical treatments. It can be noticed that the content of the AM is less than 30% (even lower than 2%). The combined thermo-mechanical processes can effectively remove the AM of RCAs.

Moreover, the performance of RAC incorporating the heated treated RCA has noticeable improvement. Figure 12 illustrates the strength improvement of RAC prepared with the microwave treated RCA. It can be noticed that the strength of RAC prepared with 100% treated RCA approaches 90% of that of conventional concrete prepared with NAs.

However, the process energy consumption of the thermo-mechanical treatments would be between 36 and 62 times higher than that of conventional recycling processes, as shown in Fig. 13 (Li et al. 2010). The heating and sorting (HS-RK) need the highest amount of energy. Compared to heat treatment processes, the improved mechanical processes (ERC, SAC, C&I) require less energy consumption. The CO_2 emission is relatively high, and more fines are produced, as shown in Fig. 14.

Table 3 Properties of RCA by thermo-mechanical treatments

Technology for RCA production	Temperature/°C	Heating time/h	Water absorption/%	AM/%
Natural Aggregate (Xuan et al. 2016)	–	–	0.80	–
Heating and sorting (Mulder et al. 2007)	700	–	–	2.0
Heating and rubbing (Shima et al. 2005)	300	0.67–1	0.80	–
Microwave and rubbing (Akbarnezhad et al. 2011)	140 Microwave 10 kW 2.45 GHz	0.033	2.8	24
Heating and milling (Prajapati et al. 2021)	500	1	0.95	4.0

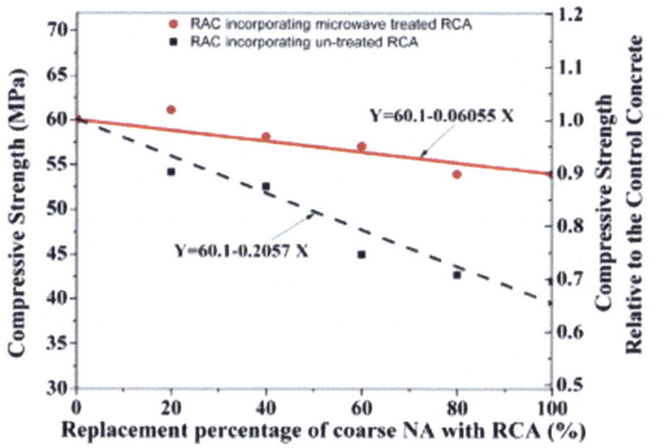

Fig. 12 Strength improvement of RAC prepared with the microwave treated RCA (Akbarnezhad et al. 2011)

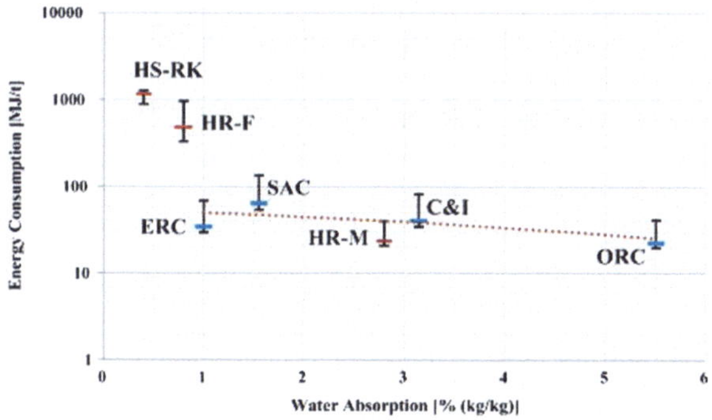

Fig. 13 Energy consumption of advanced crushing processes compared with the conventional crushing process (ORC) (Li et al. 2010)

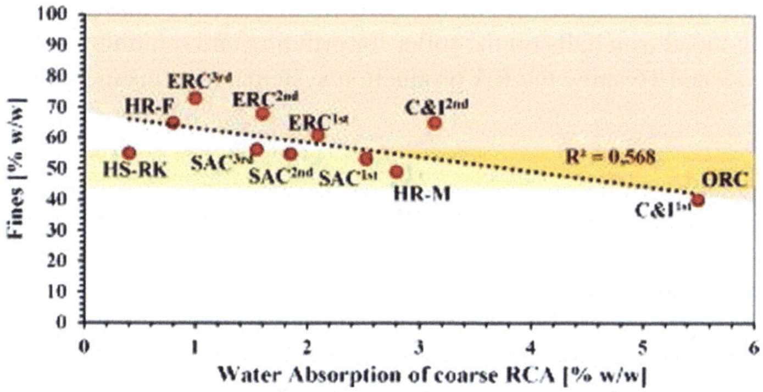

Fig. 14 Fines production of advanced crushing processes compared with conventional

3.3 Conclusions

High-quality RCA may be obtained by reducing adhered cement paste volume by improved mechanical processes and collaborative thermo-mechanical processes. However, the disadvantage of these techniques is the high energy input required and the corresponding increase in recycled fines from 40 to 100%. Choosing a proper mechanical crushing method of RCAs should be based on local conditions and requirements. Also, an integrated utilization strategy should be considered when recycling and reusing high-quality RCA and the increased generation of recycled fine.

4 Technologies to Remove AM from RA

RA includes hardened cement mortar from old concrete. The AM on RA has higher porosity, water absorption, and lower strength than NA. Therefore, two methods can be adopted to improve the physical properties of RA: removing or strengthening the AM (Al-Bayati et al. 2016; Shi et al. 2016).

4.1 Mechanical Treatment

In Japan, technologies for RA include an eccentric-shaft rotor and mechanical grinding (Sakai and Noguchi 2012). In the eccentric shaft rotor method, the crushed concrete blocks are transmitted downward between the high-speed eccentric rotating outer cylinder and the inner cylinder, and the concrete is separated into coarse aggregate and mortar by grinding. Mechanical grinding is a method of producing coarse aggregate and fine aggregate. The method is to divide the roller into small pieces with partition, and load iron balls on the roller for grinding and rotating the partitions.

Figures 15 and 16 show the RA production system using this method.

4.2 Thermo-mechanical Treatment

4.2.1 Heating and Scrubbing

By combining thermal and mechanical treatments to replace multi-mechanical abrading steps and produce a high-quality RA. Hydrated cement paste will be easy to remove due to the decomposition of paste and the decrease of mechanical strength at high temperature (Quattrone et al. 2014). Table 4 summarizes literature's data concerning the temperature range of hydrate production, losing most water and

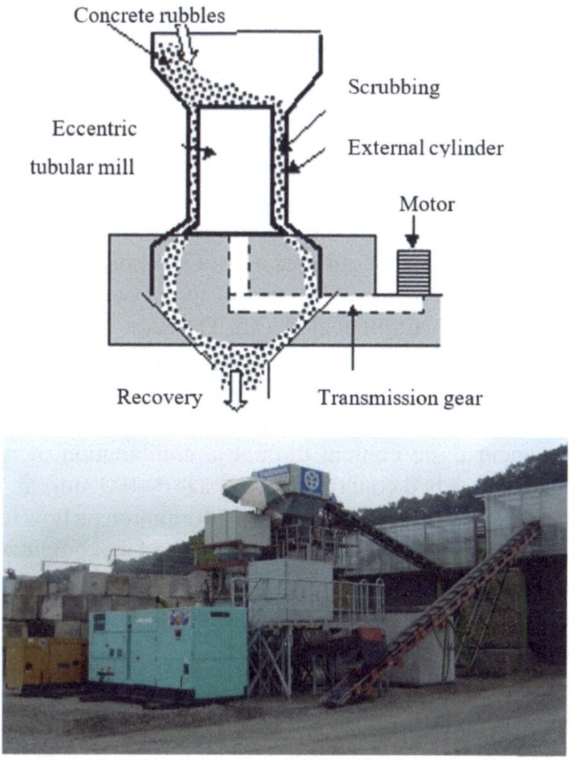

Fig. 15 RA production via eccentric-shaft rotor method (Sakai and Noguchi 2012)

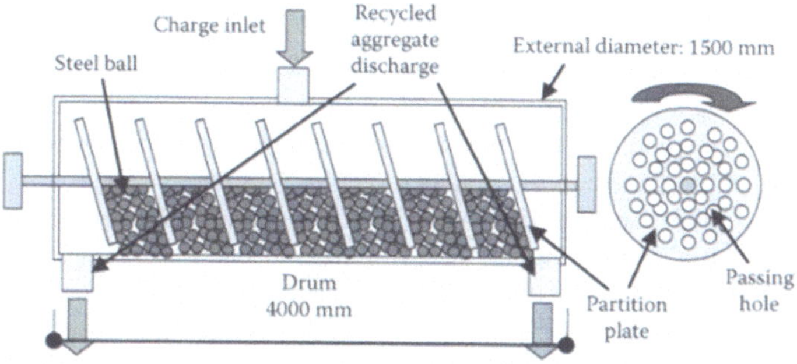

Fig. 16 Mechanical grinding method: screw mill type (Sakai and Noguchi 2012)

decomposing. Apart from physical–chemical processes connected with the liberation of water from cement paste, some cracking due to mechanical stress may also be observed. The increase in total porosity of cement paste in the temperature range 300–500 °C would be observed, which may be caused by the formation of micro-cracks (Piasta et al. 1984). Quattrone et al. (2014) summarized the literature's data concerning the thermo-mechanical treatments (Table 5).

Many studies have focused on the use of high temperature heat treatment (Gupta et al. 2012). The preliminary investigation on the scale of the laboratory showed that 450 kg of gravel (>4 mm), 350 kg of sand (<4 mm), 130 kg of cement stone (<150 μm) and 10 kg of steel bars could be recovered from 1 ton of concrete when the concrete rubble was treated at 700 °C (Larbi et al. 2000). The same authors also propose that heat treatment of RCA at about 800 °C can produce aggregates of suitable quality, whose properties are comparable to those of traditional river dredging aggregates.

Mulder, de Jong et al. 2007 described a method capable of producing RAC aggregate with a low cement paste content through a combination of mechanical and thermal treatment. The crushed concrete pieces (size $\leq$ 100 mm) are subject to 700 °C thermal treatment. Unfortunately, there is no information on how long the thermal treatment should last. However, the adhesive properties of the cement paste are broken down, and the constituents can be separated by simple sieving and air classifying (for powders). Only about 2% of the cement paste in volume is attached to these secondary aggregates. The same author also proposed that if the thermal treatment is carried out at a relatively high temperature greater than 750 °C, there will be almost no residual mortar on the aggregate. In this case, the gravel and sand shall be of the same quality as the original gravel and sand. However, the performance of recycled aggregate needs to be verified by experiments (Mulder et al. 2002).

Sui and Mueller (2012) investigated the effect of combining heat treatment at various temperatures (100–600 °C) for 30 min and mechanical treatment in a ball mill at different milling times on RCA properties. The obtained results showed that if the thermal treatment temperature of aggregate was set between 250 and 300 °C, followed by strong and sufficient mechanical treatment, the attached mortar on the surface of aggregate could be removed and the aggregate with the same performance as the original aggregate could be obtained.

Shima et al. (2005) used the "heating and rubbing method" to develop a technology that can produce high-quality aggregates from demolished concrete. Rough-crushed concrete rubble with size less than 40 mm shall be heated in a vertical kerosene furnace for 40–60 min at a temperature of about 300 °C, and then crushed and rubbed in a tube mill or a ball mill. It is necessary to remove most of the cement paste by

Table 4 Temperature range for losing most water and decomposing in hydrated production

	C-S-H and aluminates (Alarcon-Ruiz et al. 2005)	Portlandite (Alonso and Fernandez 2004)	Calcium carbonate (Taylor 1997)
Temperature range	70 ~ 400 °C	450 ~ 550 °C	Above 700 °C

Table 5 Thermo-mechanical treatments of coarse RCA (Quattrone et al. 2014)

Technology treatment	Heating temperature and heating system	Size of input material/mm	Production capacity/t/h	Heating time/h	Rubbing time/h		After treatment	Mass balance/%	AM/% kg/kg
						Density/g/cm^3	Water absorption/% kg/kg		
Heating and screening (Mulder et al. 2007)	700 °C Rotary kiln	≤ 100	–	–	–	–	–	45 coarse	2
								35 fine	
								13 powder	
								7 vapor	
Heating and rubbing (Shima et al. 2005; Kasai and Yoshio 2006)	300 °C vertical kerosene furnace	≤ 50	3–5	0.67–1	0.5–0.75	2.68^{OD}	0.8	35 coarse	–
								30 fine	
								35 powder	
Heating and rubbing (microwave) (Akbarnezhad and Ong 2010; Akbarnezhad et al. 2011)	140 °C Microwave 10 kW 2.45 GHz	≤ 30	0.65^{b}	0.033	0.05^{a}	2.46^{SD}(bulk)	2.8^{SD}	51 coarse	24^{SD}
								49 fine	
						2.43^{AD} (bulk)	3.4^{AD}	68 coarse	32^{AD}
								31 fine	

AD is air-dried; SD is saturated [a]. Test results after 100 revolutions with ten steel balls in the Los-Angeles equipment (31–33 rpm) estimate 193–181 s [b], Value estimated

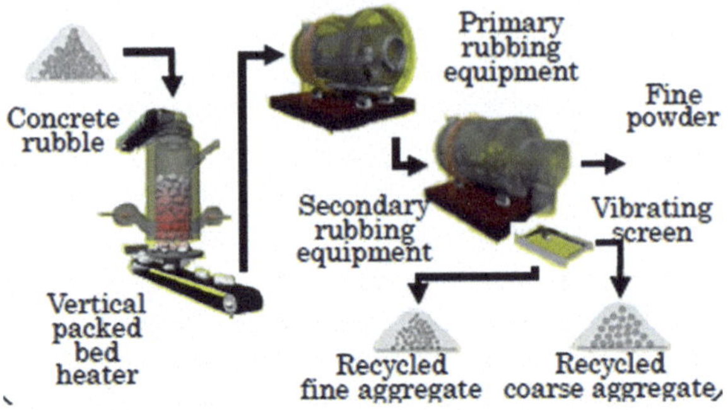

Fig. 17 Process flow of high-quality RA manufacturing process (Shima et al. 2005)

milling because of the temperature limitations of the process and the resulting partial decomposition. Due to heating, the cement paste will produce some cracks, and some of the AM will also be separated from the aggregate surface. Thus, the remaining mortar must be milled, transforming the porous cement paste into powders. Thus, it is an indispensable step to transform the porous cement paste into powder by milling. Figure 17 shows the process flow of the high-quality RA manufacturing process. The average aggregate recovery and fines generation by original concrete weight were 35% coarse aggregate, 30% fine aggregate, and 35% fines, respectively. On the basis of field testing with the portable plant developed in their research, a commercial portable plant with a capacity of 10 tons/hour was designed and constructed, which was used for two large-scale building reconstruction.

4.2.2 Use of Microwave (Selective Heating and Grinding)

Microwave heating technology is not new in building and construction applications. Industrial microwave heating technology applied to concrete materials has been investigated and developed (Benedetto and Calvi 2013). Several researchers have demonstrated its applications in the rapid curing of concrete, called "microwave decontamination of concrete" and the "microwave cutting and drilling" (Li et al. 1993; Leung and Pheeraphan 1995; Akbarnezhad and Ong 2010). The overall advantages of microwave heating can be summarized in the followings (Benedetto and Calvi 2013):

- Avoid air emissions or liquid pollutants;
- Speed heating;
- Easier to control and isolation of risk conditions;
- Thermal gradients, energy lost, prolonged heat up periods are minimal, almost negligible;
- Strict control in the automatic implementation of heating processes;
- Microwave penetration allows volumetric heating of samples;
- Direct action on polar bonds and entropy of the material to be heated.

Akbarnezhad et al. (2011) proposed a novel microwave-assisted technique, which can partially remove the mortar adhering to RCA particles and break up the lumps of mortar present in RCA to improve the quality of RCA. The technique generates high thermal stresses within the mortar by making use of the different electromagnetic properties of the adhering mortar and NAs, especially at the interface with the embedded NAs, to cause delamination. The lump of mortar also is broken up into smaller pieces because of the generated stresses. In their study, 2 kg (oven-dried weight) RCA samples were heated at the maximum power (10 kW) for one min. After heating, the samples were immediately immersed in 25 °C water to cool down. To investigate the capability of microwave-assisted RCA beneficiation technique to be used on an industrial scale, the authors simulated a deployable industrial system rather than the system configuration used in the experimental study (Fig. 18). A typical industrial microwave heating system may comprise a heating tunnel with a conveyor belt that can be easily integrated as part of the processes of a typical concrete recycling plant. A typical industrial microwave heating system consists of a heating tunnel, a microwave generator unit, a horn (antenna), and a series of microwave waveguides, where the heating tunnel has a conveyor belt that can be easily integrated as part of the process of a typical concrete recycling plant, the horn (Antenna) capable of spreading microwave power over a specific area, and the microwave waveguide is responsible for transferring the microwave power from the generator to the horn.

The same authors also compared the remaining mortar content, water absorption and particle density of RCA treated using the microwave-assisted RCA beneficiation technique obtained using other RCA beneficiation methods discussed. As can be seen in Table 6, among the single-stage processing methods, microwave heating resulted in significantly higher improvements in the properties of RCA.

Benedetto and Calvi (2013) conducted a pilot study analyzing microwave heating of aggregates and asphalt materials in the laboratory, comparing the data with conventional heating, and using numerical simulations to calibrate and validate the model

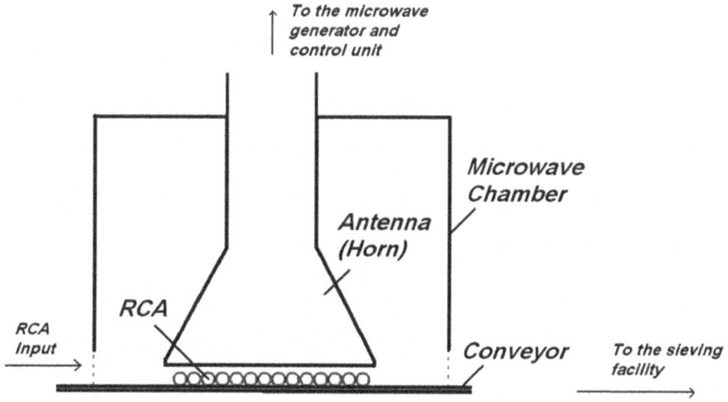

Fig. 18 The RCA beneficiation system considered for numerical simulation (Akbarnezhad 2011)

Table 6 Properties of RCA before and after treatment using various beneficiation techniques. (Akbarnezhad et al. 2011)

Beneficiation process		Process duration/h	Properties of RCA		
			24-h water absorption/%	Particle density (OD)/kg/m^3	Mortar content (%) by mass
Before beneficiation			4.2	2370	47
Single-stage processes					
Microwave heating	Pre-saturated RCA	~0.02	2.8	2460	24
Conventional heating	Air-dried RCA	~0.02	3.4	2430	32
	300 °C	2	4.1	2380	44
	500 °C	2	3.8	2390	41
Mechanical rubbing		~0.1	3.5	2410	34
Acid soaking	0.1 M sulfuric acid	24	4.1	2380	45
		120	4.1	2380	45
	0.5 M sulfuric acid	24	3.9	2390	41
		120	3.4	2420	33
	1 M sulfuric acid	24	3.5	2410	34
		120	1.6	2500	13
Combined processes					
Conventional heating and mechanical rubbing	300 °C	~2.1	3.3	2430	31
	500 °C	~2.1	2.1	2480	21
Microwave heating and mechanical rubbing	Pre-saturated RCA	~0.12	1.1	2550	7

for further application. The results are excellent, confirming several literature findings on the advantages of microwave heating compared to conventional heating, and validating a numerical simulation model that can be used to design prototypes of microwave heating devices.

Ménard et al. (2013) proposed that microwave heating or electric impulse could be used to weaken the mortar matrix and select crushing of concrete further, due to the generated mechanical stresses. Concrete samples manufactured in the laboratory were tested, which represent concrete waste from concrete-mixer trucks and concrete waste collected. The results showed that both of these techniques can weaken concrete samples and enhance the selective liberation of aggregates (that is, produce aggregates without cement paste) during crushing and grinding.

4.3 *Chemical Treatment: Acidic Treatment*

The hydration products of cement in the hardened paste can react with an acid solution and then be dissolved. Thus, the acidic solution can be used to remove the AM effectively and enhance the quality of RCA (Shi et al. 2016). Tam et al. (2007a) proposed and compared three pre-soaking treating methods for RCA. They used three different acids, which were hydrochloric acid (HCl), sulfuric acid (H_2SO_4) and phosphoric acid (H_3PO_4). After that, several authors demonstrated that pre-soaking RCA in an acidic environment could remove the AM from RCA and improve RCA's properties (Ismail and Ramli 2013; Purushothaman et al. 2015; Hasan et al. 2016). Acids used including hydrochloric acid (HCl), sulfuric acid (H_2SO_4), phosphoric acid (H_3PO_4), nitric acid ($HN\ O_3$) and acetic acid ($C_2H_4O_2$).

The whole procedure proposed by Tam et al. (2007a) are first to soak the RCA in an acidic environment at around 20 °C for 24 h and then wash with distilled water to remove the acidic solvents afterward. Before concrete mixing, 24 h water soaking of RCA is stipulated according to the specifications of the Buildings Department (2005). Figure 19 illustrates the presoak treatment procedures, while Table 7 shows the symbols used. Ismail and Ramli (2013) adapted this pre-soaking method from Tam et al. In their experiment, the first step is to add HCl to the RCA sample in a plastic container until the surface of RCA is covered. It was important to shake the container occasionally to make sure that acid in the degradation bond of mortar adhered to the original aggregate react more effectively. After immersion, the aggregates were washed with distilled water and drained, depending on the age of treatment. Most researchers adopted the procedure proposed by Tam et al. (2007a).

The main problem of the acidic treatment method is that it is not easy to determine the concentration of acid and the duration of immersion. Tam et al. (2007a) proposed that choosing the concentration of the acid solution to be 0.1 mol would remove the old cement mortar and avoid reducing the quality of the aggregate due to the suitable acid environment for the aggregate. Table 8 summarizes the properties of RCA before and after acid pre-soaking, and indicates that water absorption rates after the pre-treatments have been significantly reduced with improvements between

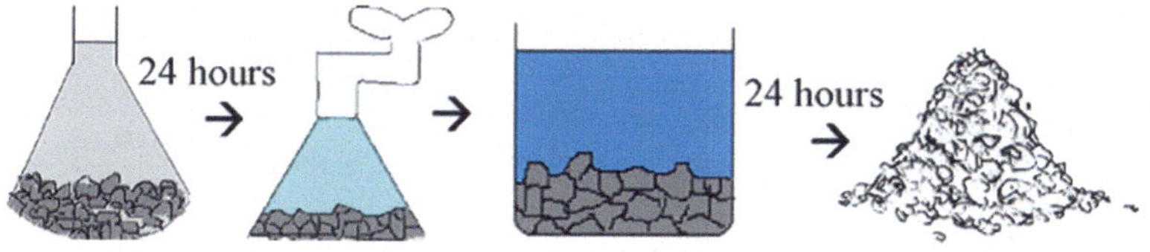

Fig. 19 Pre-soaking treatment procedures for RA (Tam et al. 2007a)

Table 7 Symbols used for representing various procedures and materials (Tam et al. 2007a)

Soaking Aggregate in Acidic Environment	Watering Aggregate with Water	Recycled Aggregate	Soaking Aggregate with Water

7.27 and 12.17%. Regarding the pH values, the pre-treated RA using ReMortarHCl, ReMortarH_2SO_4 and ReMortarH_3PO_4 are still within the alkaline group (above 8.5 pH). These show the small adversarial effect to the RA after the acidic pre-treatments. However, several authors pointed out that such a slight reduction in water absorption seems to suggest only marginal removal of mortar using the proposed acid treatment technique when considering the actual differences between the water absorptions of RCA and NA (Akbarnezhad et al. 2011; Akbarnezhad et al. 2013). Akbarnezhad et al. 2011 subsequently showed that the complete removal of mortar through acid treatment may be achieved using sulfuric acid at much higher concentrations (>2 molars) and for longer soaking durations (>5 days) than those proposed originally by Tam et al. However, the latter method proposed by Akbarnezhad et al. (2011) has the drawback of destroying NA (Table 9).

Additionally, Ismail and Ramli (2013) tested the effect of various concentrations (0.1, 0.5, and 0.8 M) of hydrochloric acid (HCl) with different immersion times (1, 3, and 7 days) on the treatment of the RCA to improve its performance. The results of the study showed that the amount of mortar loss was linearly related to the increase in acid concentration, while immersion time had no significant effect. Butler (2012) treated RCA with high concentration nitric acid (20% volume) at 85 °C for 2–3 h,

Table 8 Properties of RA before and after pre-soaking treatments (Tam et al. 2007a)

Properties of RA	Sizes of aggregate		Before pre-soaking treatment		After pre-soaking treatment
			ReMortar$_{HCl}$	ReMortar$_{H2SO4}$	ReMortar$_{H3PO4}$
Water absorption (%)	20 mm	1.65	1.45	1.48	1.53
	10 mm	2.63	2.31	2.37	2.41
Chloride content (%)	20 mm	0.0016	0.0025	0.0001	0.0001
	10 mm	0.0012	0.0056	0.0001	0.0001
Sulfate content (%)	20 mm	0.0025	0.0076	0.1090	0.0110
	10 mm	0.0025	0.0082	0.1040	0.0109
Value of pH	20 mm	10.46	9.07	8.95	8.55
	10 mm	11.63	9.34	9.35	9.33

Table 9 Types of surface improving agents (Tsujino et al. 2007)

Type	Oil (O)		Silane (S)	
Application	Release agent used in wooden form		A water-repellent agent with permeability to the concrete surface	
Main constituent	Mineral oil (Paraffin)	85–95%	Silicon analog	28–38%
	Emulsifying agent	1–5%	Emulsifying agent	Minute quantity
	Lanolin fatty acid salt	1–5%	Water	68–72%
State	Emulsion		Emulsion	

and observed that the outer layer of RCA was removed, but there were yellowish spots on the surface of RCA.

Al-Bayati 2016 investigated the effect of an acidic solution composed of hydrochloric acid (HCl) and Acetic acid ($C_2H_4O_2$) at a low concentration of 0.1 M. The results show that low-concentration acid treatment is an effective technique to improve the quality of RCA, and weak acid can more effectively reduce the effect of acid attack on RCA surface than strong acid.

Recently, pre-soaking in acidic solutions to remove the AM has attracted much research interest, and some researchers combined this method with other treatment methods to enhance the properties of RCA. Ismail and Ramli (2014) proposed a method to strengthen coarse RCA, which combines two surface treatment methods to improve the surface structure of RCA. In their study, the first step is treating coarse RCA with hydrochloric (HCl) acid at 0.5 mol (M) concentration. Then, their surface are coated with calcium metasilicate (CM) particles by impregnating them with CM solution. Hasan et al. (2016) also proposed a surface treatment method that combined pre-soaking in hydrochloric acid and coating with sodium metasilicate pentahydrate. The experimental results showed that, in general, using treated RA improved the beams' shear capacity slightly compared with natural and untreated RA.

5 Technologies to Strengthen AM of RA

5.1 *Surface Modification by Aggregate Coating*

Various surface modification treatment that can consolidate the adhering mortar layer, reduce its porosity, and improve the interfacial bond between RA and cement paste includes the use of surface-coating lime and pozzolanic materials such as silica fume (Shayan and Xu 2003; Katz 2004; Li et al. 2009; Kong et al. 2010; Shayan 2003; Spaeth and Tegguer 2013; Güneyisi et al. 2014), mineral oil (Tsujino et al. 2007; Santos et al. 2017), and polymeric material such as silane, siloxane, and polyvinyl alcohol (Tsujino et al. 2007; Spaeth and Tegguer 2013; Santos et al. 2017).

Shayan and Xu (2003) reported that 28-day strength values of 50–60 MPa could be obtained for concrete made using 100% coarse RA and 50% fine RA using silica fume. They studied various treatment methods, such as using concentrated sodium silicate solution or diluted sodium silicate solution along with different combinations of lime and silica fume and reported that the surface treatment of RA was most effective with coating with a coating slurry of silica fume. Concrete made using commercial RA treated with a slurry of silica fume or sodium silicate developed higher strength values than untreated RA in the concrete mixture containing the same amount of silica fume in the binder. Katz (2004) investigated the effect of silica fume impregnation of coarse RA on the compressive strength of RAC. Concretes at three strength levels (w/c = 0.35, 0.53, 0.77 with 28-day fc' = 27 MPa, 52 MPa, 67 MPa, respectively) served as base concrete, which was crushed using a mini jaw crusher after 28 days after strength testing to produce RA. The impregnation with a solution of silica fume was intended to add a thin layer of silica fume particles over the surface of coarse RA (RA was dried in an oven for 48 h, cooled to room temperature, soaked in a silica fume solution for 24 h, and then dried again in the oven to ensure proper penetration of silica fume particles into the surface of RA). The amount of silica fume impregnated into the surface of RA was estimated to be 0.5–0.8% of RA by wt. The silica fume treatment improved the compressive strength of new concretes made of treated coarse RA by 13–16% after 28 days over reference concrete with untreated RA.

Tsujino et al. (2007) used two different water-repelling surface treatments using paraffin oil- and silane-type agents with properties shown in Table 3, which were applied on coarse RA of two different qualities, middle quality (BSG_SSD = 2.47, water absorption = 4.8%) and low-quality (BSG-SSD = 2.46, water absorption = 5.48%). Four repetitions were required to obtain a stable coating to form an alkali metallic salt film due to the oil-type agent and a water-repellent coating due to the silane-type agent, respectively. There was concern that peeling-off and variations would reduce strength during surface modification treatments. The water absorption decreased to about 3.5% for the oil-type agent and about 1% for silane-type agents for middle- and low-quality RA. Different concrete mixes were made using OPC, middle- and low-quality coarse RA, and river sand with two different water-cement ratios of 0.6 and 0.4. The compressive strength of concretes using an oil-type agent reached a level similar to ordinary concrete depending on the water-cement ratio, as shown in Fig. 20. By contrast, considerable strength reduction was observed in all specimens treated with the silane-type agent, due to significantly weakened bonding properties. Three specimens were tested for each test level, and few quality control problems, i.e., no significant strength variations, were observed.

Santos et al. (2017) investigated the porosity of RA treated by paraffin wax and silane solution to improve RA's characteristics. Mercury Intrusion Porosimetry (MIP) was used to determine total aggregate's porosity and pore size distribution, with results shown in Table 10 and Fig. 21. The porosities of RCAs and recycled brick aggregates (RBAs) were 9 and 32 times higher than NAs, respectively. Silane treatment did not significantly reduce the porosity of RA: 33% and 6% for RCA and

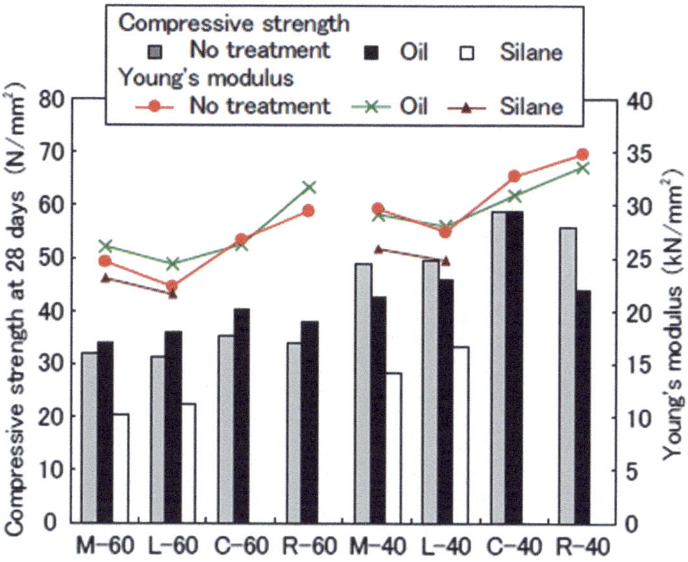

Fig. 20 Test results of compressive strength and Young's modulus (Tsujino et al. 2007): M, L, C, R—middle quality RA, low-quality RA, C—crushed stone, R—river gravel; 60, 40—w/c ratio of 60% and 40%

RBA, respectively. Paraffin treatment reduced almost 70% of the porosity of RA (Table 11).

Güneyisi et al. (2014) investigated the aggregate surface treatment method using water-soluble water–glass ($Na_2O.nSiO_2$—sodium silicate) dispersion in self-compacting concrete (SCC) with coarse RA. The coarse RA was produced from concrete cubic samples (150 × 150 × 150 mm) with 20 MPa strength using a jaw crusher and a secondary cone crusher. The RA was submerged in a water–glass for 30 min and then dried in an oven for 1 h. SCC mixes were produced with a water-binder ratio of 0.38. SEM analysis of hardened concrete demonstrated that the

Table 10 Porosity and total pore area of non-treated and treated RAs (Santos et al. 2017)

Aggregate ID	Porosity/% ml/ml	Total pore area/m^2/g
NA	1.36	0.07
RCA	9.54	3.58
RCA-Si	6.65	2.28
RCA-Pa	2.99	5.02
RBA	32.12	15.77
RBA-Si	30.98	14.93
RBA-Pa	8.84	6.94

Note RCA—Recycled coarse aggregate, RBA—recycled ceramic aggregate, Si—silane, Pa—paraffin

Table 11 Physical properties of PVA impregnated RA (Kou and Poon 2010)

Particle size (mm)		Density/kg/m^3		Water absorption/%		10% fine value/kN
PVA concentration (%)		20	10	20	10	14
6	Oven-dried	2431	2353	5.89	6.12	126
	Air-dried	2434	2356	4.25	5.84	132
8	Oven-dried	2442	2361	4.63	4.83	130
	Air-dried	2453	2367	2.98	3.17	137
10	Oven-dried	2461	2372	2.39	4.32	154
	Air-dried	2473	2385	1.62	2.38	158
12	Oven-dried	2468	2378	2.35	4.12	155
	Air-dried	2482	2389	1.57	2.27	160

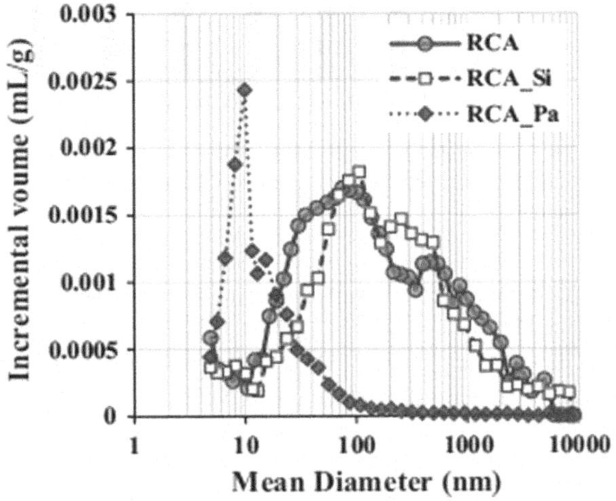

Fig. 21 Pore size distributions of RA

pre-soaking treatment of water–glass created the aggregate surface with an outer, more robust coating, which led to a decrease in RA's water absorption to about 2% as determined by water absorption results. RA treatment with water–glass also produced improved compressive and splitting tensile strength results than SCC with untreated RA.

Kou and Poon (2010) studied improving RA properties by impregnation with polyvinyl alcohol (PVA), a water-soluble polymer was used as a cement modifier (Kim et al. 1999). RA was soaked in 6, 8, 10, and 12% PVA solutions. Density, crushing value (in terms of 10% fine value), and water absorption of PVA impregnated RA were determined with results in Table 12, showing that RA properties improve with PVA impregnation. The density and 10% fine value of PVA treated RA increase with an increase in the concentration of PVA solution while water absorption significantly decreases. For the concrete mixtures, the 10% PVA treated RA were

Table 12 Physical properties of NA, RA, and 10% PVA impregnated (Kou and Poon 2010)

Property	Particle size	Aggregate type			
		Natural granite	RAC	PVA-modified (oven-dried)	PVA-modified (air-dried)
Density (kg/m^3)	20	2662	2423	2466	2472
	10	2583	2356	2378	2385
Water absorption (%)	20	0.68	6.23	2.39	1.62
	10	0.87	7.76	4.32	2.38
10% fine value (kN)	14	168	120	154	158

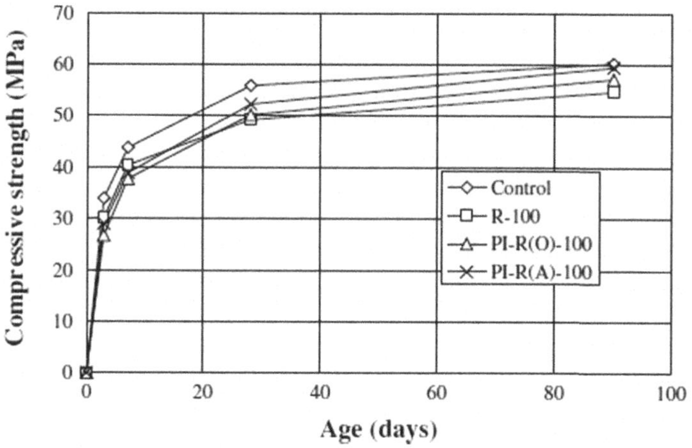

Fig. 22 Development of compressive strength of concrete (Kou and Poon 2010)

used oven-dried (PVA treated RA were dried at 60 °C for 24 h) and air-dried conditions (PVA treated RA were dried at 23 °C and 65% R.H. for 2 days). However, the compressive strength of concrete made using PVA impregnated RA was lower than that of the control concrete after 28 days. After 90 days, the compressive strength of concrete made with air-dried PVA impregnated RA was similar to the control concrete, as shown in Fig. 22.

5.2 Carbonation

Carbonation of Portland cement-based materials is the chemical reaction of calcium hydroxide and calcium silicate hydrate with carbon dioxide resulting in calcium carbonate and water, which takes place in the pores of concrete (Neville 2011). Calcium hydroxide forms calcium carbonate by reacting with CO_2 as shown in

Eqs. (1, 2). $CaCO_3$ occupies a greater volume than $Ca(OH)_2$, replacing the porosity of carbonated concrete. It is believed that CO_2 also reacts with C-S-H inducing a decrease in its C/S ratio and a concomitant loss of water, as shown in Eqs. (3, 4).

$$Ca(OH)_2 + CO_2 \rightarrow CaCO_3 + H_2O \tag{3}$$

$$C - S - H + CO_2 \rightarrow C - S - H + CaCO_3 + H_2O \tag{4}$$

The highest rate of carbonation occurs at an R.H. between 50 and 70%. Since the pores are almost filled with water, CO_2 does not penetrate well in the paste, so carbonation is low under high humidity conditions. However, under low humidity conditions, carbonation is also reduced due to the lack of water film (Mindess et al. 2003). Carbonation leads to a decrease in total porosity, while the decrease of porosity increases with increasing CO_2 concentrations (Leemann and Moro 2017). There are two major consequences of carbonation: (1) the drop of pH (the drop in hydroxide concentration in the pore solution, which destroys the passive condition of embedded reinforcement), (2) change in effective permeability due to the volume changes and microcracking caused by the chemical reactions (Johannesson and Utgenannt 2001). Recently accelerated carbonation of RA was studied for enhancement of RA properties (Kou et al. 2014; Zhan et al. 2014; Silva et al. 2015; Shi et al. 2016; Tam et al. 2016; Xiao et al. 2016). In a study by Zhan et al. 2014, hardened parent concretes with varying water-cement ratios were crushed after 90 days to produce RA with different particle sizes in the laboratory. RA with particle size ranging from 5–20 mm was produced, and the porosity of RA was determined. The CO_2 pressure in the chamber was controlled at 10 kPa while temperature and relative humidity ranged between 15–30 °C and 30%–70%, respectively. The density of RA increased while water absorption and porosity of RA were reduced after carbonation, as shown in Table 13. Factors influencing the accelerated carbonation process in curing time, particle size, and moisture contents of RA were investigated. RA with particle size ranges of 5–10 mm, 10–14 mm, and 14–20 mm was used to investigate particle size's effect on carbonation. The carbonation percentage was up to 56% for RA with 5–10 mm and was only 7% for 14–20 mm. RA with smaller particle size and a larger specific surface took up more CO_2 during the CO_2 curing process. RA with higher water absorption values attained a higher mass gain ratio and carbonation percentage. For RA of size 10–14 mm with 0.5–4 h, CO_2 curing time resulted in a higher percentage of carbonation with increased curing time: 25% after 2 h, 28.5% after 4 h. Three different moisture conditions of RA were tested: RA was (1) dried in an oven at 105 °C for 2 h, (2) stockpiled without any treatment, (3) soaked in water for 2 h with resulting moisture condition of 0.08%, 3.37%, and 5.03%, respectively. The stockpiled RA with a moisture content of 3.37% attained the highest carbonation percentage, as shown in Fig. 23. Various researchers used different CO_2 concentration (20, 100%), CO_2 chamber pressure (10, 75, 150, 500 kPa), and CO_2 cure duration (0.5–6, 12, 24, 48, 72 h). Xuan et al. 2016 studied carbonation of 5–20 mm coarse RA using chamber pressure of 0.1 bar and 5 bar, respectively, for 24 h (chamber temperature

= 25 °C, R.H. = 50%) and reported up to 16.7% reduction of water absorption, 4% increase in 10% fine value, and 25.9% reduction of crushing value. Carbonation pressure used had a lesser impact when a longer treatment duration (i.e., 24 h) was used. Zhang et al. (2015a, b, c) used CO_2 concentration of 20% and chamber temperature of 20 °C and R.H. of 60% and reported that the apparent density was increased by 4.7–5.6%, water absorption and crushing value decreased by 22.6%–28.3% and 7.6%–9.6%, respectively. The carbonation rate and strength development would increase with the increase of CO_2 concentration and pressure, but when the pressure was greater than 500 kPa, the CO_2 curing degree and compressive strength had no obvious differences.

It is noted that carbonation of RA can be an environmentally friendly method as it consumes part of CO_2 generated by calcination during cement manufacturing. This should be especially true for carbonated fine RA as CO_2 uptake is more effective with RA with smaller particle size and a larger specific surface.

Table 13 Physical properties of RA before and after CO_2 curing (Zhan et al. 2014)

	Apparent density/kg/m^3		Water absorption/%		Porosity/%	
	Mean	Std. dev.	Mean	Std. dev.	Mean	Std. dev.
(a) Before CO_2 curing						
RCA30	2639	3.7	6.58	0.6	15.66	1.3
RCA45	2623	7.1	5.98	0.2	14.30	0.5
RCA60	2620	4.8	5.95	0.2	14.23	0.4
RCA80	2602	13.7	5.81	0.3	13.84	0.8
(b) After CO_2 curing						
RCA30	2619	3.45	5.25	0.42	12.68	0.9
RCA45	2606	7.71	4.71	0.19	11.65	0.7
RCA60	2670	3.77	4.61	0.33	11.18	0.4
RCA80	2661	5.66	4.41	0.11	10.94	0.3

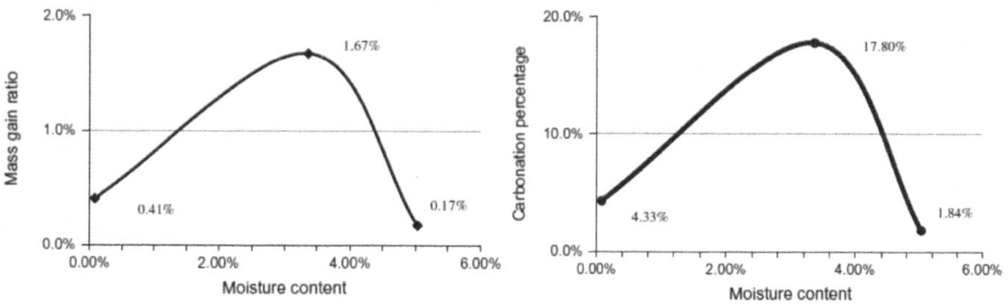

(a) Mass gain (b) Carbonation percentage

Fig. 23 Mass gain and carbonation percentage of RA versus moisture content of RA (Zhan et al. 2014)

5.3 Calcium Carbonate Bio-deposition

Calcium carbonate bio-deposition or microbial carbonate deposition (MCP) refers to the technique of using a microorganism, such as bacteria, to induce calcium carbonate crystal precipitation. The bacterially induced $CaCO_3$ precipitated on the surface and in the pores of RA can work as a barrier for the penetration of water, and hence the water absorption of the aggregates can be decreased while weight is increased. MCP use on RA has been studied by Grabiec et al. (2012), Wang et al. (2017).

Grabiec et al. (2012) first investigated and confirmed the feasibility of modifying the surface of RA by calcium carbonate bio-deposition (Fig. 24). They used Bacillus pasteurensis, and calcium chloride was used to precipitate calcium carbonate. The medium used was composed of beef extract, peptone and urea. Sporosarcina pasteurii bacteria can attract Ca ions (Ca^{2+}), which react with carbonate ions (CO_3^{2-}) derived from the hydrolysis of urea ($CO(NH_2)_2$). At the same time, ammonia ion (NH^{4+}) increases the pH value of the surrounding medium and the precipitation efficiency of calcite, as shown in Eqs. (5–7). This technique resulted in a reduction in the water absorption of RA. The observation confirmed that there were calcium carbonate crystals covered with aggregate particles under SEM.

$$\text{Sp. Cell} + Ca^{2+} \rightarrow \text{Sp.Cell-}Ca^{2+} \tag{5}$$

$$CO(NH_2)_2 \rightarrow 2NH^{4+} + CO_3^{2-} \tag{6}$$

$$\text{Sp. Cell} \; - Ca^{2+} + CO_3^{2-} \rightarrow \text{Sp. Cell} \; - CaCO_3 \tag{7}$$

The microbial calcium precipitation (MCP) on recycled aggregate (RA) treated with Sporosarcina pasteurii was influenced by pH (7.0–10.5), temperature (25°C and 35°C), calcium concentration (5.6–50.4 g/L of $CaCl_2$), and bacterial cell density (10^6–10^8 cells/mL). The highest MCP yield was observed at pH 9.5, with further

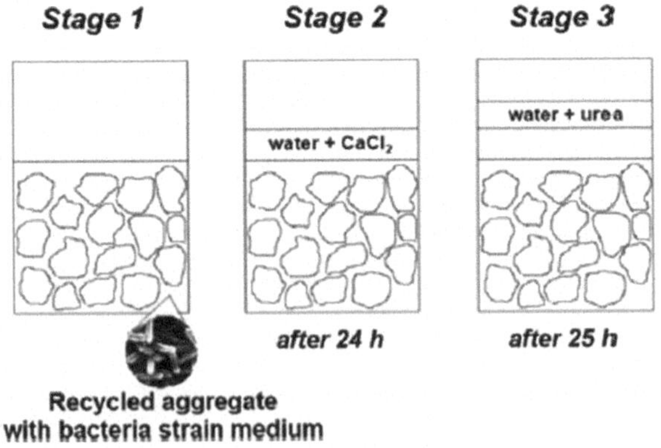

Fig. 24 Procedure of bio-deposition experiment (Grabiec et al. 2012)

enhancement noted under elevated temperatures, higher bacterial concentrations, and increased calcium availability. The feasibility of RA surface modification via MCP was confirmed by the increased weight and decreased water absorption of the treated aggregates. Wang et al. (2017) used two different types of RA, i.e., RAC aggregates (CA) and mixed aggregates (MA), which were subjected to several different MCP treatment schemes such as a spraying and three different immersion procedures. With the best method (two times immersion treatment), the weight increase (2% for CA and 2.5% for MA) and a decrease in the water absorption (1% for CA and 2% for MA) were determined. The biogenic $CaCO_3$ had an excellent cohesion and strong bond with the aggregate surface, as evidenced by the ultrasonic pulse test result and Los-Angeles abrasion resistance coefficient (30% for treated CA and 40% for treated MA).

6 Testing Methods

6.1 Particle Density and Water Absorption

According to ASTM C127-15 and GBT1743.2, the primary method to test water absorption of coarse and fine aggregate is the 24 h water absorption method.

In this method, the aggregate sample is immersed in water for 24 ± 4 h to substantially fill the pores. It is then removed from the water, the water is dried from the particle surface, and the mass is determined. Subsequently, the water displacement method is used to determine the volume of the sample. Finally, the sample is oven-dried and its mass determined. Using the mass values thus obtained and the formulae in this test method, the relative density (specific gravity) and absorption can be calculated.

$$\text{Absorption, \%} = [(B - A)/A]^{*}100 \tag{8}$$

where, A = mass of oven-dry test sample in air, g,

B = mass of the saturated surface-dry test sample in air, g.

According to BS 812-2-1995, three main methods of aggregates are described in detail, one is the wire basket method for aggregates larger than 10 mm, one is the gas jar method for aggregates 40–5 mm, and one is the pycnometer or gas jar method for aggregates of nominal size 10 mm and below.

A sample of no less than 2 kg of aggregate shall be tested for the wire basket method. The test procedure is divided into five steps.

Firstly, the prepared test sample is placed in the wire basket and immersed in the water with at least 50 mm of water above the top of the basket. Secondly, after immediately immersion, entrapped air was removed from the sample by raising the gondola 25 mm above the base of the tank and lowering it 25 times at a rate of approximately once per second. During this operation, the basket and aggregate should remain completely submerged for 24 h ± 0.5 h. If they must be transferred to another jar for weighing, jolt it 25 times as above in the new jar (mass B) before weighing.

Then take the basket and aggregate out of the water, let them drain for a few minutes, then gently pour the aggregate in the basket onto a dry cloth, then put the empty basket back into the water, jolt it 25 times and weigh it in the water (mass C). Use a cloth to gently dry the surface of the aggregate placed on the dry cloth, and when the first dry cloth does not remove more moisture, transfer it to the second dry cloth. Then unfold the second cloth to a depth not exceeding that of a stone, exposing it to the atmosphere, avoiding direct sunlight or any other heat source, until all visible water films are removed, but the aggregate still has a damp appearance. Weigh the aggregate (mass A). The final step is to place the aggregate in an oven in a shallow tray at a temperature of 105 °C ± 5° C and keep it at this temperature for 24 h ± 0.5 h. It is then removed from the oven, cooled in an airtight container, and weighed (mass D). The particle density on an oven-dried basis (ρ_{rd} in kg/m^3) is calculated from the formula:

$$\rho_{\mathrm{rd}} = \frac{\mathrm{D}}{\mathrm{A} - (\mathrm{B} - \mathrm{C})} \tag{9}$$

The particle density on a saturated and surface-dried basis (ρ_{ssd} in kg/m^3) is calculated from the formula:

$$\rho_{\mathrm{ssd}} = \frac{\mathrm{A}}{\mathrm{A} - (\mathrm{B} - \mathrm{C})} \tag{10}$$

The apparent density ($\rho_{apparent}$ in kg/m^3) is calculated from the formula:

$$\rho_{\mathrm{apparent}} = \frac{\mathrm{D}}{\mathrm{D} - (\mathrm{B} - \mathrm{C})} \tag{11}$$

The water absorption (WA_{24} as % of dry mass) is calculated from the formula:

$$\mathrm{WA}_{24} = \frac{100(\mathrm{A} - \mathrm{D})}{\mathrm{D}} \tag{12}$$

For the gas jar method, a sample of about 1 kg of the aggregate shall be used. Immerse the prepared test sample in water in the glass vessel. Shortly after soaking and again at the end of the soaking period, gently stir to remove air or air bubbles from the surface of the aggregate. This can be achieved by rapidly rotating the container clockwise and counterclockwise between the operator's hands. Overfill the container by adding water and sliding plane-ground glass to find openings and ensure there is no air in the container. Then dry the vessel on the outside and weigh it (mass B). The container is then emptied to allow the aggregate to drain while water is added to filling the container, sliding the glass disc into place as previous. Then dry the container on the outside and weigh it (mass C). The temperature difference between the water in the container during the first and second weighing shall not exceed 2 °C. Place the aggregate on a dry cloth and use it to gently dry the surface of the aggregate, transferring it to a second dry cloth when the first one does not remove any more moisture. It was then spread on a second cloth, no more than a stone deep, and exposed to the atmosphere, away from direct sunlight or any other heat source, until all visible water films were removed, but the aggregate still had a moist appearance. Weigh the aggregate (mass A). The aggregate was placed in a shallow tray in a 105 °C ± 5 °C oven for 24 h ± 0.5 h. It was then cooled in an air-tight container and weighed (mass D).The particle density on an oven-dried basis (ρ_{rd} in kg/m^3) is calculated from the formula:

$$\rho_{rd} = \frac{D}{A - (B - C)} \quad (13)$$

The particle density on a saturated and surface-dried basis (ρ_{ssd} in kg/m^3) is calculated from the formula:

$$\rho_{ssd} = \frac{A}{A - (B - C)} \quad (14)$$

The apparent density ($\rho_{apparent}$ in kg/m^3) is calculated from the formula:

$$\rho_{apparent} = \frac{D}{D - (B - C)} \quad (15)$$

The water absorption (WA_{24} as % of dry mass) is calculated from the formula:

$$WA_{24} = \frac{100(A - D)}{D} \quad (16)$$

For materials with nominal sizes of 10 to 5mm inclusive of 10 to 5mm, a sample of approximately 1 kg should be used, and if less than 5 mm, a sample of approximately 500 g should be used for the pycnometer or gas jar method. Aggregates that have been artificially heated shall not usually be used. However, according to BS 1097–6 2000, the pycnometer method shall be used for aggregate particles between 4 mm and 31.5 mm. A sampling of the aggregate shall be following EN 932–1 1997, and reduction shall be following EN 932–2 1999. The mass of the aggregate test portion shall not be less than that shown in Table 14.

Test procedures of this method are similar according to the two standards mentioned, BS 1097–6 2000 is taken as the primary reference here. Figure 25 shows the example of the pycnometer used in BS 1097–6 2000.

Immerse the prepared test portion in water at (22 ± 3) °C in the pycnometer and remove entrapped air by gently rolling and jolting the pycnometer in a tipped position. Stand the pycnometer in the water bath and keep the test portion at a temperature of (22 ± 3) °C for (24 ± 0.5) h. At the end of the soaking period, take the pycnometer

Table 14 Sampling of aggregates

Maximum size of aggregates (mm)	The minimum mass of the test portion (kg)
31.5	5
16	2
8	1

Note For other sizes, the minimum mass of the test portion may be interpolated from the masses given above

Fig. 25 Example of pycnometer. 1-Glass funnel, 2-Mark, 3-Ground section to fit the wide-neck flat bottom flask, 4-Wide-neck flat bottom flask

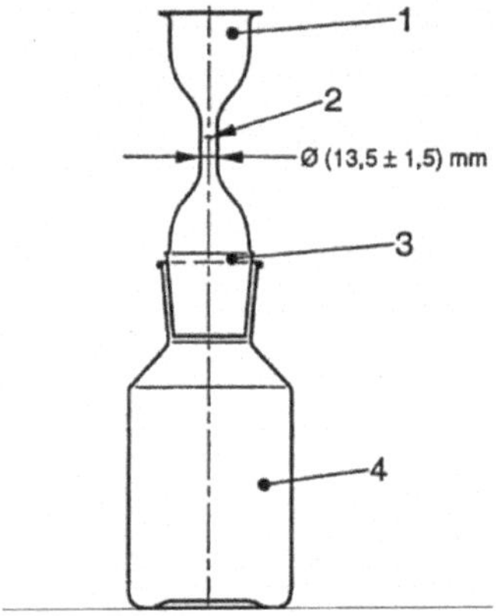

1-Glass funnel, 2-Mark, 3-Ground section to fit the wide-neck flat bottom flask, 4-Wide-neck flat bottom flask

from the water bath and remove any remaining entrapped air by gentle rolling and jolting.

Note 1 Entrapped air may also be removed by applying a vacuum.

Overfill the pycnometer by adding water and place the cover on top without trapping air in the vessel. Then dry the pycnometer on the outside and weigh it (M_2). Record the temperature of the water.

Remove the aggregate from the water and allow it to drain for a few minutes. Refill the pycnometer with water and place the cover in position as before. Then dry the pycnometer on the outside and weigh it (M_3). Record the temperature of the water.

The difference in the temperature of the water in the pycnometer during the M_2 and M_3 weighings shall not exceed 2°C.

Note 2 Instead of measuring the pycnometer volume at each test, it can be pre-calibrated. In that case, the pycnometer should be tempered in a thermostatic bath at the calibration temperature $\pm$ 0.5 °C.

Transfer the drained test portion onto one of the dry clothes. Lightly surface-dry the aggregate placed upon it and when the first cloth cannot remove more water, transfer it to the second absorbent cloth. Spread the aggregate on the second plate no more than a stone deep and expose it to the atmosphere, away from direct sunlight or any other heat source, until all visible film of water is removed but the aggregate still has a moist appearance. The saturated and surface-dry test portion is transferred to a tray and weigh the aggregate (M_1). The aggregate is dried in a ventilated oven at 110 ± 5 °C until its mass has been constant (M_4). All masses are recorded with an accuracy of 0.1% or higher of the mass of the test section.

Calculation:

Particle density on an oven-dried basis

$$\rho_{rd} = \frac{M_4}{M_1 - (M_2 - M_3)} \tag{17}$$

Particle density on a saturated and surface-dried basis

$$\rho_{ssd} = \frac{M_1}{M_1 - (M_2 - M_3)} \tag{18}$$

Water absorption after immersion for 24 h

$$WA_{24} = \frac{100*(M_1 - M_4)}{M_4} \tag{19}$$

6.2 *Porosity*

Mercury Intrusion Porosimetry can determine open porosity and particle density (Pepe et al. 2016).

According to ASTM D4044 2010, the MIP method can be described as follows: the volume of the invading pores is determined by measuring the volume of mercury entering the pore under different pressures. Single measurement includes continuous or gradual increase of pressure, and record the intrusion volume measured at different pressures.

In this method, the sample from which the test specimen is drawn shall represent the soil or rock. The test specimen shall be as large as practicable considering the test apparatus. A minimum of three specimens should be run on samples from similar material, and the pore-measuring capacity of the penetrometer limits specimen size. The small specimen size may prevent the measurement of porosity represented by relatively large cracks and fissures in the material.

6.3 Attached Mortar Content

The main factor affecting the density, porosity and water absorption of RCA is residual AM on aggregate. The density of RCA is generally lower than that of NA because the density of AM is lower than that of underlying rocks. At present, there is no standard method prescribed to determine the content of RCA mortar. There are three methods to determine the content of RCA attached mortar described in the literature.

A. Acid solution soaking treatment (Nagataki et al. 2004)

Akbarnezhad et al. (2011) proposed a method to determine the content of RCA mortar. In order to remove mortar from type of RCA thoroughly tested, they recommend that it needs to be soaked in sulfuric acid concentrations of at least 2 M for at least 5 days. Therefore, they measured the mortar content of RCA batches which was treated with various method in the study by using this procedure.

The procedure was described as the following:

After immersion in acid, the coarse aggregates were collected by washing the sample on a 4mm sieve. Then calculated the mortar content with the weight loss rate of the collected coarse aggregate.

$$M_c = \frac{W_i - W_r}{W_i} * 100\% \tag{20}$$

where M_c is the mortar content, W_i is the initial oven-dried weight of the RCA sample and W_r is the oven-dried weight of the RCA samples retained on the 4 mm sieve after soaking in the acid.

B. Production of new concrete (Hansen and Narud 1983)

This method produces a new concrete made with RA and new colored cement to easily detect the mortar surface in a sliced specimen utilizing the different colors between NA and new mortar.

C. Salicylic acid and methanol dissolution method to measure the hardened cement paste content (Zhao et al. 2013, 2018).

The experimental protocol used is as follows:

(a) A representative sample is dried at 105 °C, then ground until passing 0.2 mm sieve;
(b) 0.5 g of the dried representative sample is immersed in a solution of 14 g of salicylic acid in 80 ml of methanol and stirred for one h;
(c) The solid fraction is filtered on a glass filter (Pyrex N°4, pores: 10–16 μm) and washed four times using methanol (2–3 mm high on top of filter);
(d) The solid residue is dried in the oven at 70 °C for 30 min;
(e) The hardened cement paste content is then calculated as follows:

Table 15 Summary from previous researchers on AM content for RCA (Butler 2012)

Researcher	Method	AM content (%)
Nagataki et al. (2004)	Hydrochloric acid dissolution	30.2–55.0
Gokce et al. (2004)	Hydrochloric acid dissolution	32.4–55.7
Fathifazl (2008)	Freeze–thaw + sulfate attack	23–41
Movassaghi (2006)	Nitric acid dissolution	37.6–62.6
Liu (2011)	Image analysis	42.2–46.5
Juan and Gutiérrez (2009)	Thermal treatment	40–55
Hansen and Narud (1983)	Linear traverse method	41–43

$$CPC(\%) = \frac{M_1 - M_2}{M_1} \times 100 \tag{21}$$

where M_1 is the mass of dried material before dissolution and M_2 is the mass of dried filtrate.

D. Thermal treatment

This method consists of several cycles of soaking in water and heating the aggregate (Juan and Gutiérrez 2009). Juan and Gutiérrez (2009) compared results obtained with these three methods from literature and found the thermal method is better as it could be used for all kinds of aggregates (including limestone) and is easier to do.

In addition, computer tomography (CT) and image analysis methods also could be used to determine the amounts of attached mortar according to Pepe et al. (2016) and ASTME1570. The same author presented Table 15 summarizing the findings from previous researchers on the attached mortar content of coarse RCAs given by Butler (2012).

6.4 Crushing Value

Regarding the crushing value of aggregates, different standards described similar testing methods. (BS 812–110 1990; IS 2386 1997; GBT14685 2011 etc.). BS 812–110 1990 is taken as the main reference here (Tables 16 and 17). The testing procedure was described as follows:

Place the cylinder of the testing device on the bottom plate, and add test samples in three layers of approximately equal depth, each layer being subjected to 25 strokes from the tamping rod evenly distributed on the surface of the layer and dropping from the height of about 50 mm above the aggregate surface. The aggregate surface shall be leveled carefully and insert the plunger so that it rests horizontally on this surface. Make sure that the plunger does not block the cylinder.

Table 16 Principal dimensions of cylinder and plunger apparatus

Component	Dimensions	Nominal 150 mm internal diameter of the cylinder	Nominal 75 mm internal diameter of the cylinder (see Appendix)
Cylinder	Internal diameter, A	154 ± 0.5	78.0 ± 0.5
	Internal depth, B	125–140	70.0–85.0
	Minimum wall thickness, C	16.0	8.0
Plunger	Diameter of the piston, D	152 ± 0.5	76.0 ± 0.5
	Diameter of stem, E	$> 95 - \leq$ D	$> 45.0 - \leq$ D
	The overall length of piston plus stem, F	100 to 115	60.0 to 80.0
	Minimum depth of piston, G	Not less than 25.0	Not less than 19.0
	Diameter of hole, H	20.0 ± 0.1	10.0 ± 0.1
Baseplate	Minimum thickness, I	10	10
	Length of each side of the square, J	200–230	110–115

Table 17 Guide to the minimum mass of test portions required to obtain a suitable mass of material to determine the aggregate crushing value

Grading of the aggregate	The minimum mass of the test portion
mm	kg
AU-in aggregate 40 max. size	60
AU-in aggregate 20 max. size	45
Graded aggregate 40 to 5	40
Graded aggregate 20 to 5	25
Graded aggregate 14 to 5	15

Place the above prepared sample and plunger between the platens of the testing apparatus, and load as evenly as possible (see Note), so as to achieve the required 400 KN force within 10 min ± 30 s.

Note When there is a significant deformation during the early stages of the test, it may not be possible to maintain the loading rate's required loading rate and variations, especially at the beginning of the test. These changes should be kept to a minimum, with the primary objective being to complete the test in a total time of 10 min ± 30 s.

Load is released and crushed material is removed by holding the cylinder to a clean tray of known mass and striking the outside of the cylinder with a rubber mallet until the particles are sufficiently perturbed to allow the sample mass to fall freely on the tray.

Note that if this fails to remove the compacted aggregate, other methods may be used but not to cause further crushing of the particles.

Calculation of aggregate crushing value (ACV):

$$\mathrm{ACV} = \left(\frac{\mathrm{M}_1}{\mathrm{M}_2}\right) * 100\% \tag{22}$$

where, M_1 is the mass of the test specimen (in gms); M_2 is the mass of the material passing the 2.36 mm test sieve.

Calculate the average of the two results to the nearest integer. Averages are reported as aggregate crush values unless individual results differ by more than 0.07 times the mean. In this case, the test was repeated with the other two samples, the median of the four results are calculated to the nearest integer, and the median is reported as the aggregate crush value.

Note First exclude the highest and lowest values, and then calculate the average of the intermediate results to calculate the median of the four results Fig.26.

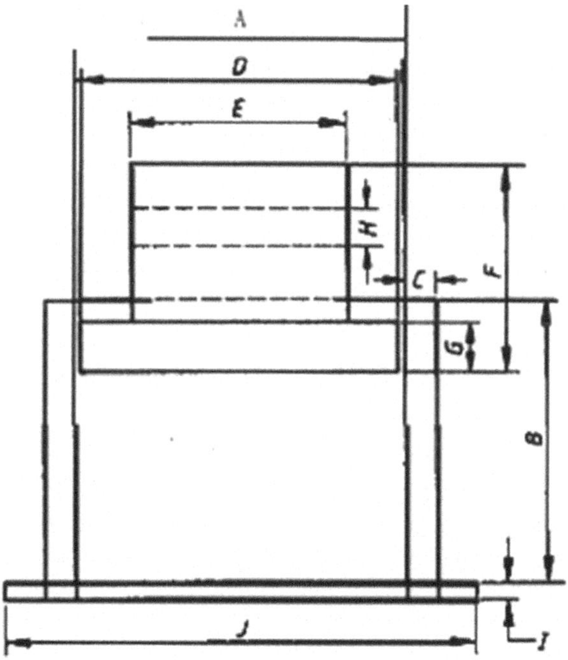

Fig. 26 Outline form of cylinder and plunger apparatus for the aggregate crushing value

7 Enhancement of RCA with the Use of Nano-silica Particles

7.1 Technical Background of the Incorporation of Nano-silica Particles

Nanotechnology, nano materials have been regarded as the materials that possess excellent application prospects in the twenty-first century (Jing-xian and Li-jiu 2004). Nanomaterials of high reactivity and advanced properties have attracted increasing research interests in improving cement-based materials' properties. Nanomaterials can be defined as particles with particle sizes between 1 and 100 nm. The nano-particle exists in the atom cluster's transition areas to the macroscopic matter with a high specific surface ratio. Because of the ultra-fine particle size, these particles have some beneficial physical and chemical properties (Zhang 2001). Almost ten years have passed since the first use of nanomaterials in the concrete industry. Many research types have been carried out on the effects of nano particles on the matrix of cement-based materials. Till now, several nanoparticles have been studied for incorporating in cement and concrete, including nano-silica (NS) (Kong et al. 2015; Li et al. 2015; Shah et al. 2016), nano-limestone (NL) (Li et al. 2015, Li et al. 2016), nano-metallic oxide (Farzadnia et al. 2013; Pan et al. 2015; Polat et al. 2015; Zhang et al. 2015a, b, c), graphene oxide (Kai et al. 2014; Pan et al. 2015; Lin et al. 2016), carbon nanotubes (Tyson et al. 2011; Metaxa, Seo et al. 2012). Among all these nanomaterials, NS particle is considered to have the most extensive concrete applications because of its superior pozzolanic activity in concrete than others (Zhang et al. 2017). A transmission electron micrograph of NS is shown in Fig. 27.

7.2 Microstructure of ITZs in RAC Prepared with Nano-silica Particles

RCA is a three-phase material that contains the old raw aggregate and the adhesive cement mortar as well as the old ITZ between the two materials, as shown in Fig. 28. Thus, when RAC contains RCA, there are at least two ITZ, namely the old ITZ between the old aggregate and the adhesive old mortar, and the new ITZ between the old and new mortar as shown in (Zhang et al. 2015a, b, c). ITZ is often a weak point for cement-based materials (Ollivier et al. 1995; Prokopski and Halbiniak 2000; Hosseini et al. 2011) with a large number of CH crystals and relatively high concentration of pores.

In recent years, some researchers have made trials to apply NS particles to improve RAC properties. Hosseini et al. (2011) added low percentages of NS particles (1.5 and 3% of the cement contents) to RAC prepared with 100% RCA, and examined the slump, compressive strength, water absorption, and microstructure of the modified

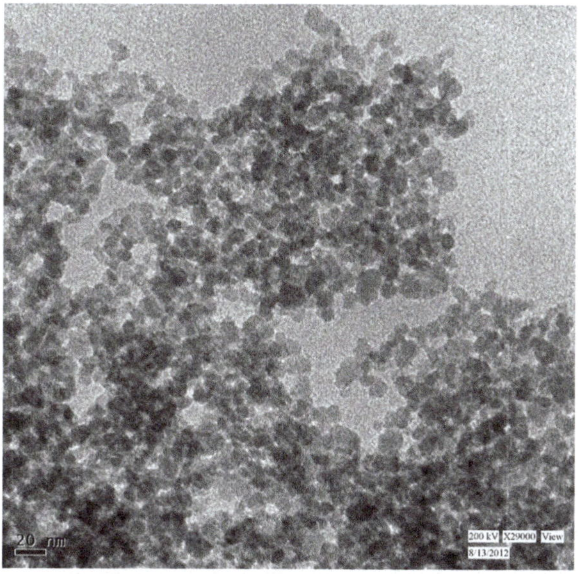

Fig. 27 Transmission electron micrograph of NS (Mukharjee and Barai 2014a, b)

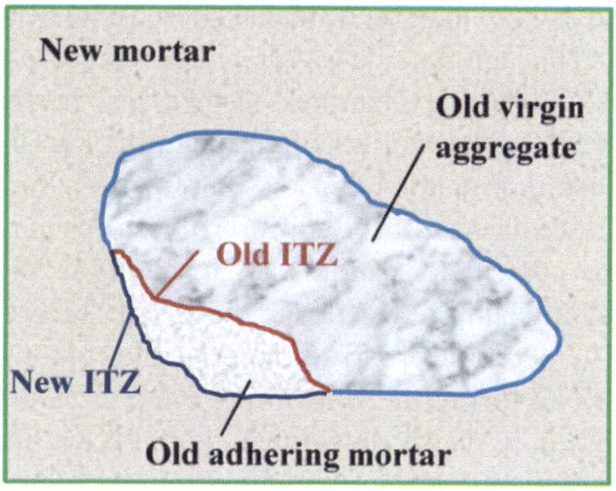

Fig. 28 Schematic diagram of RAC's microstructure and the definition of the old and new ITZs (Zhang et al. 2015a, b, c)

RAC. The test results showed that by using low dosages of silica nanoparticles at contents of up to 3%, the mechanical behavior and RAC permeability could be improved. By comparing the ITZs of RAC with the addition of NS and those of the control RAC (See Fig. 29a and b), by adding NS particles, the ITZs of RAC became denser and more uniform, and even tiny voids had been eliminated. These nanoparticles react with portlandite, and the produced dense C-S-H gel was the main reason for the more uniform and denser ITZs in RAC.

In another study carried out by Mukharjee and Barai (2014a, b), the compressive strength of natural aggregate concrete (NAC) and RAC prepared with 100% RA, with or without NS incorporation, were tested and compared. The test results showed that by adding 3% NS to RAC, the 7-day compressive strength of RAC was increased

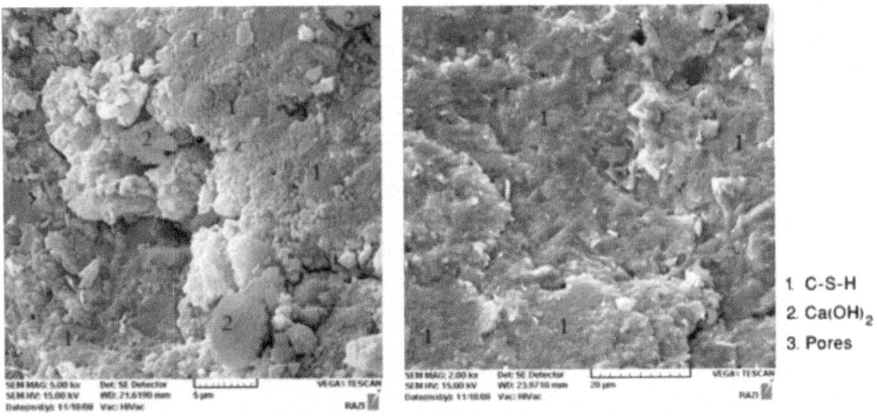

(a) the control RAC without NS particles (b) RAC incorporating 3% NS particles

Fig. 29 The ITZ of RAC

by about 15%, which was similar to the compressive strength of NAC without NS at the same age. To reveal the mechanism for such enhancement, Mukharjee and Barai (2014a, b) also used Backscattered SEM (BSEM) images to study porosity changes and the amount of un-hydrated cement across the ITZs in the modified and the control RAC. The microhardness test was also performed to obtain the mechanical properties of ITZs. The analysis of the quantity of un-hydrated cement calculated across the ITZ zone in Fig. 30a indicated that the un-hydrated cement content was higher for RAC and reduced with the addition of NS, this may be because the presence of NS increases the cement hydration rate. In addition, the porosity across the ITZ of RAC was significantly reduced, and these values were similar to those of NAC (see Figs. 30b). The addition of NS to the concrete mixture (NAC and RAC) reduced the porosity of ITZ due to the void-filling effect of nanoparticles. The Vickers microhardness of ITZ in RAC was lower than that of NAC, indicating that there are more microcracks and micropores.

However, the addition of 3% NS in the concrete enhanced the microhardness of ITZ of RAC to a great extent (See Fig. 30c), which was due to the reduction of porosity across the ITZ.

7.3 *Mechanical and Durability of RAC Prepared with Nano-silica Particles*

The test results by Mukharjee and Barai (2014a, b) showed that the addition of NS enhanced the compressive strength of RAC, and with the incorporation of 3% NS, the 28-day compressive strength was similar to that of the control NAC as shown in Fig. 31a. Also, although both the splitting tensile strength and flexural strength of

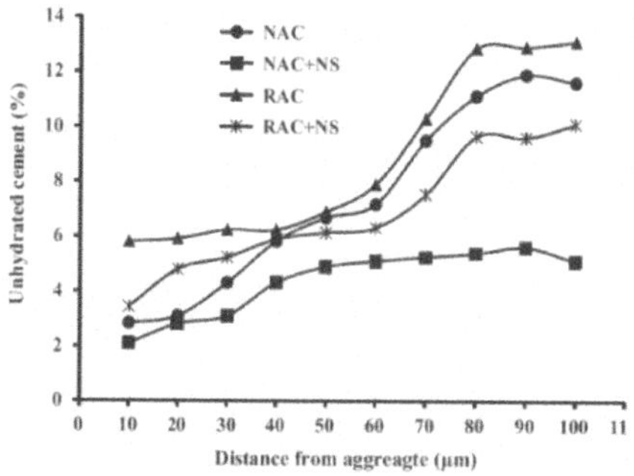

(a) Amount of un-hydrated cement particles

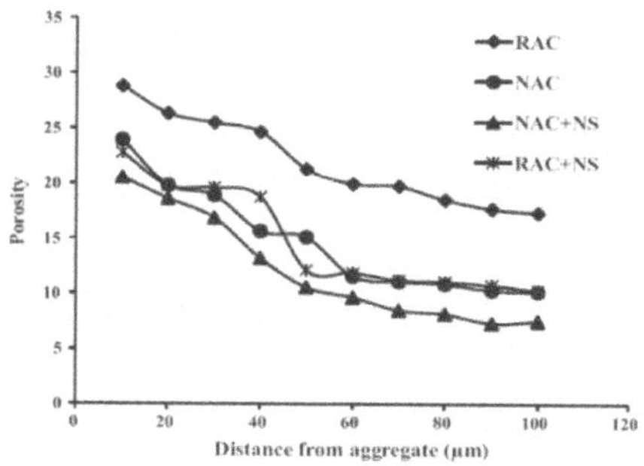

(b) Porosity

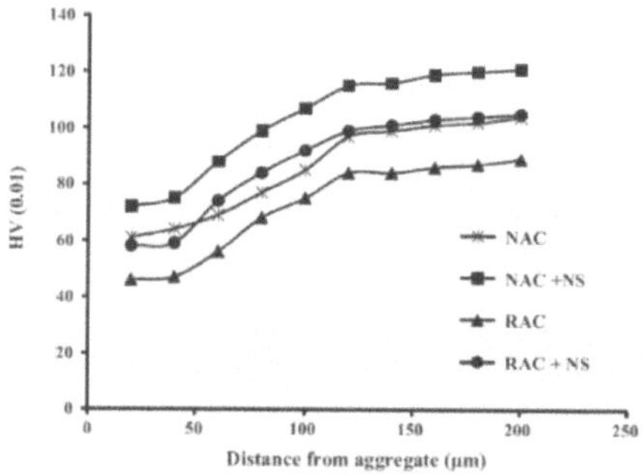

(c) Vickers microhardness with distance away from aggregate

Fig. 30 Microstructural analysis across ITZs in RAC modified by NS particles (Mukharjee and Barai 2014a, b)

RAC prepared with 100% RCA were inferior to that of NAC, the addition of NS may improve these tensile properties, as shown in Fig. 31b and c. Mukharjee and Barai (2014a, b) also proved that the modulus of elasticity and density of RAC was also improved with NS particles' incorporation, and the volume of voids and the water absorption were reduced.

Li et al. (2017) compared the effects of NS particles and NL particles on the crack propagation and microstructure properties of RAC. The resulted show that NS was more effective in improving the microstructure properties and mechanical strength of RAC than NL. The NS incorporation reduced the porosity and water absorption of RAC, as shown in Table 18 and Fig. 32.

In a study carried out by Zhang et al. (2015a, b, c), NS was used to prepare a slurry for RCA's surface treatment before the RAC production. In this study, two strengthening slurries incorporating different nanoparticles were prepared and employed to treat RCA's surface. The first slurry mainly contained NS and NL, labeled as the "nSi + nCa" slurry. Simultaneously, the other was prepared with a reduced dosage of NS

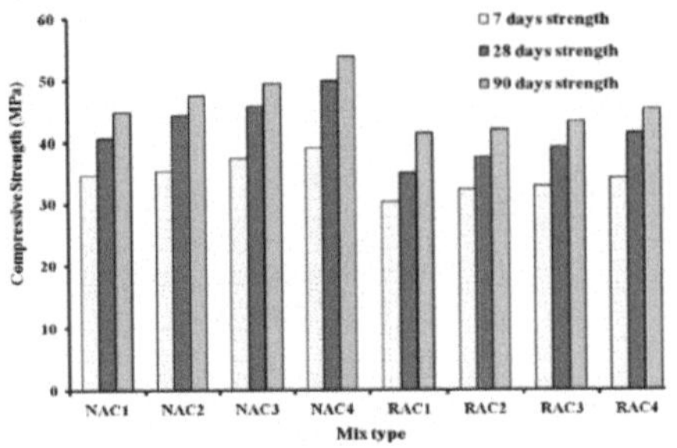

(a) Compressive strength

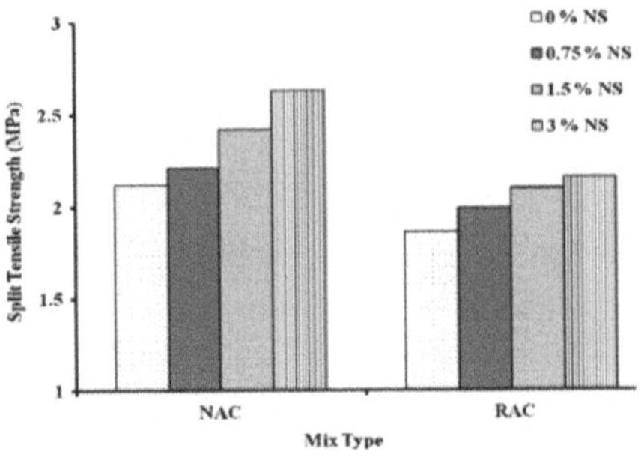

(b) Splitting tensile strength

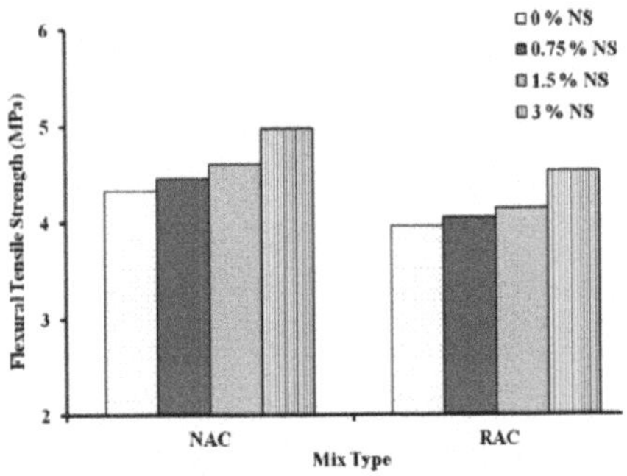

(c) Flexural tensile strength

Fig. 31 Mechanical properties of RAC with the incorporation of NS particles (Mukharjee and Barai 2014a, b)

Table 18 Pore size distribution of the modified RAC by NS and NL particles (Li et al. 2017)

Specimens	Total porosity/%		Average pore diameter/nm			Pore size distribution/%
			0–10 nm	10–50 nm	50–100 nm	>1000 nm
RAC	12.00	48.34	7.2	9.2	43.2	40.4
NL-RAC	8.82	71.65	2.8	10.8	37.6	48.8
NS-RAC	9.06	52.64	7.7	4.8	60.4	27.1

but was added with more cement as a substitute, considering the nanoparticles' high price, which was labeled as the "Cement + nSi" slurry.

It was found that although the old ITZ in RAC might not be strengthen due to the inability of nanomaterials or cement particles to penetrate far into the old mortar, the performance of the new ITZ in RAC was significantly enhanced (see Fig. 33). The old mortar, at the same time, was likely surface strengthened, which led to a reduction of the water absorption of RCA and thereby improving the workability

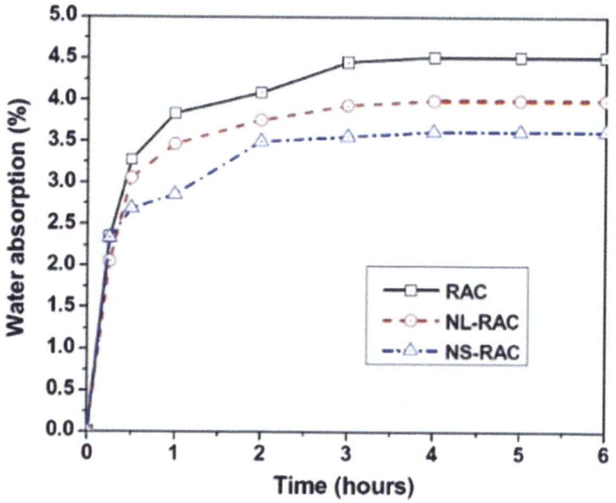

Fig. 32 Water absorption of the modified RAC by NS and NL particles (Li et al. 2017)

of the corresponding concrete, RAC1 (modified by the "nSi + nCa" slurry) and RAC2 (modified by the "Cement + nSi" slurry). Both the significant improvement in the new ITZs and the surface strengthening of the old mortar resulted in higher compressive strength and better resistance to Cl^- diffusion of RAC, as shown in Fig. 34a and b.

7.4 Summary

Although some achievements have been made in modifying RAC properties and its structural performance by using nanoparticles, especially by using NS particles, there are still several problems that need further studies.

The first problem lies in the adverse effects of nanoparticles on concrete workability, which has already been widely reported. RAC is known to have low flowability because of RCA's introduction with much higher water absorption than the NCA. Hence, if nanoparticles with a very high specific surface area are to be used in RAC, effective and efficient measures should be taken to eliminate their adverse effects on further reducing the flowability of RAC.

Moreover, even though NS addition's beneficial effects have been reported, NS particles and other nanoparticles' content should be well controlled. At high NS concentrations, increased autogenous shrinkage due to self-drying leads to a higher probability of cracking. High concentrations of superplasticizer and water should be added, and proper curing methods must be used to avoid this effect. In addition, nanomaterials are mainly used in functional or high-performance concrete, mainly due to their high cost. Therefore, if nanomaterials are widely used for RAC modification, their dosage must be controlled to make sure the economic efficiency of the target RAC project.

Fig. 33 Probability distribution of the modulus of the old and new ITZs in the modified RAC: and (Zhang et al. 2015a, b, c)

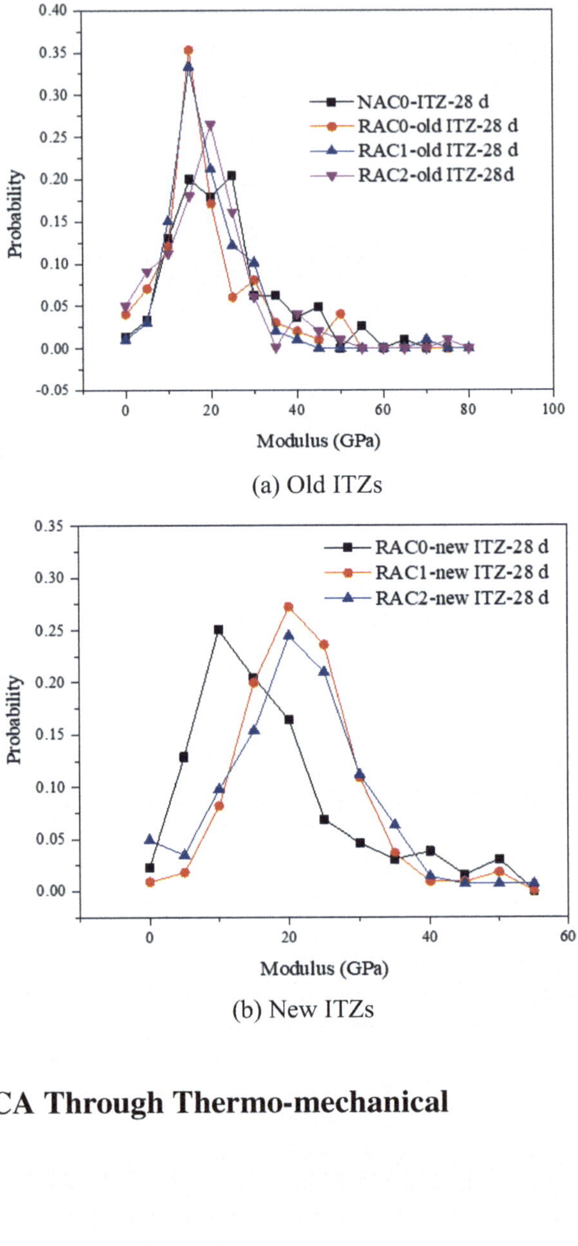

8 Enhancement of RCA Through Thermo-mechanical Beneficiation

8.1 Technical Basis

The working principle of thermo-mechanical beneficiation is based on the weakening of attached mortar (AM) by dehydrating the cement hydration products; differential thermal stresses due to heating and cooling also enable the weakening of AM, after which it can be more easily detached/debonded from the aggregate surface by virtue

Fig. 34 Properties of the modified RAC (Zhang et al. 2015a, b, c)

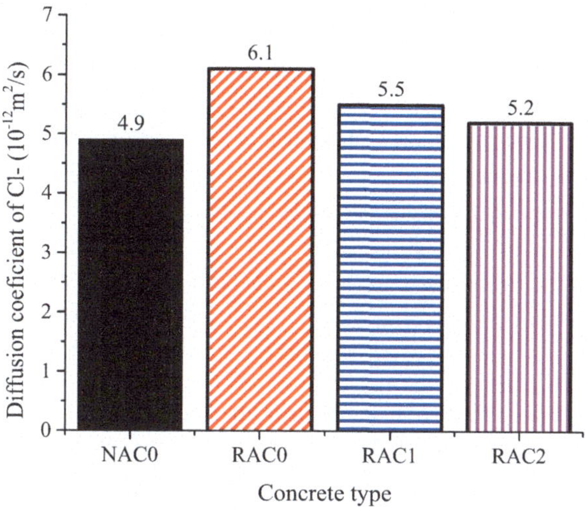

(a) Compressive strength

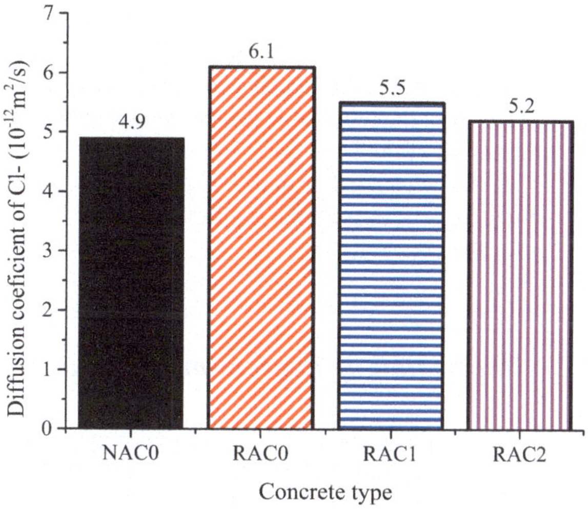

(b) Diffusion coefficient of Cl^-

of mechanical scrubbing, say, through ball milling. The efficiency of the beneficiation process is mostly governed by the heating temperature and duration; see the effect of temperature on various cement hydrated compounds in Table 4. It is to be noted that the temperature is to never exceed 750ºC so as to prevent the conversion of the carbonate phases of the AM release of CO_2, in addition to limiting the energy requirement. In the case of granite aggregates, it is preferable that the heating temperature is kept lower than 570ºC to avoid the phase transition of α-SiO_2 into β-SiO_2 (Ahn et al. 2001).

Table 19 Quality parameters and acceptance criteria for RCA (Prajapati et al. 2021)

Parameter	Limits	
Parameters for efficiency of process		
Ty: Total yield of recyclable products	Ty $\geq$ 90%	
CF: Concrete fines	CF $\geq$ 20%	
Parameters for quality of recycled products		
	Coarse aggregate (granite)	Fine aggregates
Wabs: Coefficient of water absorption (%)	Wabs $\leq$ 2%	Wabs $\leq$ 5%
Dd: Dry density (kg/m^3)	Dd $\geq$ 2500	Dd $\geq$ 2200

The effectiveness of the thermo-mechanical beneficiation depends on the maximum size of the concrete chunks (feed size), among other factors. Considering that larger feed sizes would increase the heating time required and smaller feed sizes would increase the preparatory crushing energy, a maximum feed size of about 50 mm is considered appropriate when the maximum aggregate size in the parent concrete is about 20 mm (Prajapati et al. 2021). The overall efficiency of the beneficiation could be assessed considering the parameters and their limits given in Table 19, where the quality of the RCA is expressed in terms of the coefficient of water absorption and dry density, in addition to the process efficiency parameters—the total yield of the recyclable products (i.e., all particles smaller than or equal to the maximum size of the aggregates in the parent concrete) and the quantity of concrete fines (i.e., particles smaller than 300 μm). These limits are in accordance with most recommendations, except that the Japanese code has a stricter water absorption limit of 2% for the coarse RCA.

8.2 *Illustration of Thermo-mechanical Beneficiation*

In a study at IIT Madras (Prajapati et al. 2021) concrete chunks, from demolition waste, with a maximum size of 50 mm were heated at various temperatures over 120 min in an electric furnace. Following the heating, the chunks were cooled at ambient temperature for 24 h and then scrubbed using a ball mill; the milling was done with ten steel balls of 45 mm diameter for 500 revolutions. Heating at 500°C for 60 min was found to be the most reasonable treatment in line with the limits given in Table 19. The physical and mechanical properties coarse and fine RCA thus obtained are compared with other commercially-available aggregates, in India, in Tables 20 and 21. The properties of beneficiated coarse RCA (see Table 20) were found comparable to the pristine granite aggregates; similar results were obtained by other researchers as well (Purushothaman et al. 2014; Shima et al. 2005; Sui and Mueller 2012). The AM content of beneficiated RCA was found to be 4%, which is much lower than the crushed and industrially-produced RCA. The water absorption by the beneficiated fine RCA was close to 6%, which is higher than the target value but 40% lower than that of the crushed fine RCA. Similarly, the dry density was not as high as that of river sand but close to that of crushed sand. The concrete fines (CF) obtained had a high amount of calcium bearing phases, suggesting that it can be used for replacing the binder in concrete (Florea et al. 2014; Shima et al. 2005) or cement clinker production.

To verify the feasibility of using the fine and coarse RCA, three concrete mixes were prepared, namely (i) NAC: reference concrete using pozzolanic portland cement cement and 0.45 w/c with pristine granite and river sand (ii) CRAC: NAC with the pristine aggregates were replaced with coarse RCA, and (iii) FRCA: NAC with the river sand was replaced with fine RCA. The 28-day cube compressive strengths of the concretes are shown in Table 22. Though there was a slight reduction in the strength compared to the reference mix, it was much higher than expected (Evangelista and De Brito 2014; Khatib 2005), indicating that the beneficiated RCA could be used for high-quality applications.

Table 20 Comparison of the beneficiated coarse RCA with other coarse aggregates (Prajapati et al. 2021)

Aggregate type: coarse/mm		Dry density/kg/m³	Water absorption/%	Bulk density/kg/m³	Crushing value	Impact value	Attached mortar/%
Beneficiated RCA	10	2790	1.18	1555	23.7	12.5	4.0
	20	2870	0.95				
Pristine granite aggregates		2740	0.40	1540	21.1	19.2	–
Jaw-crushed RCA	10	2230	5.92	1205	31.2	29.3	42.5
	20	2400	4.81				
Industrially-produced RCA	10	2290	5.43	1425	33.8	40.8	29.6
	20	2520	5.72				
Aggregates extracted by acid washing the concrete waste	10	2720	0.73	1530	13.5	15.6	0
	20	2780	0.40				

Table 21 Comparison of the beneficiated fine RCA with other fine aggregates (Prajapati et al. 2021)

Aggregate type: Fine	Dry density/kg/m³	Water absorption/%	Bulk density/kg/m³
Beneficiated RCA	2350	5.91	1480
River sand	2610	0.65	1630
Pristine crushed sand	2450	2.95	1515
Jaw-crushed RCA	2040	9.80	1450
Industrially-produced RCA	2260	10.75	1305

Table 22 Compressive strength of concrete (mean values ± standard deviations) (Prajapati et al. 2021)

Concrete	NAC	CRAC	FRAC
Cube strength (MPa)	47.5 ± 1.7	44.7 ± 0.5	41.0 ± 1.4

9 Enhancements of RCA with Microbial Induced Carbonation Precipitation

9.1 Technical Background of Microbial Induced Carbonation Precipitation

In nature, calcium carbonate precipitation is a straightforward chemical process affected by four key factors: the concentration of calcium, dissolved inorganic carbon, the pH, and the availability of nucleation sites (Hammes, Verstraete et al. 2002). Microbial induced carbonation precipitation (MICP), one of the mineralization processes widely existing in nature, refers to the phenomenon of using a microorganism, especially bacteria, to induce calcium carbonate precipitation. With given conditions, such as temperature, pH, and so on, most bacteria can induce calcium carbon precipitation (Boquet et al. 1973). There are many different mechanisms for MICP, such as microbial photosynthesis, microbial sulfate reduction, microbial nitrogen cycle, etc., as shown in Table 23. Although the processes for different bacteria differ, the primary role has been ascribed to their ability to create an alkaline environment through various physiological activities and act as sites of nucleation or calcium enrichment (De Muynck et al. 2010).

Table 23 The mechanisms of MICP for different types

Mechanisms	Equations
Microbial photosynthesis (Castanier et al. 1999; Hammes et al. 2002)	$CO_2 + H_2O \rightarrow (CH_2O) + O_2$
	$2HCO_3^- \rightarrow CO_2 + CO_3^{2-} + H_2O$
	$CO_3^{2-} + H_2O \rightarrow HCO_3^- + OH^-$
	$Ca^{2+} + HCO_3^- + OH^- \rightarrow CaCO_3 \downarrow + H_2O$
Microbial sulfate reduction (Wright 1999; Van et al. 2003)	$CaSO_3 \cdot 2H_2O \rightarrow Ca^{2+} + SO_4^{2-} + H_2O$
	$2(CH_2O) + SO_4^{2-} \rightarrow HS^- + HCO_3^- + CO_2 + H_2$
	$Ca^{2+} + HCO_3^- + OH^- \rightarrow CaCO_3 \downarrow + H_2O$
Microbial nitrogen cycle (Fujita et al. 2000)	$CO(NH_2)_2 + H_2O \rightarrow CO_3^{2-} + 2NH_4+$
	$Ca^{2+} + HCO_3^- + OH^- \rightarrow CaCO_3 \downarrow + H_2O$

MICP has been applied to the biodegradation of organic pollutants, removing metal ions and investigating its environment-friendly character to improve construction materials' properties. As early as 1990, Adolphe used MICP to protect ornamental stone and applied for a patent. The patent concept laid the foundation for subsequent studies by different research groups to investigate different approaches for treating limestone. As an extension of research on limestone restoration, De Muynck et al. (2008), and De Muynck et al. (2008) research on the surface treatment of cementitious materials with different porosities by employing *B.sphaericus*. The results showed that the treated cementitious materials' durability, incredibly high porosity, was improved. It was consistent with the treated limestone. Similarly, Chunxiang et al. (2009) investigated the bio-deposition as a surface treatment for protecting the concrete and proved that the properties of treated concrete have improved. Furthermore, the technology of MICP has been applied to repair the ancient masonries (Shen and Cheng 2008).

The above research proved that MICP could be applied to different construction materials and form a layer to protect them. MICP was used as a binder-based material to repair the cracks in building materials and enhance material strength. In particular, based on repairing the crack by MICP, Jonkers (2007) proposed making the self-healing concrete with bacteria as self-healing agents. Also, Luo, Qian et al. 2015 investigated the factors affecting the repair of self-healing concrete and found that bacteria are good at repairing small cracks.

Based on the above modification of construction materials, MICP technology can also modify RCA to enhance the weak layer of old mortar attached to the original aggregates and repair the micro-crack in the aggregate.

9.2 *Optimization of Modifying RCA Methods*

Grabiec et al. (2012) first investigated and confirmed the feasibility of MICP to modify RCA's surface. Four kinds of RCA were produced experimentally with particle sizes ranging from 6/8 mm and 12/16 mm from crushing laboratory prepared parent concretes with w/c ratios of 0.45 and 0.70, respectively. The procedure of bio-deposition was carried out, shown in Fig. 24.

After modifying, the carbonate precipitation was observed on RCA's surface under SEM, as shown in Fig. 35. Simultaneously, the test results of water absorption showed that the water absorption of aggregate had been reduced to varying degrees.

Subsequently, different research groups studied the methods of modifying aggregate. Qiu et al. (2014) investigated the condition factors, mainly related to bacteria growth, influencing MICP on RCA treatment. Parametric studies were conducted to determine the effects of pH (7.0–10.5), temperature (25 °C and 35 °C), calcium content (5.6–50.4 g/L of $CaCl_2$), and bacteria concentration (10^6–10^8 cells/ml) on MICP. The test results showed the MICP peaks at pH = 9.5 and increased with higher temperature, bacteria concentration, or calcium concentration. Also, the rate,

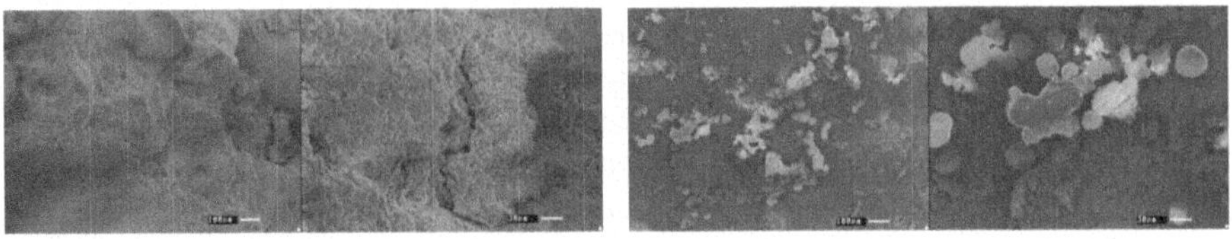

Fig. 35 Scanning electron micrograph of RCA and Scanning electron micrograph of RCA after modification (Qiu et al. 2014)

amount, and type of MICP on RCA could be modified through proper control of culturing and bio-deposition conditions.

Considering that many conditions would influence modified aggregate, the methods of modification should be investigated further. Researchers have optimized them in different ways. Wang et al. (2017) focused on the optimization of the bio-deposition treatment procedure. In the test, it was found that the immersion method was better than the spraying method. Then, four immersion treatments were conducted: (1) RCA were first immersed in the culture containing bacteria without calcium for 24 h, and then RCA were immersed in a deposition medium for three days. This is method 1; (2) In the first immersion, the medium's bacterial content was three times higher than in method 1; (3) The treatment steps were the same as in method 1. The only difference was that the deposition medium contained yeast extract (5 g/L); (4) The treatment repeated the method 1 two times. The test results showed that the best method is to use the immersion method and modify it twice. The bio-deposition treatment was more efficient to treat materials with high porosity and large pores.

Zeng et al. (2019) investigated the different bio-deposition treatments with cubic RA, as shown in Fig. 36, and three bio-deposition methods were designed as shown in Fig. 37. Also, the rotation treatment was conducted, as shown in Fig. 38. The test results showed that calcium should be added in batches, and RCA should be kept static during the modification process. Moreover, Feng et al. (2020) focused on the length of modification time and recommended that a single modification time be about seven days.

In addition to optimizing methods, research groups have also tried to optimize modification by changing bacterial species. Wu et al. (2018) chose to employ Bacillus pseudofirmus (DSM8715) instead of ureolytic bacteria to modify RCA to eliminate ammonia's adverse effects on the modification environment.

9.3 Properties of Products Prepared with Modified RCA

After modification, the properties of RCA have been improved. Simultaneously, it is still necessary to test the properties of the cementitious materials prepared with the modified RCA for practical application.

Fig. 36 Cubic recycled aggregate (Zeng et al. 2019)

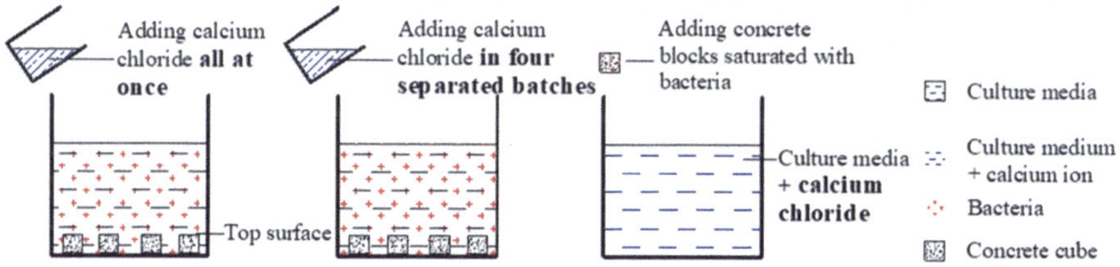

Fig. 37 Three bio-deposition methods (Zeng et al. 2019)

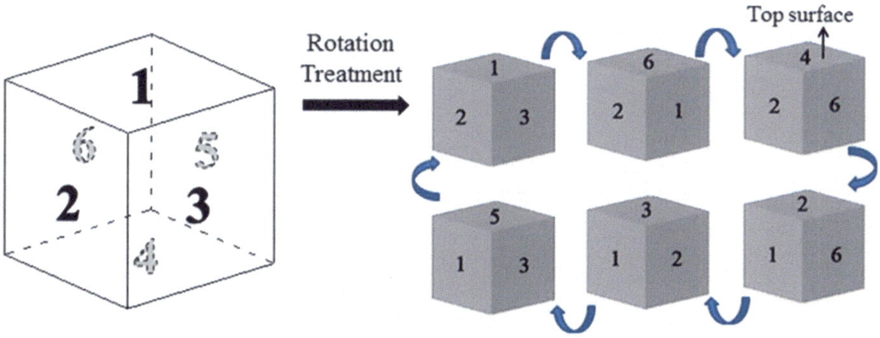

Fig. 38 Schematic diagrams of the rotation treatment (Zeng et al. 2019)

Wang et al. (2017) investigated NAC's properties, RAC and RAC prepared with modified RCA. In the experiment, two different RCA kinds were compared, crushed concrete aggregate and MA. The MA consisted of crushed concrete, masonry, glass, wood, etc. As shown in Fig. 39, the results indicated that the concrete made with bio-deposition treated aggregates showed a significant improvement in concrete properties compared to those made with untreated aggregates. Zhao et al. 2019 investigated the relationship between microstructure and properties. The ITZs of concrete were measured to obtain the microhardness. Two types of ITZs were measured, including ITZ_1 between the original NA and the new mortar, and ITZ_2 between the old and the new mortar, as shown in Fig. 40. Figure 41 indicated that the original microhardness of ITZ_2 was lower than that of ITZ_1. After the MICP modification, the microhardness of ITZs, especially ITZ_2, increased. The enhancement in microhardness of ITZs, between new and old mortars explained the improvement of compressive strength at the micro-level.

Feng et al. (2020) carried out mechanical tests on the mortar cast by modified RCA and found that the mortar properties were significantly improved, especially the flexural properties were twice as good as those of the unmodified mortar. However, there is still a big gap between the mortar cast by modified RCA and the mortar cast by NA. Wu et al. (2018) also obtained similar conclusions for strength properties. What is more, the experimental results show that the workability of fresh mortars cast by modified RCA is also improved, as shown in Fig. 42.

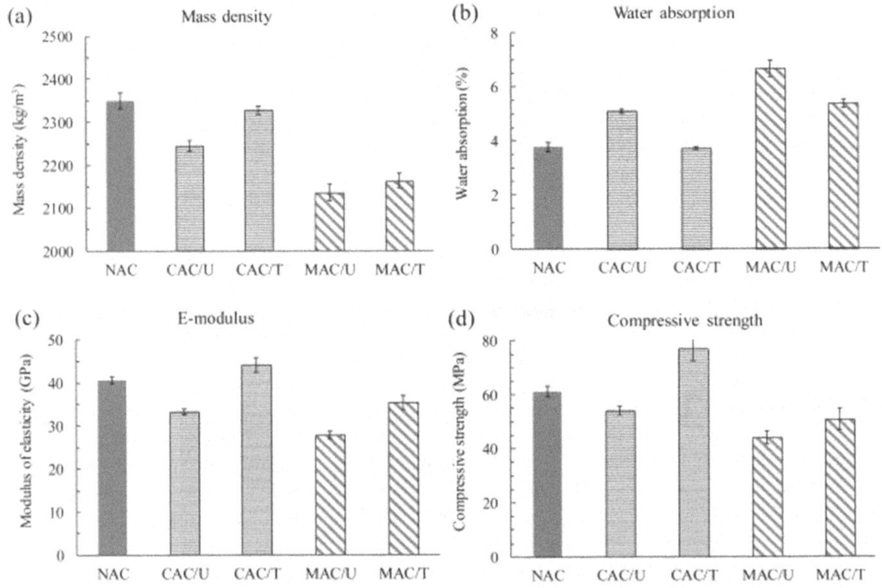

Fig. 39 Properties of concrete made with NA, untreated CA and MA (CA/U and MA/U), treated CA and MA (CA/T and MA/T) (Wang et al. 2017)

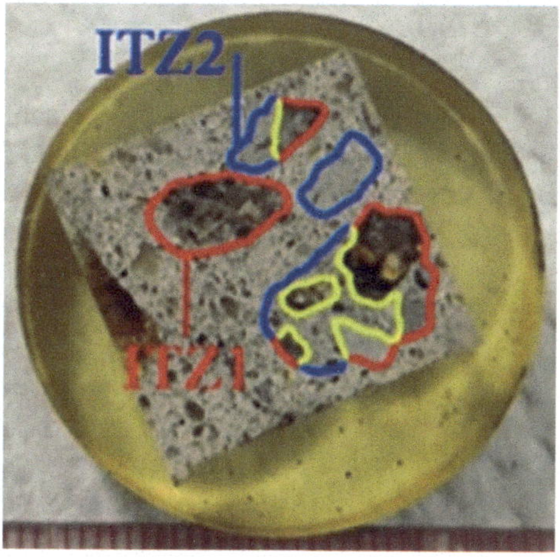

Fig. 40 The schematic diagrams of different types of ITZs (Zeng et al. 2019)

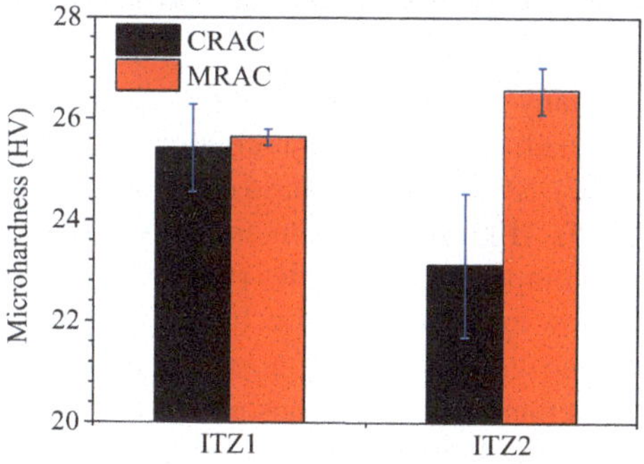

Fig. 41 The mean values of microhardness of two types of ITZs (Zeng et al. 2019)

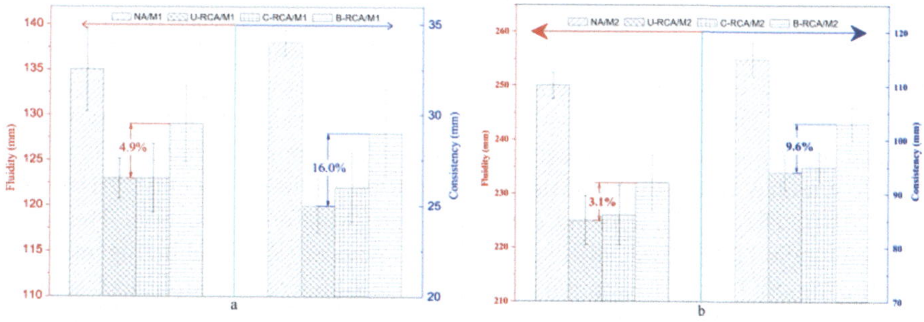

Fig. 42 Workability of fresh mortars prepared with 0.35 w/c ratio (left) and 0.5 w/c ratios (right) (Wu et al. 2018)

9.4 Summary

Modifying RCA with MICP is a sustainable and environmentally friendly method that can improve RCA's properties. Although some achievements, such as the modified RAC's micro and basic materials properties, have been obtained. Several problems still exist, including the efficiency of modification, the systematic modification process, the cost of modification, and the side effects of the residual chloride-ion on the modified RCA, which should be studied further.

10 Modification of SCC by Ultra-fine Recycled Powder from Construction and Demolition Waste

10.1 Technical Background

For many years, SCC has been studied for its strength, volume stability, and longevity (Wu et al. 2015). SCC was created because it outperformed normal vibrated concrete (NVC). SCC is a two-phase composite material, according to Topcu and Kocataşkin (1995). High powder content, superplasticizers, and VMA are required to achieve self-compacting properties (Gaimster and Dixon 2003). Increase SCC paste volume to allow aggregates to flow better and consolidate under their weight. Growing paste volume will impact SCC costs and the environment. This requires 1.58 billion tons of supplemental cementitious materials (SCMs) like fly ash, which is only 800 million tons globally available (Lin et al. 2010).

Many inert filler, pozzolanic, and cementitious additives have been studied. In the cement and concrete industries, powders such as marble (Güneyisi et al. 2009), ceramic waste (Ferrara 2019), granite (Sadek 2016), dimensional stone (Rana et al. 2016) and glass (Liu 2011) are well-established. CDW is one major concern with 2 billion tons of CDW in 2017 (Wang et. al. 2019) Trash generated by CDW produces recycled fine aggregates (RFA) and dust powder (0.16) mm) (Vázquez 2012). Due to its low added value, dust powder is rarely used in fillers (Li et al. 2016; Lin et al. 2010). They are now being researched for Chinese construction. The most common method is to physically crush them into powder (RP). Stabilizing and then amorphizing -SiO_2 improves activity. Grinding waste concrete amorphizes $Ca(OH)_2$ and C-S-H.

RP from CDW was expected to help improve the long-term characteristics and microstructure of hydration products (Shui et al. 2008; Xiao et al. 2018). Topič and Prošek (2017) found that substituting RP for 15% cement increased hydration heat and improved concrete flexural strength. Xiao et al. 2018 and Lu et al. 2014a found that substituting RA for cement improved mechanical and tensile properties. Using RP and SCMs instead of cement also addresses the issue of fine fraction disposal. SCMs such as RP, silica fume (SF) and FA are chemically similar to cement (Fig. 43).

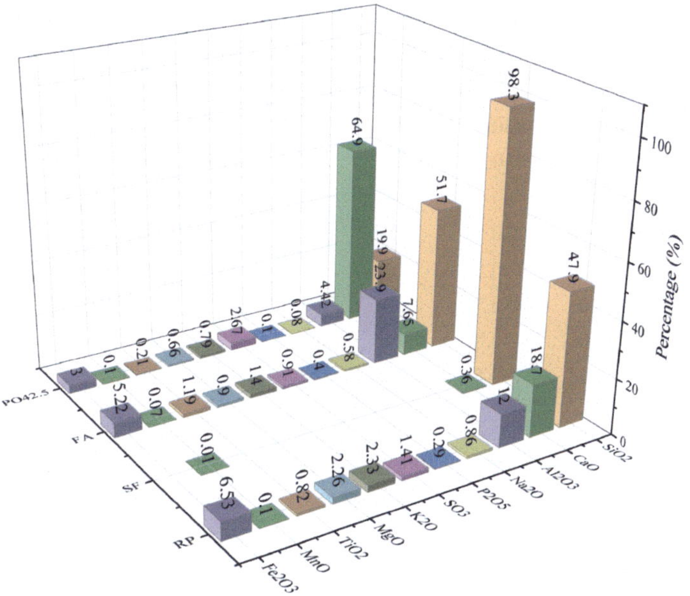

Fig. 43 Chemical compositions

RP's primary oxide components are similar to fly ash. Surrounding it are sand and un-hydrated cement particles.

10.2 Optimization of Modifying SCC with RP

In a study by (Duan et al. 2020), RP was produced from the CDW, utilizing the cone cavity mill. This ultra-fine powder showed better performance in many studies (Ma et al. 2019). Most of the time, the constituents determine the composite's performance. Establishing the link between the characteristic material and the performance parameters is important research. So, the mechanical performance along with chloride transport and NDt testing such as X-CT was used to determine the influence of the RP on the performance of SCC. The mix proportion used in this study is shown in Tables 24 and 25.

10.3 Properties of SCC with Ultra-Fine RP

Table 24 shows the trend of fresh properties for SCC mixed with RCA and RP. SCC with 10% RP decreased by 0.66, 3.04, 9, and 14.56%, while SCC with 20% RP

Table 24 Mix proportions of the experimental program in this study (kg/m^3) (Duan et al. 2020)

Mix	Notation	Cement	FA	SF	RP	NFA	NCA	RCA
Control	Control	430.5	184.5	0	0	660	780	0
Series I	C0RP10	430.5	61.5	61.5	61.5	660	780	0
	C25RP10	430.5	61.5	61.5	61.5	660	585	195
	C50RP10	430.5	61.5	61.5	61.5	660	390	390
	C100RP10	430.5	61.5	61.5	61.5	660	0	780
Series II	C0RP20	430.5	0	61.5	123	660	780	0
	C25RP20	430.5	0	61.5	123	660	585	195
	C50RP20	430.5	0	61.5	123	660	390	390
	C100RP20	430.5	0	61.5	123	660	0	780

Table 25 Fresh properties of SCC (modified from Duan et al. 2020)

Mix	Notations	Dia (mm)	T-500 (sec)	V-Funnel (sec)	L-Box (h2/h1)
Control	Control	755	1.76	4.12	0.96
Series I	C0RP10	750	1.8	4.12	0.96
	C0RP20	730	1.9	4.67	0.93
	C25RP10	732	1.96	4.25	0.93
	C25RP20	643	3.4	5.15	0.89
Series II	C50RP10	687	2.1	4.54	0.89
	C50RP20	630	3.1	5.65	0.86
	C100RP20	610	3.9	6.12	0.79
	C100RP10	645	2.5	5.12	0.87

decreased by 3.3, 14.83, 16.55, and 19.2%. For example, when using RCA at 100%, T500 increased by 38.89%, while the V-funnel result increased by 24.27%. The high water absorption of RCA affects SCC fresh properties. Second, increased friction and surface area due to increased RCA can affect interlocking behavior, decreasing flowability (Montero and Laserna 2017, Behera et al. 2019).

According to Wang et al. (2015) and Kim et al. (2012), increased surface area of powder content can reduce SCC's fresh properties. Figure 44 compares the mechanical properties of SCC prepared with RCA and RP at various curing ages.

Both SCC with and without RCA/RP had increasing compressive strength with age. SCC compressive strength gradually decreases with increasing RCA/RP replacement ratios, and this decrease is more pronounced when the RCA replacement ratio exceeds 50%, regardless of the RP dosage. The reason may be that RP and SF can replace FA quickly and effectively.

The greatest reduction was seen with SCC with 100% RCA and 10% RP. Less than 50% RCA reduced the rate of reduction. The initial stage of hydration RP has low pozzolanic activity, so adding RP can increase strength. RP can boost strength

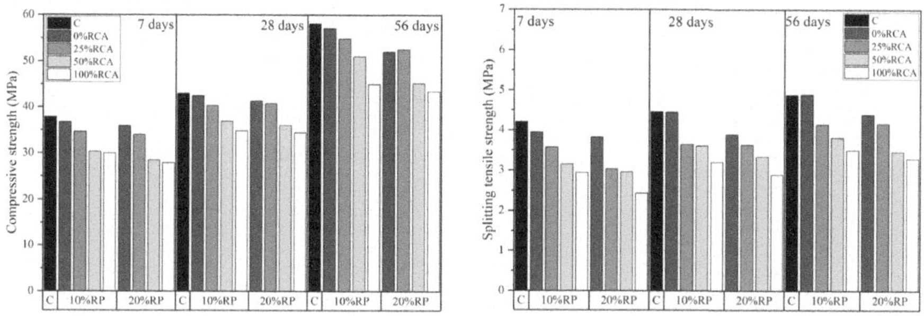

Fig. 44 Mechanical properties of SCC with RCA and RP (modified from Duan et al. 2020)

by increasing reactivity. Pawluczuk (2017) suggested mixing 20% RP into concrete for good results. The residual cement (alite, belite) and aggregates (quartz, anorthite) in RP increase concrete strength. The chloride concentration (Fig. 45) was slightly higher in SCC with 10% RP than in control SCC with 50% RCA. The chloride content of SCC with 20% RP increases with RCA replacement. Too much RP can cause particle aggregates and reduce interfacial energy between cementitious particles and dispersion medium, while too little can cause particle aggregation (Zhang et al. 2016). Agglomeration may slow the SCC reaction, affecting chloride ingress.

X-Ray CT was used to study the durability and the microstructure of SCC. X-Ray CT showed that the particulates, attached mortar, and RCA/RP secondary hydration products all affect concrete defect content. Pore sphericity is shown in Fig. 46. A scale of 0 to 1 represents the perfect circular shape of pores. All samples with a higher number of smaller pores showed an exponential behavior. SCC with 20% RP had more pores per slice than SCC with 10% RP, even though the sphericity was higher. In the cement matrix, fine particles agglomerate, causing higher porosity (Zhang et al. 2016). Figure 47 depicts micropore distribution in concrete. The AM

Fig. 45 Chloride concentration (Duan et al. 2020)

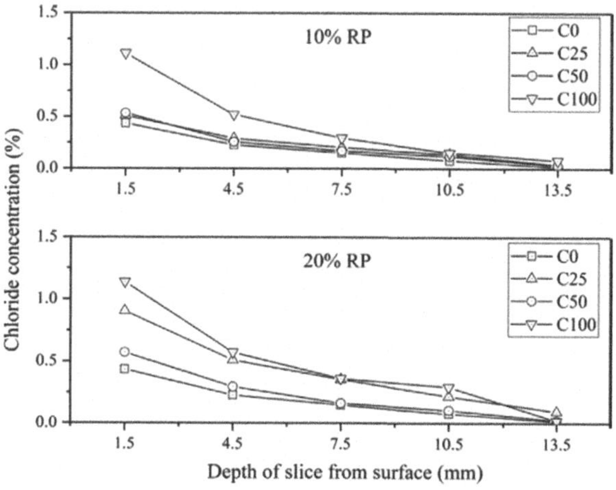

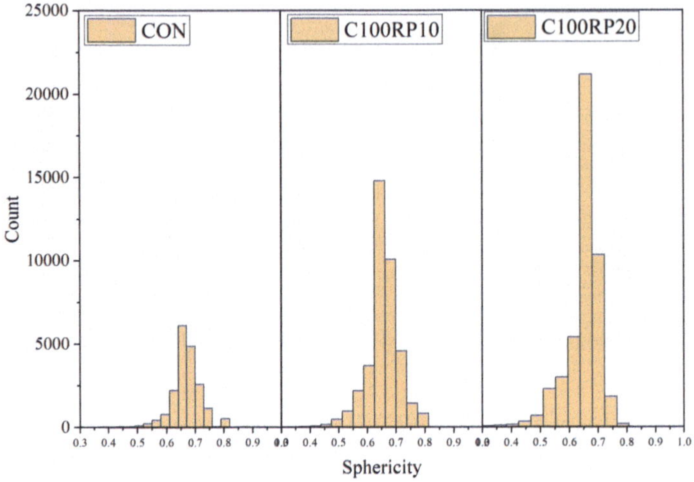

Fig. 46 Sphericity of pores (Duan et al. 2020)

on aggregates reduces micropores compared to RCA concrete. A higher amount of RP may cause particle aggregation, as well as the filler effect.

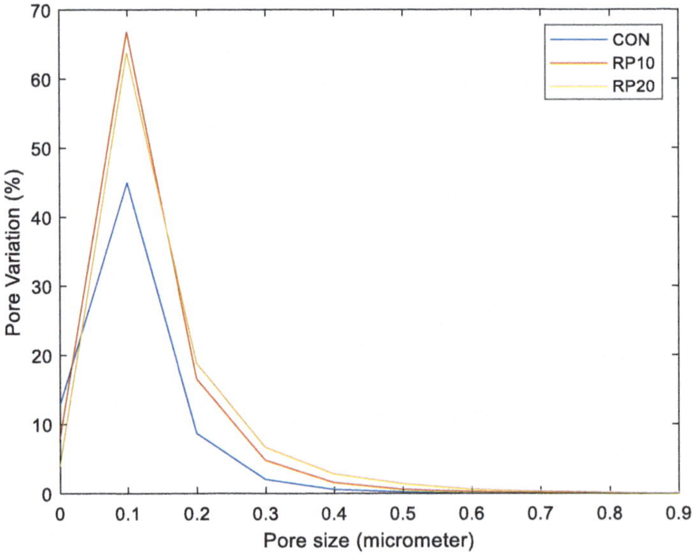

Fig. 47 Pore variation in concrete (in %)

10.4 Summary

Various studies on the fresh and hardened characteristics of SCC with RP and RCA led to the following conclusions:

The SCC mixtures tested in this study functioned well even when all FA was replaced by RP.

In the fresh state, there is more interlocking between the particles, but there is also more agglomeration due to the increased amount of RP.

SCC compressive and splitting tensile strength decreased as RCA replacement increased, however the rate of drop was slower when RCA replacement was less than 50%.

However, when RP's alternate use was increased from 10 to 20%, a small increase in compressive strength was found at 28 days.

The chloride concentration in SCC with 10% RP was substantially greater than in control SCC when RCA was fully utilized, but was quite low when RCA was utilized just 50%.

These findings suggest that more RP may lead to particle aggregation and increased porosity in concrete.

References

Ahn, J.W., Kim, H.S., Han, G.C.: Recovery of aggregates from waste concrete by heating and grinding. Geosystem Eng. **4**, 123–129 (2001). https://doi.org/10.1080/12269328.2001.10541178

Akbarnezhad, A., Ong, K.: Microwave decontamination of concrete. Mag. Concr. Res. **62**(12), 879–885 (2010)

Akbarnezhad, A., Ong, K.C.G., Zhang, M.H., Tam, C.T., Foo, T.W.J.: Microwave-assisted beneficiation of recycled concrete aggregates. Constr. Build. Mater. **25**(8), 3469–3479 (2011)

Akbarnezhad, A., Ong, K.C.G., Tam, C.T., Zhang, M.H.: Effects of the parent concrete properties and crushing procedure on the properties of coarse recycled concrete aggregates. J. Mater. Civ. Eng. **25**(12), 1795–1802 (2013)

Alarcon-Ruiz, L., Platret, G., Massieu, E., Ehrlacher, A.: The use of thermal analysis in assessing the effect of temperature on a cement paste. Cem. Concr. Res. **35**(3), 609–613 (2005)

Al-Bayati, H.K.A., Das, P.K., Tighe, S.L., Baaj, H.: Evaluation of various treatment methods for enhancing the physical and morphological properties of coarse recycled concrete aggregate. Constr. Build. Mater. **112**, 284–298 (2016)

Alonso, C., Fernandez, L.: Dehydration and rehydration processes of cement paste exposed to high temperature environments. J. Mater. Sci. **39**(9), 3015–3024. (2004)

Benedetto, A., Calvi, A.: A pilot study on microwave heating for production and recycling of road pavement materials. Constr. Build. Mater. **44**, 351–359 (2013)

Berger, R.L., Young, J.F., Leung, K.: Acceleration of hydration of calcium silicates by carbon dioxide treatment. Nat. Phys. Sci. **240**(97), 16–18 (1972)

Bierlich, K.B.: Manufacture of Portland cement products. Google Patents (1969)

Boquet, E., Boronat, A., Ramos-Cormenzana, A.J.N.: Production of calcite (calcium carbonate) crystals by soil bacteria is a general phenomenon. **246**(5434): 527–529 (1973)

Buildings Department (2005). https://www.info.gov.hk/bd/english/index.html.

Bukowski, J.M., Berger, R.L.: Reactivity and strength development of CO_2 activated non-hydraulic calcium silicates **9**(1): 57–68 (1979)

Butler, L.: Evaluation of recycled concrete aggregate performance in structural concrete (2012)

Castanier, S., Le Métayer-Levrel, G., Perthuisot, J.-P.J.S.G. : Ca-carbonates precipitation and limestone genesis—the microbiogeologist point of view **126**(1–4): 9–23 (1999)

Chunxiang, Q., Jianyun, W., Ruixing, W., Liang, C.J.M.S.: Corrosion protection of cement-based building materials by surface deposition of CaCO3 by bacillus pasteurii **29**(4): 1273–1280 (2009)

De Juan, M.S., Gutiérrez, P.A.: Study on the influence of attached mortar content on the properties of recycled concrete aggregate. Constr. Build. Mater. **23**(2), 872–877 (2009)

De Muynck, W., Debrouwer, D., De Belie, N., Verstraete, W.J.C., Research C.: Bacterial carbonate precipitation improves the durability of cementitious materials. **38**(7): 1005–1014 (2008)

Duan, A., Singh, A., Xiao, J., Hou, S.: Combined use of recycled powder and recycled coarse aggregate derived from construction and demolition waste in self-compacting concrete. Constr. Build. Mater. **254**, 119-323 (2020). https://doi.org/10.1016/j.conbuildmat.2020.119323

Evangelista, L., De Brito, J.: Concrete with fine recycled aggregates: a review. Eur. J. Environ. Civ. Eng. **18**, 129–172 (2014)

Farzadnia, N., Abang Ali, A.A., Demirboga, R., Anwar, M.P.: Characterization of high strength mortars with nano Titania at elevated temperatures. Constr. Build. Mater. **43**(Complete): 469–479 (2013)

Fathifazl, G.:. Structural performance of steel reinforced recycled concrete members, Carleton University (2008)

Feng, Z., Zhao, Y., Zeng, W., Lu, Z., Shah, S.P.J.C., Materials, B.: Using microbial carbonate precipitation to improve the properties of recycled fine aggregate and mortar. **230**: 116949 (2020)

Ferrara, L., Deegan, P., Pattarini, A., Sonebi, M., Taylor, S.: Recycling ceramic waste powder: effects its grain-size distribution on fresh and hardened properties of cement pastes/mortars formulated from SCC mixes. J. Sustain. Cem. Mater. **8**, 145–160 (2019). https://doi.org/10.1080/21650373.2018.1564396

Florea, M.V.A., Ning, Z., Brouwers, H.J.H.: Activation of liberated concrete fines and their application in mortars. Constr. Build. Mater. **50**, 1–12 (2014). https://doi.org/10.1016/j.conbuildmat.2013.09.012

Fujita, Y., Ferris, F.G., Lawson, R.D., Colwell, F.S., Smith, R.W.J.G.J.: Subscribed content calcium carbonate precipitation by ureolytic subsurface bacteria **17**(4): 305–318 (2000)

Gaimster, R., Dixon, N. (2003). https://doi.org/10.1016/B978-075065686-3/50295-0

Gokce, A., Nagataki, S., Saeki, T., Hisada, M.: Freezing and thawing resistance of air-entrained concrete incorporating recycled coarse aggregate: the role of air content in demolished concrete. Cem. Concr. Res. **34**(5), 799–806 (2004)

Goodbrake, C.J., Young, J.F., Berger, R.L.: Reaction of beta-dicalcium silicate and tricalcium silicate with carbon dioxide and water vapor. J. Am. Ceram. Soc. **62**(3–4), 168–171 (1979)

Grabiec, A.M., Klama, J., Zawal, D., Krupa, D.: Modification of recycled concrete aggregate by calcium carbonate biodeposition. Constr. Build. Mater. **34**, 145–150 (2012)

Güneyisi, E., Gesoğlu, M., Özbay, E.: Effects of marble powder and slag on the properties of self compacting mortars. Mater. Struct. Constr. **42**, 813–826 (2009). https://doi.org/10.1617/s11527-008-9426-2

Güneyisi, E., Gesoğlu, M., Algın, Z., Yazıcı, H.: Effect of surface treatment methods on the properties of self-compacting concrete with recycled aggregates. Constr. Build. Mater. **64**, 172–183 (2014)

Gupta, A., Ghosh, S., Mandal, S.: Coated recycled aggregate concrete exposed to elevated temperature. Glob. J. Res. Eng. **12**(3-E) (2012)

Hammes, F., Verstraete, W.: Key roles of pH and calcium metabolism in microbial carbonate precipitation **1**(1): 3–7 (2002)

Han, J., Liang, Y., Sun, W., Liu, W., Wang, S.: Microstructure modification of carbonated cement paste with six kinds of modern microscopic instruments. J. Mater. Civ. Eng. (2014)

Hansen, T.C., Narud, H.: Strength of recycled concrete made from crushed concrete coarse aggregate. Concr. Int. **5**(1), 79–83 (1983)

Hasan, A., Ayob, A., Razali, M.E., Alias, S., Ahmad, A.G., Ali, D.S.H., Khan, A.A., Mannan, M.A., Hipolito, C.N., Sutan, N.M., Othman, A.-K.H., Kabit, M.R., Wahab, N.A.: Engineering Behavior of Concrete with Recycled Aggregate. In: MATEC Web of Conferences 87 (2016)

Hermawan, D., Hata, T., Umemura, K., Kawai, S., Kuroki, Y.: Rapid production of high-strength cement-bonded particleboard using gaseous or supercritical carbon dioxide. J. Wood Sci. **47**(4), 294–300 (2001)

Hosseini, P., Booshehrian, A., Madari, A.: Developing concrete recycling strategies by utilization of nano-SiO_2particles **2**(3): 347–355 (2011)

Houst, Y.F., Wittmann, F.H.: Influence of porosity and water content on the diffusivity of CO_2 and O_2 through hydrated cement paste (1994)

Ismail, S., Ramli, M.: Engineering properties of treated recycled concrete aggregate (RCA) for structural applications. Constr. Build. Mater. **44**, 464–476 (2013)

Ismail, S., Ramli, M.: Mechanical strength and drying shrinkage properties of concrete containing treated coarse recycled concrete aggregates. Constr. Build. Mater. **68**, 726–739 (2014)

Jing-Xian, W., Li-Jiu, W.: Advances in the applied research of nano-material in concrete. Concrete (2004)

Johannesson, B., Utgenannt, P.: Microstructural changes caused by carbonation of cement mortar. Cem. Concr. Res. **31**(6), 925–931 (2001)

Jonkers, H.M.: Self healing concrete: a biological approach, pp. 195–204. Springer, Self healing materials (2007)

Kai, G., Zhu, P., Korayem, A.H., Ling, Q., Dan, L., Collins, F., Wang, C.M., Wen, H.D.: Reinforcing effects of graphene oxide on Portland cement paste. J. Mater. Civ. Eng. **27**(2), A4014010 (2014)

Kasai and Yoshio: Development and subjects of recycled aggregate concrete in Japan. Key Eng. Mater. **302–303**, 288–300 (2006)

Katz, A.: Treatments for the improvement of recycled aggregate. J. Mater. Civ. Eng. **16**(6), 597–603 (2004)

Kauer, J., Freeman, R.: Effect of carbon dioxide on fresh concrete. J. Proc. (1955)

Khatib, J.M.: Properties of concrete incorporating fine recycled aggregate. Cem. Concr. Res. **35**(4), 763–769 (2005)

Kim, J.-H., Robertson, R.E., Naaman, A.E.: Structure and properties of poly (vinyl alcohol)-modified mortar and concrete. Cem. Concr. Res. **29**(3), 407–415 (1999)

Kim et al. (2012). https://doi.org/10.1016/j.conbuildmat.2011.11.042

Klemm, W.A., Berger, R.L.: Accelerated curing of cementitious systems by carbon dioxide: part I. Portland Cem **2**(5): 567–576 (1972)

Knopf, F.C., Roy, A., Samrow, H.A., Dooley, K.M.: High-pressure molding and carbonation of cementitious materials. Ind. Eng. Chem. Res. **38**(7), 2641–2649 (1999)

Kong, D., Lei, T., Zheng, J., Ma, C., Jiang, J., Jiang, J.: Effect and mechanism of surface-coating pozzalanics materials around aggregate on properties and ITZ microstructure of recycled aggregate concrete. Constr. Build. Mater. **24**(5), 701–708 (2010)

Kong, D., Corr, D.J., Hou, P., Yang, Y., Shah, S.P.: Influence of colloidal silica sol on fresh properties of cement paste as compared to nano-silica powder with agglomerates in micron-scale. Cement and Concrete Composites, **63**, 30–41 (2015)

Kou, S.-C., Poon, C.-S.: Properties of concrete prepared with PVA-impregnated recycled concrete aggregates. Cement Concr. Compos. **32**(8), 649–654 (2010)

Kou, S.C., Zhan, B.J., Poon, C.-S.: Use of a CO_2 curing step to improve the properties of concrete prepared with recycled aggregates. Cement Concr. Compos. **45**, 22–28 (2014)

Larbi, J., Heijnen, W., Brouwer, J., Mulder, E.: Preliminary laboratory investigation of thermal treated recycled concrete aggregate for general use in concrete. Waste materials in construction-WASCON. Pergamon, England, 129–139 (2000)

Leemann, A., Moro, F.: Carbonation of concrete: the role of CO_2 concentration, relative humidity and CO_2 buffer capacity. Mater. Struct. **50**(1), 30 (2017)
Leung, C.K., Pheeraphan, T.: Microwave curing of Portland cement concrete: experimental results and feasibility for practical applications. Constr. Build. Mater. **9**(2), 67–73 (1995)
Li, W., Ebadian, M., White, T., Grubb, R.: Heat transfer within a concrete slab applying the microwave decontamination process (1993)
Li, J., Xiao, H., Zhou, Y.: Influence of coating recycled aggregate surface with pozzolanic powder on properties of recycled aggregate concrete. Constr. Build. Mater. **23**(3), 1287–1291 (2009)
Li, Q., Quan, H., Qin, Y.: Performance and application technology of recycled concrete. China Building Materials Press. Beijing, China, 167–168 (2010)
Li, W., Huang, Z., Cao, F., Sun, Z., Shah, S.P.: Effects of nano-silica and nano-limestone on flowability and mechanical properties of ultra-high-performance concrete matrix. Constr. Build. Mater. **95**, 366–374 (2015)
Li, L., Xiao, J., Poon, C.S.: Dynamic compressive behavior of recycled aggregate concrete. Mater. Struct.struct. **49**(11), 4451–4462 (2016)
Li, W., Luo, Z., Long, C., Wu, C., Duan, W.H., Shah, S.P.: Effects of nanoparticle on the dynamic behaviors of recycled aggregate concrete under impact loading. Materials & Design **112**, 58–66 (2016)
Li, W., Long, C., Tam, V.W.Y., Poon, C.-S., Duan, W.H.: Effects of nano-particles on failure process and microstructural properties of recycled aggregate concrete. Constr. Build. Mater. **142**(Complete): 42–50 (2017)
Lin, K.L., Chen, B.Y., Chiou, C.S., Cheng, A.: Waste brick's potential for use as a pozzolan in blended Portland cement. Waste Manage. Res. **28**(7), 647–652 (2010). https://doi.org/10.1177/0734242X09355853
Lin, C., Wei, W., Hu, Y.H.: Catalytic behavior of graphene oxide for cement hydration process. J. Phys. Chem. Solids **89**, 128–133 (2016)
Van Lith, Y., Warthmann, R., Vasconcelos, C., Mckenzie, J.A.J.G.: Sulphate-reducing bacteria induce low-temperature Ca-dolomite and high Mg-calcite formation, **1**(1): 71–79 (2003)
Liu, M.: Incorporating ground glass in self-compacting concrete. Constr. Build. Mater. **25**(2), 919–925 (2011)
Lotfi, S., Rem, P.: Recycling of end of life concrete fines into hardened cement and clean sand. J. Environ. Prot. **07**, 934–950 (2016)
Lotfi, S., Eggimann, M., Wagner, E., Mróz, R., Deja, J.: Performance of recycled aggregate concrete based on a new concrete recycling technology. Constr. Build. Mater. **95**, 243–256 (2015)
Luo, M., Qian, C.-X., Li, R.-Y.J.C., Materials, B.: Factors affecting crack repairing capacity of bacteria-based self-healing concrete **87**: 1–7 (2015)
Ma, Z., Li, W., Wu, H., Cao, C.: Chloride permeability of concrete mixed with activity recycled powder obtained from C&D waste. Constr. Build. Mater. **199**, 652–663 (2019). https://doi.org/10.1016/j.conbuildmat.2018.12.065
Ménard, Y., Bru, K., Touzé, S., Lemoign, A., Poirier, J.E., Ruffié, G., Bonnaudin, F., Von Der Weid, F.: Innovative process routes for a high-quality concrete recycling. Waste Manage. **33**(6), 1561–1565 (2013)
Metaxa, Z.S., Seo, J.-W.T., Konsta-Gdoutos, M.S., Hersam, M.C., Shah, S.P.: Highly concentrated carbon nanotube admixture for nano-fiber reinforced cementitious materials. Cement Concr. Compos. **34**(5), 612–617 (2012)
Movassaghi, R.: Durability of reinforced concrete incorporating recycled concrete as aggregate (RCA). University of Waterloo (2006)
Mindess, S., Young, F., Darwin, D.: Concrete 2nd edn. Technical documents (2003)
Mukharjee, B.B., Barai, S.V.: Influence of incorporation of nano-silica and recycled aggregates on compressive strength and microstructure of concrete. Constr. Build. Mater. **71**, 570–578 (2014a)
Mukharjee, B.B., Barai, S.V.: Influence of Nano-Silica on the properties of recycled aggregate concrete. Constr. Build. Mater. **55**(2), 29–37 (2014b)

Mulder, E., de Jong, T.P., Feenstra, L.: Closed cycle construction: an integrated process for the separation and reuse of CDW. Waste Manage. **27**(10), 1408–1415 (2007)

Mulder, E., Blaakmeer, J., Nijland, T., Tamboer, L.: A closed material cycle for concrete, as part of an integrated process for the reuse of the total flow of CDW. In: Challenges of concrete construction: volume 5, sustainable concrete construction: proceedings of the international conference held at the university of Dundee, Scotland, UK on 9–11 Sept 2002. Thomas Telford Publishing (2002)

De Muynck, W., Cox, K., De Belie, N., Verstraete, W.J.C., Materials, B.: Bacterial carbonate precipitation as an alternative surface treatment for concrete. **22**(5): 875–885 (2008)

De Muynck, W., De Belie, N., Verstraete, W.J.E.E.: Microbial carbonate precipitation in construction materials: a review. **36**(2): 118–136 (2010)

Nagataki, S., Gokce, A., Saeki, T., Hisada, M.: Assessment of recycling process induced damage sensitivity of recycled concrete aggregates. Cem. Concr. Res. **34**(6), 965–971 (2004)

Neville, A.M.: Properties of concrete. UK, Longman, London (2011)

Ollivier, J.P., Maso, J.C., Bourdette, B.: Interfacial transition zone in concrete. Adv. Cem. Based Mater. **2**(1), 30–38 (1995)

Pan, Z., He, L., Qiu, L., Korayem, A.H., Li, G., Zhu, J.W., Collins, F., Li, D., Duan, W.H., Wang, M.C.: Mechanical properties and microstructure of a graphene oxide–cement composite. Cement Concr. Compos. **58**, 140–147 (2015)

Pawluczuk, E.: Recycled concrete powder as partial cement replacement in fine-grained concrete. International Multidisciplinary Scientific GeoConference: SGEM, **17**, 169–176 (2017)

Pepe, M., Toledo Filho, R.D., Koenders, E.A.B., Martinelli, E.: A novel mix design methodology for recycled aggregate concrete. Constr. Build. Mater. **122**, 362–372 (2016)

Piasta, J., Sawicz, Z., Rudzinski, L.: Changes in the structure of hardened cement paste due to high temperature. Matériaux Constr **17**(4), 291–296 (1984)

Polat, R., Demirboğa, R., Khushefati, W.H.: Effects of nano and micro size of CaO and MgO, nano-clay and expanded perlite aggregate on the autogenous shrinkage of mortar. Constr. Build. Mater. **81**, 268–275 (2015)

Prajapati, R., Gettu, R., Singh, S.: Thermomechanical beneficiation of recycled concrete aggregates (RCA). Constr. Build. Mater. **310**, 125200 (2021). https://doi.org/10.1016/j.conbuildmat.2021.125200

Prokopski, G., Halbiniak, J.: Interfacial transition zone in cementitious materials. Cem. Concr. Res. **30**(4), 579–583 (2000)

Purnell, P., Short, N.R., Page, C.L.: Super-critical carbonation of glass-fibre reinforced cement. Part 1: mechanical testing and chemical analysis. Compos. A Appl. Sci. Manuf. **32**(12), 1777–1787 (2001)

Purushothaman, R., Amirthavalli, R.R., Karan, L.: Influence of treatment methods on the strength and performance characteristics of recycled aggregate concrete. J. Mater. Civ. Eng. **27**(5), 04014168 (2015)

Qiu, J., Tng, D.Q.S., Yang, E.-H.: Surface treatment of recycled concrete aggregates through microbial carbonate precipitation. Constr. Build. Mater. **57**, 144–150 (2014)

Quattrone, M., Angulo, S.C., John, V.M.: Energy and CO_2 from high performance recycled aggregate production. Resour. Conserv. Recycl. **90**, 21–33 (2014)

Rana, A., Kalla, P., Verma, H.K., Mohnot, J.K.: Recycling of dimensional stone waste in concrete: A review. J. Clean. Prod. **135**, 312–331 (2016). https://doi.org/10.1016/j.jclepro.2016.06.126

Rimmelé, G., Barlet-Gouédard, V., Porcherie, O., Goffé, B., Brunet, F.: Heterogeneous porosity distribution in Portland cement exposed to CO_2-rich fluids. Cem. Concr. Res. **38**(8–9), 1038–1048 (2008)

Sadek, D.M., El-Attar, M.M., Ali, H.A.: Reusing of marble and granite powders in self-compacting concrete for sustainable development. J. Clean. Prod. **121**, 19–32 (2016). https://doi.org/10.1016/j.jclepro.2016.02.044

Sakai, K., Noguchi, T.: The sustainable use of concrete. CRC Press (2012)

Santos, W.F., Quattrone, M., John, V.M., Angulo, S.C.: Roughness, wettability and water absorption of water-repellent treated recycled aggregates. Constr. Build. Mater. **146**, 502–513 (2017)

Shah, S.P., Hou, P., Konsta-Gdoutos, M.S.: Nano-modification of cementitious material: toward a stronger and durable concrete. J. Sustain. Cem. Based Mater. 1–22 (2016)

Shayan, A., Xu, A.M.: Performance and properties of structural concrete made with recycled concrete aggregate. ACI Mater. J. **100**(5), 371–380 (2003)

Shen, J., Cheng, X.: Laboratory investigation on restoration of Chinese ancient masonry buildings using microbial carbonate precipitation. In: 1st International Conference on Bio-Geo-Civil Eng., Netherlands (2008)

Shi, C., Li, Y., Zhang, J., Li, W., Chong, L., Xie, Z.: Performance enhancement of recycled concrete aggregate—a review. J. Clean. Prod. **112**, 466–472 (2016)

Shima, H., Tateyashiki, H., Matsuhashi, R., Yoshida, Y.: An advanced concrete recycling technology and its applicability assessment through input-output analysis. J. Adv. Concr. Technol. **3**(1), 53–67 (2005)

Shui, Z., Xuan, D., Wan, H., Cao, B.: Rehydration reactivity of recycled mortar from concrete waste experienced to thermal treatment. Constr. Build. Mater. **22**, 1723–1729 (2008)

Silva, R.V., Neves, R., Brito, J.D., Dhir, R.K.: Carbonation behavior of recycled aggregate concrete. Cement Concr. Compos. **62**, 22–32 (2015)

Sorochkin, M.A., Shchurov, A.F., Safonov, I.A.: Study of the possibility of using CO_2 for accelerating the hardening of products made from portland cement. J. Appl. Chem. USSR **48**, 1271–1274 (1975)

Spaeth, V., Tegguer, A.D.: Improvement of recycled concrete aggregate properties by polymer treatments. Int. J. Sustain. Built Environ. **2**(2), 143–152 (2013)

Stuart, E.B.: Method for producing a hard paving or surfacing material. Google Patents (1970)

Sui, Y., Mueller, A.: Development of thermo-mechanical treatment for recycling of used concrete. Mater. Struct. **45**(10), 1487–1495 (2012)

Tam, V.W.Y., Tam, C.M.: Assessment of durability of recycled aggregate concrete produced by two-stage mixing approach. J. Mater. Sci. **42**(10), 3592–3602 (2007)

Tam, V.W., Tam, C.M., Le, K.N.: Removal of cement mortar remains from recycled aggregate using pre-soaking approaches. Resour. Conserv. Recycl. **50**(1), 82–101 (2007a)

Tam, V.W.Y., Le, K.N., Wang, J., Wang, X.: Development of a sustainable checklist in construction. Proc. Inst. Civ. Eng. Waste Resour. Manag. **169**(4), 166–180 (2016)

Taylor, H.F.: Cement chemistry **2**(459). London: Thomas Telford. (1997)

Topcu, I.B., Kocataşkin, F.: A two-phase composite materials approach to the workability of concrete. Cement and Concrete Composites, **17**(4), 319–325 (1995)

Topič, J., Prošek, Z.: Hydration process and mechanical properties of cement paste with recycled concrete powder and silica sand powder. Acta Polytech. CTU Proc. **13**, p. 125, (2017). https://doi.org/10.14311/APP.2017.13.0125

Tsujino, M., Noguchi, T., Tamura, M., Kanematsu, M., Maruyama, I.: Application of conventionally recycled coarse aggregate to concrete structure by surface modification treatment. J. Adv. Concr. Technol. **5**(1), 13–25 (2007)

Tyson, B.M., Abu Al-Rub, R.K., Yazdanbakhsh, A., Grasley, Z.: Carbon nanotubes and carbon nanofibers for enhancing the mechanical properties of nanocomposite cementitious materials. J. Mater. Civ. Eng. **23**(7), 1028–1035 (2011)

Vázquez, E.: Progress of recycling in the built environment: final report of the RILEM Technical Committee 217-PRE, 8 Springer Science & Business Media (2012)

Wagh, A., Singh, D., Knox, L.: Rapid setting of Portland cement by greenhouse carbon dioxide capture. Argonne National Lab., IL (United States) (1994)

Wang, N., Zhang, K.F., Yao, Y., Cui, Q.Y., Huang, P.Z., Deng, T.M.: Performance of recycled pulverized fodder of construction waste with different particle size. Bull. Chinese Ceram. Soc., **34**, 277–279 (2015)

Wang, J., Vandevyvere, B., Vanhessche, S., Schoon, J., Boon, N., De Belie, N.: Microbial carbonate precipitation for the improvement of quality of recycled aggregates. J. Clean. Prod. **156**, 355–366 (2017)

Wang, J., Wu, H., Tam, V.W.Y., Zuo, J.: Considering life-cycle environmental impacts and society's willingness for optimizing construction and demolition waste management fee: An empirical study of China. J. Clean. Prod. **206**, 1004–1014 (2019). https://doi.org/10.1016/j.jclepro.2018.09.170

Wright, D.T.J.S.G.: The role of sulphate-reducing bacteria and cyanobacteria in dolomite formation in distal ephemeral lakes of the Coorong region. South Australia **126**(1–4): 147–157 (1999)

Wu, L., Shi, C., Wu, Z., Lv, K.: A review on mixture design methods for self-compacting concrete. Constr. Build. Mater. **84**, 387–398 (2015). https://doi.org/10.1016/j.conbuildmat.2015.03.079

Wu, C.-R., Zhu, Y.-G., Zhang, X.-T., Kou, S.-C.J.C., Composites C.: Improving the properties of recycled concrete aggregate with bio-deposition approach **94**: 248–254 (2018)

Xiao, J., Ma, Z., Ding, T.: Reclamation chain of waste concrete: a case study of Shanghai. Waste Manag. **48**, 334–343 (2016)

Xiao, J., Ma, Z., Sui, T., Akbarnezhad, A., Duan, Z.: Mechanical properties of concrete mixed with recycled powder produced from construction and demolition waste. J. Clean. Prod. **188**, 720–731 (2018). https://doi.org/10.1016/j.jclepro.2018.03.277

Xuan, D., Zhan, B., Poon, C.S.: Assessment of mechanical properties of concrete incorporating carbonated recycled concrete aggregates. Cement Concr. Compos. **65**, 67–74 (2016)

Yonezawa, T., Kamiyama, Y., Yanagibashi, K., Kojima, M., Arakawa, K., Yamada, M.: A study on a technology for producing high-quality recycled coarse aggregate. J. Soc. Mater. Sci. Jpn. **50**, 835–842 (2001)

Zeng, W., Zhao, Y., Poon, C.S., Feng, Z., Lu, Z., Shah, S.P.J.C., Materials, B.: Using microbial carbonate precipitation to improve the properties of recycled aggregate **228**: 116743 (2019)

Zhan, B., Poon, C.S., Liu, Q., Kou, S., Shi, C.: Experimental study on CO2 curing for enhancement of recycled aggregate properties. Constr. Build. Mater. **67**, 3–7 (2013)

Zhan, B., Poon, C.S., Liu, Q., Kou, S., Shi, C.: Experimental study on CO2 curing for enhancement of recycled aggregate properties. Constr. Build. Mater. **67**, 3–7 (2014)

Zhang, H., Zhao, Y.: Performance of recycled concrete beams under sustained loads coupled with chloride-ion (Cl−) ingress. Constr. Build. Mater. **128**, 96–107 (2016)

Zhang, J., Shi, C., Li, Y., Pan, X., Poon, C.-S., Xie, Z.: Influence of carbonated recycled concrete aggregate on properties of cement mortar. Constr. Build. Mater. **98**, 1–7 (2015a)

Zhang, R., Cheng, X., Hou, P., Ye, Z.: Influences of nano-TiO_2 on the properties of cement-based materials: hydration and drying shrinkage. Constr. Build. Mater. **81**, 35–41 (2015b)

Zhang, H., Zhao, Y., Meng, T., Shah, S.P.: Surface treatment on recycled coarse aggregates with nanomaterials. J. Mater. Civ. Eng. **28**(2), 04015094 (2015c)

Zhang, Y., Kong, X., Gao, L., Bai, Y.: Characterization of the mesostructural organization of cement particles in fresh cement paste. Constr. Build. Mater. **124**, 1038–1050 (2016). https://doi.org/10.1016/j.conbuildmat.2016.08.143

Zhang, P., Wan, J., Wang, K., Li, Q.: Influence of nano-SiO_2 on properties of fresh and hardened high performance concrete: a state-of-the-art review. Constr. Build. Mater. **148**, 648–658 (2017)

Zhao, Z., Remond, S., Damidot, D., Xu, W.: Influence of hardened cement paste content on the water absorption of fine recycled concrete aggregates. J. Sustain. Cem. Based Mater. **2**(3–4), 186–203 (2013)

Zhao, Y., Dong, J., Ding, H., Jin, W.: Shape of corrosion-induced cracks in recycled aggregate concrete. Corros. Sci. **98**, 310–317 (2015)

Zhao, Z., Remond, S., Damidot, D., Courard, L., Michel, F.: Improving properties of recycled concrete aggregates by accelerated carbonation. Proc. Inst. Civ. Eng. Constr. Mater. **171**(3), 126–132 (2018)

Zhao, Z., Courard, L., Groslambert, S., Jehin, T., Leonard, A., Xiao, J.: Use of recycled concrete aggregates from pre-cast block for the production of new building blocks: an industrial scale study. Resour. Conserv. Recycl. **157**, 104786 (2020)

Zhang, L.: Nanomaterials and nanotechnology in China: current status of application and opportunities for commercialization. Mater. Rev. (2001)

Mix Proportion Design Method of RAC and Properties of Fresh RA

Romildo Dias Toledo Filho and Wengui Li

1 Optimal Mix Approach for RAC

Environmental control is an increasingly pressing concern in the construction industry. Natural resources are consumed in its day-to-day operations, and a large amount of waste is generated (Tam 2008; Yuan and Shen 2011). Construction activities impose significant environmental impacts over the entire construction life cycle (Polster et al. 1996; Morledge and Jackson 2001; Tam et al. 2012). Waste management in the construction industry has not been successfully controlled, and it is challenging to initiate improvement. It has been thought that the re-use and recycling of construction materials will effectively reduce impacts on limited landfill spaces and improve waste management.

Concrete is an essential, mass-produced material in the construction industry. However, much effort has been made to recycle and conserve precious natural resources. Repeated recycling can be suitable for concrete, as is the case for steel and aluminum (Noguchi and Tamura 2001; Tam et al. 2010). An effective method would be using RCA in the production of RAC. RCA is a particle of stones attached with old cement mortar generated by crushing demolished concrete waste. RAC is created by mixing RCA with other natural ingredients, including cement, water, fine aggregate and other materials. As concrete is composed only of cementitious materials, and powders generated during the production of RCA can be reprocessed as cement resources, repeated recycling is possible. This also enables concrete to be recycled

R. D. Toledo Filho
Civil Engineering Program, COPPE, Federal University of Rio de Janeiro, Rio de Janeiro, Brazil

W. Li (✉)
School of Civil and Environmental Engineering, The University of New South Wales, Sydney, NSW, Australia
e-mail: wengui.li@unsw.edu.au

J. Xiao et al. (eds.), *Structural Behaviour and Innovation of Recycled Aggregate Concrete*, RILEM State-of-the-Art Reports 41,
https://doi.org/10.1007/978-3-031-84136-1_3

in a fully closed system, thus improving the environment by reducing landfill and concrete waste.

The benefits to the environment of using RAC include economic aspects, reduction of environmental impacts and saving on resources (Tech Data Sheet 1998, Carneiro 2000, Tam et al. 2010). Although RCA reduces the consumption of limited natural resources and thereby saves cost, there are also shortcomings. For example, weaker interfacial behavior between aggregate and cement paste and high portions of attached cement mortar leads to low concrete quality (Xiao et al. 2012; Dong et al. 2013; Tam et al. 2014). To facilitate waste management by using RCA, it is vital to investigate how these shortcomings can be overcome.

1.1 Research Methodologies

RCA samples collected from south-eastern Australia's centralized recycling plant were used for investigating the properties of RAC. Table 1 summarizes the properties and standards used for the RCA samples.

Experimental work on RAC from different mix designs, including replacement ratios of RCA from 0 to 100%, water-to-cement ratios at 0.35, 0.40, 0.45, 0.50, 0.55 and 0.60, and aggregate to cement ratios at 3.0, 3.5, 4.0, 4.5, 5.0, 5.5 and 6.0, were examined. Details of all mix designs used for the RAC experimental work are shown in Table 2. In total 44 mixes were conducted in this paper.

The RAC mixing was first charged with about half coarse aggregate, fine aggregate, cement, and finally, the remaining coarse aggregate. According to the Australian Standard, water was then immediately added after starting the operation for two

Table 1 Properties and standards used for the RCA samples

Properties types	Properties	Standards used
Uncompacted bulk density	$1.22t/m^3$	AS 1141.4 2014
Compacted bulk density	$1.34t/m^3$	AS 1141.4 2014
Particle density	$2.52t/m^3$	AS 1141.6.1 2014
Water absorption	2.68%	AS 1141.6.1 2014
Aggregate crushing value	22.31%	AS 1141.22 2014
Wet strength	186.90kN	AS 1141.22 2014
Dry strength	155.34kN	AS 1141.22 2014
Weak particles	0.27%	AS 1141.32 2014
Flakiness index	9.70	AS 1141.15 2014
Particle shape, by proportional caliper	0.88	AS 1141.14 2014
Chloride content	0.0180%	AS 1141.20 2014
Sulfate content	0.0468%	AS 1141.20 2014

Table 2 RAC mix designs used for the experimental work

Water-to-cement ratio, R_{wc}	Aggregate to cement ratio, R_{ac}	RCA replacement ratio (%), R_{rca}	Cement (kg), Q_c	RCA 10 mm (kg), Q_{rca10}	NA 10 mm (kg), Q_{nca10}	RCA 20 mm(kg), Q_{nca20}	NA 20 mm (kg), Q_{nca20}	Sand (kg), Q_s	Water (kg), Q_w
0.30	4.5	0	400.00	0.00	360.00	0.00	720.00	720.00	120.00
		30	400.00	108.00	252.00	216.00	504.00	720.00	120.00
		100	400.00	360.00	0.00	720.00	0.00	720.00	120.00
0.35		0	400.00	0.00	360.00	0.00	720.00	720.00	140.00
		30	400.00	108.00	252.00	216.00	504.00	720.00	140.00
		100	400.00	360.00	0.00	720.00	0.00	720.00	140.00
0.40		0	400.00	0.00	360.00	0.00	720.00	720.00	160.00
		30	400.00	108.00	252.00	216.00	504.00	720.00	160.00
		100	400.00	360.00	0.00	720.00	0.00	720.00	160.00
0.45		0	400.00	0.00	360.00	0.00	720.00	720.00	180.00
		10	400.00	36.00	324.00	72.00	648.00	720.00	180.00
		20	400.00	72.00	288.00	144.00	576.00	720.00	180.00
		30	400.00	108.00	252.00	216.00	504.00	720.00	180.00
		40	400.00	144.00	216.00	288.00	432.00	720.00	180.00
		50	400.00	180.00	180.00	360.00	360.00	720.00	180.00
		60	400.00	216.00	144.00	432.00	288.00	720.00	180.00
		70	400.00	252.00	108.00	504.00	216.00	720.00	180.00
		80	400.00	288.00	72.00	576.00	144.00	720.00	180.00
		90	400.00	324.00	36.00	648.00	72.00	720.00	180.00
		100	400.00	360.00	0.00	720.00	0.00	720.00	180.00

(continued)

Table 2 (continued)

Water-to-cement ratio, R_{wc}	Aggregate to cement ratio, R_{ac}	RCA replacement ratio (%), R_{rca}	Cement (kg), Q_c	RCA 10 mm (kg), Q_{rca10}	NA 10 mm (kg), Q_{nca10}	RCA 20 mm(kg), Q_{nca20}	NA 20 mm (kg), Q_{nca20}	Sand (kg), Q_s	Water (kg), Q_w
0.50		0	400.00	0.00	360.00	0.00	720.00	720.00	200.00
		30	400.00	108.00	252.00	216.00	504.00	720.00	200.00
		100	400.00	360.00	0.00	720.00	0.00	720.00	200.00
0.55		0	400.00	0.00	360.00	0.00	720.00	720.00	220.00
		30	400.00	108.00	252.00	216.00	504.00	720.00	220.00
		100	400.00	360.00	0.00	720.00	0.00	720.00	220.00
0.60		0	400.00	0.00	360.00	0.00	720.00	720.00	240.00
		30	400.00	108.00	252.00	216.00	504.00	720.00	240.00
		100	400.00	360.00	0.00	720.00	0.00	720.00	240.00
0.45	3.0	0	600.00	0.00	360.00	0.00	720.00	720.00	180.00
		30	600.00	108.00	252.00	216.00	504.00	720.00	180.00
		100	600.00	360.00	0.00	720.00	0.00	720.00	180.00
	3.5	0	514.29	0.00	360.00	0.00	720.00	720.00	180.00
		30	514.29	108.00	252.00	216.00	504.00	720.00	180.00
		100	514.29	360.00	0.00	720.00	0.00	720.00	180.00
	4.0	0	450.00	0.00	360.00	0.00	720.00	720.00	180.00
		30	450.00	108.00	252.00	216.00	504.00	720.00	180.00
		100	450.00	360.00	0.00	720.00	0.00	720.00	180.00
	5.0	0	360.00	0.00	360.00	0.00	720.00	720.00	180.00
		30	360.00	108.00	252.00	216.00	504.00	720.00	180.00
		100	360.00	360.00	0.00	720.00	0.00	720.00	180.00
	5.5	0	327.27	0.00	360.00	0.00	720.00	720.00	180.00

(continued)

Table 2 (continued)

Water-to-cement ratio, R_{wc}	Aggregate to cement ratio, R_{ac}	RCA replacement ratio (%), R_{rca}	Cement (kg), Q_c	RCA 10 mm (kg), Q_{rca10}	NA 10 mm (kg), Q_{nca10}	RCA 20 mm(kg), Q_{nca20}	NA 20 mm (kg), Q_{nca20}	Sand (kg), Q_s	Water (kg), Q_w
		30	327.27	108.00	252.00	216.00	504.00	720.00	180.00
		100	327.27	360.00	0.00	720.00	0.00	720.00	180.00
	6.0	0	300.00	0.00	360.00	0.00	720.00	720.00	180.00
		30	300.00	108.00	252.00	216.00	504.00	720.00	180.00
		100	300.00	360.00	0.00	720.00	0.00	720.00	180.00

minutes (AS1012 1993). All mixes are controlled with a slump value between 70 and 100 mm without the addition of any additives.

Three major properties of RAC were under investigation: compressive strength, tensile strength, and flexural strength. The 28-day compressive, tensile and flexural strengths from the different mix designs for the RAC were conducted based on the Australian Standard (ASTM C293-00 2001, AS 1391 2007, AS 1012.9 2014). The average of three samples of the 28-day compressive, tensile and flexural strength tests was calculated. The characteristic compressive ($f_c{'}$), tensile ($f_{ct}{'}$) and flexural ($f_{cf}{'}$) strengths of RAC were used for the development of prediction equations for the optimal mix approach. Equation (1) shows the calculation method for the characteristic compressive strength from the average compressive strength (ACI Committee 311 1992). The same calculation method is used for the characteristic tensile and flexural strengths.

$$f_c' = f_{cr} - 1.65\sigma \quad (1)$$

where $f_c{'}$ is the characteristic compressive strength; f_{cr} is the average compressive strength, and σ is the standard deviation with a 95% confidence level.

Data collected from the experimental work were analyzed using the Statistical Package for Social Sciences Version 18.0 for Windows. Before we conduct correlation and regression analysis of the experimental work data, a sensitivity analysis was undertaken. Sensitivity analysis is the study of how the uncertainty in the output of a mathematical model or system (numerical or otherwise) can be apportioned to different sources of uncertainty in its input (Saltelli et al. 2008), which can be measured by Eq. (1) (Levy and Rockafellar 1994). All four available regression analysis methods, including Stepwise, Remove, Backward and Forward, were used. All regression analysis in this paper is assumed as linear.

$$G(p) = \left[x \in C| - F(p, x) \in M(x)\right] \quad (2)$$

To verify the accuracy of the developed equations from the development of optimal mix approach, twelve different mix designs for the RAC were conducted for verification test (see Table 3).

2 The Existing Normal Aggregate Concrete Mix Design

For NAC, the common method of expressing the proportions of ingredients of a concrete mix is in the terms of parts or ratios of cement and fine (sand) and coarse aggregate. For example, a concrete mix of 1:2:4 is the ratio of cement: fine aggregate: coarse aggregate. The proportions can be either by volume or by mass. The water-to-cement ratio is usually expressed in mass (AS 2001).

The general concrete mix design for industrial usage is proportional for various purposes (Engineer Forum 2011): (i) ratios of 1:1:2 and 1:2.2:2.4 for high-strength

Table 3 RAC verification mix designs

Water to-cement ratio, R_{wc}	Aggregate to cement ratio, R_{ac}	RCA replacement ratio (%), R_{rca}	Cement (kg), Q_c	RCA 10 mm (kg), Q_{rca10}	NA 10 mm (kg), Q_{nca10}	RCA 20 mm (kg), Q_{nca20}	NA 20 mm (kg), Q_{nca20}	Sand (kg), Q_s	Water (kg), Q_w
0.40	3.5	10	473.51	33.18	298.35	66.35	596.47	662.82	189.31
		50	473.51	165.65	165.65	331.53	331.53	662.82	189.31
		90	473.51	298.35	33.18	596.47	66.35	662.82	189.31
	5.5	10	336.17	36.89	332.92	74.01	665.84	739.62	134.56
		50	336.17	184.90	184.90	369.81	369.81	739.62	134.56
		90	336.17	332.92	36.89	665.84	74.01	739.62	134.56
0.55	3.5	10	459.36	32.25	289.54	64.26	578.84	643.10	252.65
		50	459.36	160.78	160.78	321.55	321.55	643.10	252.65
		90	459.36	289.54	32.25	578.84	64.26	643.10	252.65
	5.5	10	328.98	36.19	325.73	72.38	651.69	724.07	180.96
		50	328.98	180.96	180.96	361.92	361.92	724.07	180.96
		90	328.98	325.73	36.19	651.69	72.38	724.07	180.96

concrete; (ii) ratios of 1:1.5:3 and 1:2:4 for general concrete and reinforced concrete, respectively; and (iii) ratios of 1:3:6 and 1:4:8 for foundation and mass concrete work respectively. To achieve the target strength, a suitable water-to-cement ratio is needed for each proportion of cement, fine and coarse aggregate.

Although this guideline in AS 3600 2001 provides the ratios required for different concrete applications, this guideline is not suitable for RAC as the strength can be significantly affected by RCA replacement ratios. Therefore, it is necessary to develop an optimal mix approach considering RCA properties for the use of RAC in the industry.

Table 4 summarizes the results of compressive, tensile and flexural strengths of the RAC samples from the experimental work in this paper. The relationship between different parameters of the RAC, including the RCA replacement ratio, amount of water used, amount of cement used, amount of sand used, water-to-cement ratio and aggregate to cement ratio, and compressive, tensile and flexural strengths are linear. The implication is that the higher RCA replacement ratio will deteriorate the concrete strength.

Linear regression analysis is used for the development of the prediction equations for the RAC optimal mix approach. Table 5 shows the correlation among the characteristic strengths (f_c, f_{ct} and f_{cf}) and different RCA replacements (R_{rca}), water-to-cement (R_{wc}) and aggregate to cement (R_{ac}) ratios, and quantities of 10 mm and 20 mm natural (Q_{nca10} and Q_{nca20} respectively) and 10 mm and 20 mm recycled coarse aggregate (Q_{rca10} and Q_{rca20} respectively), water (Q_w), cement (Q_c), and sand (Q_s). The correlation between $f_{c'}$ and Q_{rca10}, Q_{nca10}, Q_{rca20}, Q_{nca20}, Q_w, Q_c, Q_s, R_{rca}, R_{wc}, and R_{ac} are 0.538, 0.512, 0.538, 0.512, 0.660, 0.646, 0.344, 0.528, 0.804 and 0.580 respectively. The significant values between the characteristic strengths and RCA parameters are less than 0.05, which means that all correlations are significant

Table 4 Characteristic of compressive, tensile and flexural strength

Water-to-cement ratios	Aggregate to-cement ratios	RCA replacement ratio (%)	Average compressive strength (MPa)	Average tensile strength (MPa)	Average flexural strength (MPa)	Characteristic compressive strength, $f_c{}'$ (MPa)	Characteristic Tensile strength, $f_{ct}{}'$ (MPa)	Characteristic flexural strength, $f_{cf}{}'$ (MPa)
0.30	4.5	0	71.0	5.9	6.6	69.8	5.2	6.3
		30	63.9	5.2	6.1	63.6	4.8	5.8
		100	49.2	4.8	4.1	47.5	4.5	4.0
0.35		0	68.0	5.6	6.2	66.3	4.9	5.9
		30	56.2	4.9	5.5	54.0	4.6	5.2
		100	46.0	4.5	4.0	44.7	3.4	3.9
0.40		0	62.0	5.3	5.8	61.5	4.7	5.3
		30	53.7	4.9	5.0	52.1	4.5	4.5
		100	41.9	4.2	3.8	39.5	4.0	3.3
0.40		0	56.8	5.1	5.6	53.1	4.9	4.7
		10	55.4	5.0	5.0	54.5	4.8	4.8
		20	55.1	4.9	4.9	48.7	4.1	4.3
		30	52.4	4.7	4.6	49.1	4.4	4.0
		40	52.0	4.6	4.4	50.9	4.2	3.7
		50	50.0	4.5	4.1	42.2	4.2	3.8
		60	48.0	4.6	4.4	46.4	4.3	4.2
		70	46.8	4.4	4.2	43.7	3.8	3.8
		80	47.2	4.3	4.2	43.9	3.7	3.7
		90	44.2	4.1	4.1	41.1	4.0	3.5
		100	39.1	3.8	3.8	34.9	3.4	3.7

(continued)

Table 4 (continued)

Water-to-cement ratios	Aggregate to-cement ratios	RCA replacement ratio (%)	Average compressive strength (MPa)	Average tensile strength (MPa)	Average flexural strength (MPa)	Characteristic compressive strength, $f_c{'}$ (MPa)	Characteristic Tensile strength, $f_{ct}{'}$ (MPa)	Characteristic flexural strength, $f_{cf}{'}$ (MPa)
0.50		0	46.0	4.6	4.9	45.3	4.2	4.7
		30	43.4	4.4	4.3	41.2	3.9	3.9
		100	32.0	3.4	3.5	30.4	3.1	2.9
0.55		0	40.0	4.1	4.4	37.2	4.0	4.0
		30	37.2	3.9	4.0	34.7	3.3	3.7
		100	29.0	3.1	3.1	27.4	2.9	2.9
0.60		0	32.2	3.8	4.2	30.4	3.5	3.7
		30	32.3	3.6	3.6	30.6	2.7	3.1
		100	25.3	3.0	3.0	21.8	2.8	2.8
0.45	3.0	0	80.4	6.2	6.7	78.7	6.1	6.6
		30	74.0	5.6	5.8	72.7	4.8	5.1
		100	55.5	4.5	4.8	54.5	4.3	3.9
	3.5	0	73.1	5.8	6.2	72.2	5.0	6.0
		30	65.9	5.2	5.3	63.6	4.7	5.0
		100	51.8	4.3	4.2	50.2	4.1	4.0
	4.0	0	66.1	5.4	5.9	64.4	5.0	5.5
		30	57.1	5.0	4.8	56.2	4.6	4.5
		100	45.6	4.1	3.9	44.1	3.7	3.7

(continued)

Table 4 (continued)

Water-to-cement ratios	Aggregate to-cement ratios	RCA replacement ratio (%)	Average compressive strength (MPa)	Average tensile strength (MPa)	Average flexural strength (MPa)	Characteristic compressive strength, $f_c{}'$ (MPa)	Characteristic Tensile strength, $f_{ct}{}'$ (MPa)	Characteristic flexural strength, $f_{cf}{}'$ (MPa)
	5.0	0	52.2	4.9	4.9	49.5	4.6	4.6
		30	47.6	4.6	4.3	47.2	4.5	3.7
		100	34.4	3.6	3.4	33.4	3.4	3.3
	5.5	0	50.4	4.6	4.4	49.3	4.0	3.4
		30	42.9	4.3	3.9	40.5	3.2	3.6
		100	32.3	3.4	3.1	31.9	3.2	2.7
	6.0	0	42.9	4.3	4.0	41.5	3.9	3.1
		30	34.9	3.7	3.6	34.1	3.3	3.2
		100	27.1	2.9	2.8	24.3	2.3	2.1

at the 95% confidence level. The sensibility analysis results for the data for RAC is 0.918, which shows that the data is suitable for analysis and the development of the optimal mix approach.

There are four methods: Stepwise, Remove, Backward and Forward, used in the linear regression analysis. Prediction equations are developed if the correlation significant (sig. or α) of the models is less than 0.05. Table 6 summarizes the acceptance models for the characteristic compressive strength with their coefficients, R^2 values and correlation significant, which leads to the development of Eqs. (3) and (4). Where f_c' is the characteristic compressive strength; R_{wc} is the water-to-cement ratio; Q_{rca10} is the quantity of 10 mm RCA by proportion (in %); Q_w is the quantity of water by proportion (in %); Q_s is the quantity of sand by proportion (in %); R_{rca} is the RCA replacement ratio; and R_{ac} is the aggregate to cement ratio.

$$f_c' = 105.118-(111.411 \times R_{wc})-(1.161 \times Q_{rca10}) \tag{3}$$

$$\begin{aligned} f_c' &= 324,870 + (69.32 \times Q_w) - (26.148Q_s) \\ &\quad - (0.175 \times R_{rca}) - (132,236 \times R_{wc}) + (131,455 \times R_{ac}) \end{aligned} \tag{4}$$

Tables 7 and 8 summarize the acceptance models for the characteristic tensile and flexural strengths, respectively, with their coefficients, R^2 values and correlation significance, which leads to the development of Eqs (5 and 6) respectively.

$$f_{ct}' = 7.192-(5.892 \times R_{wc}) - (0.01 \times R_{rca}) \tag{5}$$

$$f_{cf}' = 8.246-(7.648 \text{ x } R_{wc})-(0.015 \text{ x } R_{rca}) \tag{6}$$

where f_{ct}' is the characteristic tensile strength; and f_{ct}' is the characteristic flexural strength.

Equations (3–6) are provided to predict mix designs' target characteristic compressive, tensile and flexural strengths. The independent variables for Eq. (3)are water-to-cement ratio and quantity of 10 mm RCA by proportion. The independent variables for Eq. (5) and (6) are water-to-cement ratio and RCA replacement ratios. Equation (3) is re-analyzed using the same independent variables as in Eqs. (5 and 6) in the linear regression analysis with the four methods, Eq. (1) is deviated based on the acceptance model in Table 9.

This study also found that there are strong correlations among the characteristic compressive, tensile and flexural strengths (Table 10). Linear regression analysis is used to develop prediction equations with similar procedures as above. Tables 11, 12 and 13 summarize the acceptance models for setting different characteristic strengths as the dependent variables with their coefficients, R^2 values and correlation significant, which leads to the development of Eqs. (8)–(9).

$$f_c' = 105.509-(112,283 \times R_{wc})-(0.175 \times R_{rca}) \tag{7}$$

Table 5 Correlation between characteristic strengths and quantity of mix proportions

		f_c'	f_{ct}'	f_{cf}'	Q_{rca10}	Q_{nca10}	Q_{rca20}	Q_{nca20}	Q_w	Q_c	Q_s	R_{rca}	R_{wc}	R_{ac}
f_c'	Pearson correlation	1	0.926**	0.915**	−0.538**	0.512**	−0.538**	0.512**	−0.660**	0.646**	−0.344**	−0.528**	−0.804**	−0.580**
	Sig. (1-tailed)		0.000	0.000	0.000	0.000	0.000	0.000	0.000	0.000	0.009	0.000	0.000	0.000
f_{ct}'	Pearson correlation	0.926**	1	0.892**	−0.557**	0.537**	−0.557**	0.537**	−0.628**	0.566**	−0.278*	−0.550**	−0.756**	−0.529**
	Sig. (1-tailed)	0.000		0.000	0.000	0.000	0.000	0.000	0.000	0.000	0.029	0.000	0.000	0.000
f_{cf}'	Pearson correlation	0.915**	0.892**	1	−0.616**	0.594**	−0.616**	0.594**	−0.575**	0.565**	−0.303*	−0.610**	−0.730**	−0.551**
	Sig. (1-tailed)	0.000	0.000		0.000	0.000	0.000	0.000	0.000	0.000	0.019	0.000	0.000	0.000
Q_{rca10}	Pearson correlation	−0.538**	−0.557**	−0.616**	1	−0.997**	1.000**	−0.997**	−0.005	−0.033	0.036	0.999**	0.012	0.029
	Sig. (1-tailed)	0.000	0.000	0.000		0.000	0.000	0.000	0.486	0.412	0.404	0.000	0.469	0.423
Q_{nca10}	Pearson correlation	0.512**	0.537**	0.594**	−0.997**	1	−0.997**	1.000**	−0.010	−0.033	0.038	−0.999**	0.015	0.038
	Sig. (1-tailed)	0.000	0.000	0.000	0.000		0.000	0.000	0.474	0.414	0.400	0.000	0.459	0.400
Q_{rca20}	Pearson correlation	−0.538**	−0.557**	−0.616**	1.000**	−0.997**	1	−0.997**	−0.005	−0.033	0.036	0.999**	0.012	0.029
	Sig. (1-tailed)	0.000	0.000	0.000	0.000	0.000		0.000	0.486	0.412	0.405	0.000	0.469	0.423

(continued)

Table 5 (continued)

		f_c'	f_{ct}'	f_{cf}'	Q_{rca10}	Q_{nca10}	Q_{rca20}	Q_{nca20}	Q_w	Q_c	Q_s	R_{rca}	R_{wc}	R_{ac}
Q_{nca20}	Pearson correlation	0.512**	0.537**	0.594**	−0.997**	1.000**	−0.997**	1	−0.010	−0.033	0.038	−0.999**	0.016	0.038
	Sig. (1-tailed)	0.000	0.000	0.000	0.000	0.000	0.000		0.475	0.414	0.400	0.000	0.459	0.400
Q_w	Pearson correlation	−0.660**	−0.628**	−0.575**	−0.005	−0.010	−0.005	−0.010	1	−0.272*	−0.198	0.001	0.827**	0.188
	Sig. (1-tailed)	0.000	0.000	0.000	0.486	0.474	0.486	0.475		0.032	0.091	0.497	0.000	0.103
Q_c	Pearson correlation	0.646**	0.566**	0.565**	−0.033	−0.033	−0.033	−0.033	−0.272*	1	−0.889**	−0.005	−0.748**	−0.974**
	Sig. (1-tailed)	0.000	0.000	0.000	0.412	0.414	0.412	0.414	0.032		0.000	0.487	0.000	0.000
Q_s	Pearson correlation	−0.344**	−0.278*	−0.303*	0.036	0.038	0.036	0.038	−0.198	−0.889**	1	0.004	0.369**	0.903**
	Sig. (1-tailed)	0.009	0.029	0.019	0.404	0.400	0.405	0.400	0.091	0.000		0.489	0.005	0.000
R_{rca}	Pearson correlation	−0.528**	−0.550**	−0.610**	0.999**	−0.999**	0.999**	−0.999**	0.001	−0.005	0.004	1	0.000	0.000
	Sig. (1-tailed)	0.000	0.000	0.000	0.000	0.000	0.000	0.000	0.497	0.487	0.489		0.500	0.500
R_{wc}	Pearson correlation	−0.804**	-0.756**	−0.730**	0.012	0.015	0.012	0.016	0.827**	−0.748**	0.369**	0.000	1	0.707**
	Sig. (1-tailed)	0.000	0.000	0.000	0.469	0.459	0.469	0.459	0.000	0.000	0.005	0.500		0.000

(continued)

Table 5 (continued)

		f_c'	f_{ct}'	f_{cf}'	Q_{rca10}	Q_{nca10}	Q_{rca20}	Q_{nca20}	Q_w	Q_c	Q_s	R_{rca}	R_{wc}	R_{ac}
R_{ac}	Pearson correlation	−0.580**	−0.529**	−0.551**	0.029	0.038	0.029	0.038	0.188	−0.974**	0.903**	0.000	0.707**	1
	Sig. (1-tailed)	0.000	0.000	0.000	0.423	0.400	0.423	0.400	0.103	0.000	0.000	0.500	0.000	

Note ** Correlation is significant at the 0.01 level (1-tailed). * Correlation is significant at the 0.05 level (1-tailed)

Table 6 Proposed models' coefficients for $f_c^{'}$ as a dependent variable

Model		Unstandardized coefficients		Standardized coefficients	T	Sig	R	R^2	Std. error of the estimate	Analysis method
		B	Std. Error	Beta						
1	(Constant)	105.118	2.703		38.888	0.000	0.962	0.926	3.72075	Stepwise
	R_{wc}	−111.411	5.742	−0.798	−19.404	0.000				
	Q_{rca10}	−1.161	0.090	−0.529	−12.857	0.000				
2	(Constant)	324.870	62.780		5.175	0.000	0.980	0.961	2.79422	Backward
	Q_w	69.320	21.048	5.563	3.294	0.002				
	Q_s	−26.148	4.891	−1.762	−5.346	0.000				
	R_{rca}	−0.175	0.010	−0.528	−17.093	0.000				
	R_{wc}	−1321.236	349.498	−9.460	−3.780	0.000				
	R_{ac}	131.455	34.063	6.655	3.859	0.000				

Table 7 Proposed model's coefficients for f_{ct}' as a dependent variable

Model		Unstandardized coefficients		Standardized coefficients	t	Sig.	R	R^2	Std. the error of the estimate	Method
		B	Std. error	Beta						
1						0.00	0.935	0.874	0.27090	Stepwise, forward
	(Constant)	7.192	0.197		36.451	0.000				
	R_{wc}	− 5.892	0.418	− 0.756	− 14.095	0.000				
	R_{rca}	− 0.010	0.001	− 0.550	− 10.259	0.000				

Table 8 Proposed model's coefficients for f_{cf}' as a dependent variable

Model		Unstandardized coefficients		Standardized coefficients	T	Sig.	R	R^2	Std. error of the estimate	Method
		B	Std. error	Beta						
1						0.00	.951	.905	.31636	Stepwise, forward
	(Constant)	8.246	.230		35.788	.000				
	R_{wc}	− 7.648	.488	−.730	− 15.667	.000				
	R_{rca}	−.015	.001	−.610	− 13.091	.000				

Table 9 Proposed model's coefficients for f_{ct}' as a dependent variable

Model		Unstandardized coefficients		Standardized coefficients	t	Sig	R	R^2	Std. error of the estimate	Method
		B	Std. error	Beta						
1						0.00	0.962	0.925	3.72974	Stepwise,forward
	(Constant)	105.509	2.716		38.841	0.000				
	R_{wc}	−112.283	5.755	−0.804	−19.510	0.000				
	R_{rca}	−0.175	0.014	−0.528	−12.818	0.000				

Table 10 Correlation among the characteristic compressive, tensile and flexural strength

		f_c'	f_{ct}'	f_{cf}'
f_c'	Pearson correlation	1	0.926**	0.915**
	Sig. (1-tailed)		0.000	0.000
f_{ct}'	Pearson correlation	0.926**	1	0.892**
	Sig. (1-tailed)	0.000		0.000
f_{cf}'	Pearson correlation	0.915**	0.892**	1
	Sig. (1-tailed)	0.000	0.000	

** Correlation is significant at the 0.01 level (1-tailed)

$$f_{ct}' = 1.647 + \left(0.052 \times f_c'\right) \tag{8}$$

$$f_{cf}' = 0.888 + \left(0.069 \times f_c'\right) \tag{9}$$

$$f_{ct}\prime = 1.346 + \left(0.664 \times f_{cf}\prime\right) \tag{10}$$

2.1 *Verification for the Developed Equations*

To verify the accuracy of the developed equations, twelve different mix designs for the RAC were conducted. The average and characteristic compressive (f_c'), tensile (f_{ct}') and flexural (f_{cf}') strengths of RAC in the twelve mix designs for verifying the prediction equations are shown in Table 14.

Table 15 shows the accuracy of Eqs. (3), (4), and (7) for the characteristic compressive strength. Error percentages from different mix designs from the verification data set for Eqs. (3) and (7) are all less than 10%, which is an acceptable level. It is ranged between 0.75% and 8.26%, and 0.44% and 8.41% respectively. However, error percentages from different mix designs from the verification data set for Eq. (4) are very high, ranging between 10.93% and 156.91%. Therefore, Eqs. (3) and (7) are suggested to predict characteristic compressive strength. Tables 16 and 17 shows the accuracy of Eqs. (5) and (6) for the characteristic tensile and flexural strengths, respectively. The error percentages for both equations are less than 10%, ranging between 0.62% and 6.52%, and between 0.60% and 9.89%, respectively.

Table 18 shows the accuracy of Eqs. (8), (9) and (10). The error percentages from different mix designs from the verification data set are reasonable, ranging between 0.20% and 5.95%, between 0.08% and 15.58%, and between 0.15% and 12.25%, respectively. This proves the accuracy of the developed equations in the optimal mix approach.

Table 11 Proposed model's coefficients for f_{ct}' as a dependent variable

Model		Unstandardized coefficients		Standardized coefficients	t	Sig.	R	R^2	Std. error of the estimate	Method
		B	Std. error	Beta						
1						0.00	0.926	0.858	0.28401	Stepwise, Remove, backward, forward
	(Constant)	1.647	0.154		10.728	0.000				
	$f_{c'}$	0.052	0.003	0.926	16.478	0.000				

Table 12 Proposed model's coefficients for f_{cf}' as a dependent variable

Model		Unstandardized coefficients		Standardized coefficients	t	Sig.	R	R^2	Std. error of the estimate	Method
		B	Std. error	Beta						
1						0.00	0.915[a]	0.838	0.40772	Stepwise, remove, backward, forward
	(Constant)	0.888	0.220		4.026	0.000				
	$f_{c'}$	0.069	0.005	0.915	15.246	0.000				

Table 13 Coefficients for f_{ct}' as a dependent variable

Model		Unstandardized coefficients		Standardized coefficients	t	Sig	R	R^2	Std. error of the estimate	Method
		B	Std. error	Beta						
1							0.796	0.792	0.34014	Stepwise, remove, backward, forward
	(Constant)	1.346	0.212		6.337	0.000				
	f_{cf}'	0.664	0.050	0.892	13.254	0.000				

Table 14 Characteristic compressive, tensile and flexural strength of the verification data set

Water-to-cement ratio	Aggregate to-cement ratio	RCA replacement ratio (%)	Characteristic compressive strength, f_c' (MPa)	Characteristic tensile strength, f_{ct}' (MPa)	Characteristic flexural strength, f_{cf}' (MPa)
0.40	3.5	10	61.5	4.8	5.1
		50	54.7	4.2	4.5
		90	44.7	3.9	4.0
	5.5	10	60.0	4.5	4.8
		50	51.7	4.3	4.3
		90	43.4	4.0	3.9
0.55	3.5	10	40.7	3.8	3.9
		50	37.8	3.4	3.5
		90	29.8	3.2	2.6
	5.5	10	43.9	3.7	3.8
		50	34.3	3.2	3.3
		90	30.0	3.2	2.8

The developed equations in the optimal mix approach can estimate the suitable mix designs for structural applications or the highest possible strength of RAC. Only required to input the RCA replacement ratios and water-to-cement ratios of RAC, estimated characteristic compressive, tensile and flexural strengths can be predicted. Alternatively, the required water-to-cement ratios can be predicted by inputting the RAC and RCA replacement ratios' required strength.

This is the first time developing equations for predicting RAC strength from RCA replacement ratios and water-to-cement ratios. The developed equations can significantly help the industry use RCA to achieve the highest possible strength for RAC, which can potentially be used for structural applications.

It needs to state that the developed equations are developed from the local materials, of which it can likely apply to the set of concreting materials; particular RA is highly variable from location to location.

2.2 Conclusion

The paper developed an optimal mix approach for RAC with seven accurate prediction equations using linear regression analysis. It has also been proved that the developed equations are reliable. Using the developed equations, characteristic compressive, tensile and flexural strengths by providing mix designs' parameters or one of the character strengths can be predicted with their designed mix designs. The developed equations can also estimate the requirements for concrete mix designs to achieve the required strength. This paper can significantly help the industry use RCA to achieve

Table 15 Verification of Eqs. (3, 4, and 7) for the characteristic compressive strength

Water-to-cement ratio	Aggregate to cement ratio	RCA replacement ratio (%)	f_c' (MPa)				Error (%)		
			Verification data	Equation (3)	Equation (4)	Equation (7)	Equation (5)	Equation (3)	Equation (4)
0.40	3.5	10	61.5	58.9	77.3	58.5	4.19	25.72	4.91
		50	54.7	52.3	69.8	51.5	4.49	27.55	5.97
		90	44.7	45.6	62.3	44.5	2.18	39.58	0.44
	5.5	10	60.0	58.7	89.2	58.5	2.22	48.58	2.64
		50	51.7	51.3	82.7	51.5	0.75	60.04	0.45
		90	43.4	43.9	76.2	44.5	1.04	75.49	2.33
0.55	3.5	10	40.7	42.2	92.1	41.6	3.81	126.50	2.29
		50	37.8	35.8	84.4	34.6	5.28	123.34	8.41
		90	29.8	29.4	76.7	27.6	1.65	156.91	7.48
	3.5	10	43.9	42.0	48.7	41.6	4.28	10.93	5.23
		50	34.3	34.8	42.0	34.6	1.48	22.56	0.97
		90	30.0	27.5	35.3	27.6	8.26	17.65	8.02

Table 16 Verification of Eq. (5) for the characteristic tensile strength

Water-to-cement ratios	Aggregate to cement ratio	RCA replacement ratio (%)	fct' (in MPa)		Error (in %)
			Verification data	Eq	
0.40	3.5	10	4.8	4.7	1.56
		50	4.2	4.3	2.24
		90	3.9	3.9	1.95
	5.5	10	4.5	4.7	4.53
		50	4.3	4.3	1.05
		90	4.0	3.9	2.11
0.55	3.5	10	3.8	3.9	2.16
		50	3.4	3.5	0.62
		90	3.2	3.1	3.74
	5.5	10	3.7	3.9	3.53
		50	3.2	3.5	6.52
		90	3.2	3.1	4.35

Table 17 Verification of Eq. (6) for the characteristic flexural strength

Water to-cement ratio	Aggregate to cement ratio	RCA replacement ratio (%)	f_{cf}' (in MPa)		Error (in %)
			Verification data	Eq	
0.40	3.5	10	5.1	5.2	1.75
		50	4.5	4.4	1.45
		90	4.0	3.6	7.89
	5.5	10	4.8	4.6	4.76
		50	4.3	4.2	2.90
		90	3.9	3.7	3.89
0.55	3.5	10	3.9	4.0	1.51
		50	3.5	3.2	9.33
		90	2.6	2.3	9.89
	5.5	10	3.8	3.7	2.60
		50	3.3	3.2	0.60
		90	2.8	2.7	1.11

the highest possible strength for RAC, which can potentially be used in structural applications.

Table 18 Verification of Eqs. (8, 9 and 10) for the characteristic tensile and flexural strength

Water-to-cement ratio	Aggregate to-cement ratio	RCA replacement ratio (%)	f_{ct}' (in MPa)			Error (in %)		f_{cf}' (in MPa)		Error from Eqs. 9 (in %)
			Verification data	Eqs. 8	Eqs. 10	Eqs. 8	Eqs. 10	Verification data	Eqs. 9	
0.40	3.5	10	4.8	4.8	4.7	0.70	4.99	5.1	5.1	0.97
		50	4.2	4.5	4.3	5.95	4.58	4.5	4.6	4.43
		90	3.9	4.0	4.1	2.82	4.12	4.0	4.0	0.48
	5.5	10	4.5	4.8	4.6	5.28	5.02	4.8	5.0	4.18
		50	4.3	4.3	4.2	1.04	4.37	4.3	4.5	3.77
		90	4.0	3.9	3.9	2.84	3.89	3.9	3.9	0.08
0.55	3.5	10	3.8	3.8	4.0	0.20	3.95	3.9	3.7	6.22
		50	3.4	3.6	3.7	5.31	3.51	3.5	3.5	0.74
		90	3.2	3.2	3.3	0.91	2.79	2.6	3.0	15.58
	5.5	10	3.7	3.9	3.9	5.65	3.83	3.8	3.9	1.98
		50	3.2	3.4	3.5	5.85	3.14	3.3	3.3	0.20
		90	3.2	3.2	3.2	0.57	2.55	2.8	3.0	6.84

3 Mixture Proportioning for RAC by Using Compressible Packing Model

3.1 Introduction

The RCAs are composed of NAs and attached mortar, and, as a consequence, they generally present a higher porosity than ordinary NAs (Faella et al. 2016). The higher porosity of RCAs can affect the resulting properties of RAC. In this context, the mix design procedure available in the literature for natural concrete mixture cannot be applied for the case of RAC.

Amario et al. (2017) presented a study in which the optimal mixing procedure for RAC was investigated and then a possible extension of the Compressible Packing Model (CPM) (De Larrard 2014) in the case of RAC was proposed. The main advantage of CPM is the possibility of considering the intrinsic characteristics of each component and determining the desired properties for the concrete. Initially, an experimental study was conducted to determine the RCAs' optimum water absorption compensation content to be considered during concrete mixing. Thus, the mixing ratio of RAC was developed considering different levels of RA content for three classes of compressive strength (25, 45 and 65 MPa). Finally, some of the designed mixtures were selected for experimental validation. In this phase, concrete mixtures were produced for mechanical and durability tests to evaluate the possible extension of the CPM for the mixture proportion of structural RAC.

3.2 Some Considerations About the Application of the CPM for RAC Mixtures

The concrete mixtures were designed with the CPM for describing the packing density achieved by a granular mixture. This model (De Larrard 2014) considers that all the grains interact with each other influencing the overall density of the mixture. The packing density of the granular mixtures is governed by the shape, surface texture and size of the aggregate particles. For this reason, the packing density of each aggregate must be measured experimentally to use this model (De Larrard 2014).

In addition, the CPM considers the properties of the aggregates to predict the compressive strength of the concrete. For this, the method proposes the following analytical formulation:

$$f_c = \frac{p \cdot f_{cm}}{q \cdot f_{cm} + 1} \tag{12}$$

where f_{cm} is the compressive strength of the cement matrix (in MPa) and f_c is the resulting concrete strength (in MPa). The p and q are two empirical constants depending on the aggregates, in which p represents the bond between the aggregate and the paste, while q identifies the intrinsic strength of the aggregates. In order to determine these constants for each aggregate, it is necessary to produce two concrete mixtures of different strengths (i.e., a concrete with low strength and a high-strength concrete).

The compressible packing theory has already been successfully applied for predicting and controlling the rheological, mechanical and durability performance of several families of concrete mixtures (Sedran and De Larrard 1999; De Larrard and Sedran 2002). The CPM assumes that the aggregates are employed in dry conditions during the mixing procedure, and their water absorption capacity at 24 h is considered. Therefore, the RCAs do not absorb the whole amount of water considered for saturation due to other granular materials. Consequently, part of the additional water is free within the fresh mixture, and, consequently, the water-to-cement ratio was affected by the presence of RCAs within the concrete. Consequently, the concrete performance at both fresh and hardened states was affected (Pepe 2015).

Amario et al. (2017), to evaluate the role of this additional water for RCAs, designed several mixtures: thus, the evolution of the temperature development, the corresponding workability and compressive strength of these mixtures were measured.

The results showed that the best initial moisture condition for RCAs is fully dried condition and with an addition of 50% of their water absorption capacity (considered for saturation during mixing). The free water available in the mixture is not affected by this amount of additional water (Pepe 2015).

3.3 Materials and Methods

The binder used in their study (Amario et al. 2017) was a high initial strength Portland cement and, in addition, a polycarboxylate superplasticizer was used in all mixes for workability control. Natural sand with a nominal diameter smaller than 4.75 mm was used as fine natural aggregate (NFA), and granite type stone with a nominal diameter from 9.5 mm down to 4.75 mm was used as NCA. The RCAs were obtained from the debris of reinforced concrete beams produced in a laboratory. The compressive strength of the original waste concrete was 33–36 MPa, and the water-to-cement ratio was 0.60. During the production process of the RCAs, the original concrete was broken into smaller pieces with a stone crusher, and then the debris was double crushed using a jaw crusher. The RA was then subjected to sun drying and sieving. In the final step, the RCAs were homogenized by the longitudinal blending bed technique.

3.4 *Mixture Proportioning for Normal and High-Strength RAC Mixtures*

Simulations of mixtures composition (Table 19) were performed using the BetonLab Pro 3 software in which the CPM is implemented. Three classes of compressive strength were considered: 25, 45 and 65 MPa. Additionally, all the concrete mixtures were designed for a slump value ranging between 100 and 150 mm. The mixture proportioning procedure was optimized by maximizing the compactness of the dry granular mixture to obtain the desired compressive strength by minimizing the overall cement amount.

Table 19 Mixture proportioning of RAC with CPM

Concrete mixture	Composition (kg/m^3)						Properties		
	NCA	RCA	NFA	Cement	SP	Total water	% of RCA	Effective w/c	Free water (l/m^3)
C25-01,100	609.7	381.7	842.5	267.2	2.7	193.2	39	0.64	170.0
C25-00	1005.2	0	842.0	267.2	1.1	185.1	0	0.64	170.0
C25-0120	803.0	195.6	844.4	266.4	2.7	188.7	20	0.64	170.0
C25-2040	609.7	381.7	842.5	267.2	2.7	193.2	39	0.64	170.0
C25-4060	600.6	390.1	842.0	267.6	2.7	193.6	40	0.64	170.3
C25-6080	395.4	577.8	831.5	277.8	2.8	203.9	60	0.63	176.3
C25-80,100	197.0	767.7	828.7	278.5	2.8	208.4	80	0.63	176.2
C25-100	0	935.9	808.1	299.9	3.0	225.5	100	0.63	189.8
C45-01,100	496.8	318.3	917.3	385.2	3.9	184.0	40	0.43	165.0
C45-00	949.1	0	795.0	391.3	1.5	183.9	0	0.43	170.0
C45-0120	570.8	159.9	913.9	391.5	3.9	179.1	22	0.42	165.0
C45-2040	496.8	318.3	917.3	385.2	3.9	184.0	40	0.43	165.0
C45-4060	494.0	320.9	917.2	385.3	3.9	184.1	40	0.43	165.1
C45-6080	325.9	476.2	907.4	397.3	5.2	189.3	60	0.42	167.5
C45-80,100	162.4	632.9	904.5	400.5	5.0	193.3	80	0.42	167.6
C45-100	0	786.4	899.1	406.2	4.6	198.9	100	0.42	169.4
C65-01,100	610.6	56.4	913.9	537.3	5.4	184.7	9	0.33	175.0
C65-00	835.8	0	700.1	591.9	3.2	201.0	0	0.32	190.0
C65-0120	610.6	56.4	913.9	537.3	5.4	184.7	9	0.33	175.0
C65-2040	530.5	129.2	906.5	547.2	5.5	188.3	20	0.32	177.0
C65-4060	397.7	258.3	906.1	548.9	5.5	189.4	40	0.32	175.0
C65-6080	260.6	380.8	890.4	570.2	5.7	196.6	60	0.32	179.6
C65-80,100	129.4	504.2	884.3	579.1	5.8	199.5	80	0.31	179.7
C65-100	0	608.9	854.3	619.3	6.2	211.7	100	0.31	190.0

For each strength class, several mixtures proportioning were performed considering the optimal variation of the RCA content. Moreover, ten mixtures proportioned with CPM were chosen to validate the methodology applied in this study (the ones highlighted in Table 19). Thus, the optimum mixtures for each class were chosen for this. Additionally, mixtures with RCA content close to 20, 40 and 60% were also selected if these mixtures have not already been chosen to be the optimal mixture.

3.5 *Results and Analysis*

3.5.1 Workability

The results of workability and mechanical properties for the concrete mixtures are presented in Table 20.

As in the case of ordinary concrete, the RAC workability is mainly governed by the aggregate shape, grain size distribution, and the amount of free water available in the mixture. The slump values presented a tendency to increase as the RCA content increases: this can be attributed to the more rounded shape observed for RCA due to the crushing process performed for producing them but also due to the lower amount of fine particles in comparison with the NA (Amario et al. 2017).

Finally, it is worth highlighting that the desired slump (i.e., ranging between 100 and 150 mm) was not obtained in all cases: this demonstrates that, probably, further input parameters should be considered in the CPM for predicting and achieving the desired workability.

First, from the experimental data reported in Table 20, it emerges that the desired compressive strength was achieved for all the concrete mixture class, demonstrating

Table 20 Workability and mechanical properties for the selected mixtures designed with CPM (Amario et al. 2017)

Concrete mixture	Slump (mm)	f_c (MPa)	Strain at maximum stress (με)	E (GPa)	$F_{t,sp}$ (MPa)	$f_{t,sp}/f_c$ (%)
C25-01,100	60	25.4	2397	19.7	2.79	11.0
C25-0120	45	25.2	2180	20.0	2.30	9.0
C25-6080	70	23.5	2282	19.5	2.08	8.9
C45-01,100	45	46.0	2560	26.6	3.79	8.2
C45-0120	30	45.8	2495	26.9	3.88	8.5
C45-6080	55	46.9	2736	27.0	3.77	8.0
C65-01,100	55	68.1	3205	31.1	5.01	7.4
C65-2040	80	63.9	2620	33.7	3.90	6.1
C65-4060	85	62.5	2691	32.4	4.77	7.6
C65-6080	95	62.4	2876	31.4	3.76	6.0

that the methodology adopted in this study can be successfully applied for designing structural RAC mixtures (at least for controlling their concrete compressive strength). Moreover, also in terms of elastic modulus, the concrete performance seems to be unaffected by the presence of RCA. In fact, these values seem to be mainly influenced by the concrete strength. Then, by analyzing the strain at maximum compressive stress values, it emerges that, as a general trend, this value tends to increase by increasing the RCA amount. These results confirm that the elastic modulus and the peak strain increase by increasing the compressive strength. Finally, based on the tensile-to-compressive strength ratio values, it is possible to note that a lower $f_{t,sp}/f_c$ values is registered as the compressive strength class increases.

3.5.2 Mechanical Behavior

The compressive stress–strain curves at 28 days are presented in Fig. 1. The mechanical properties were not affected by increasing the RCA content in each mixture regardless of strength class. In addition, the experimental results of compressive strength showed a high correlation with the values predicted by the CPM: this demonstrates that this method can be successfully applied for the mixture proportioning of RAC for both normal and high-strength classes.

3.5.3 Physical Properties and Durability Indexes

The physical properties and the related durability indexes experimentally measured for the selected mixtures are summarized in Table 21.

The analysis of the results in terms of concrete density highlights that there are two main factors influencing these physical properties: the paste-to-aggregate ratio and the amount of RCA within the mixture. Then, the experimental results of the water absorption tests (by immersion) highlight that the void index (and the total water absorption) is mainly governed by the paste characteristic than the type of coarse aggregates. Similar conclusions can be drawn from the results of the water absorption tests (by capillarity), also in this case the water absorption values are directly related to the concrete strength class (i.e., the effective water-to-cement ratio): lower strength lead to having higher values of water absorption.

Moreover, the results of nitrogen permeability highlight that, as in the case of water absorption, the increasing of concrete strength causes a decrease in permeability.

3.5.4 Conclusions

The study proposed by Amario et al. 2017 analyzed the feasibility of using the CPM for the rational proportioning of RAC mixture and the following consideration can be remarked: the optimal procedure to produce RAC without affecting the water-to-cement ratio is to use the coarse RCAs in dried condition and add (during

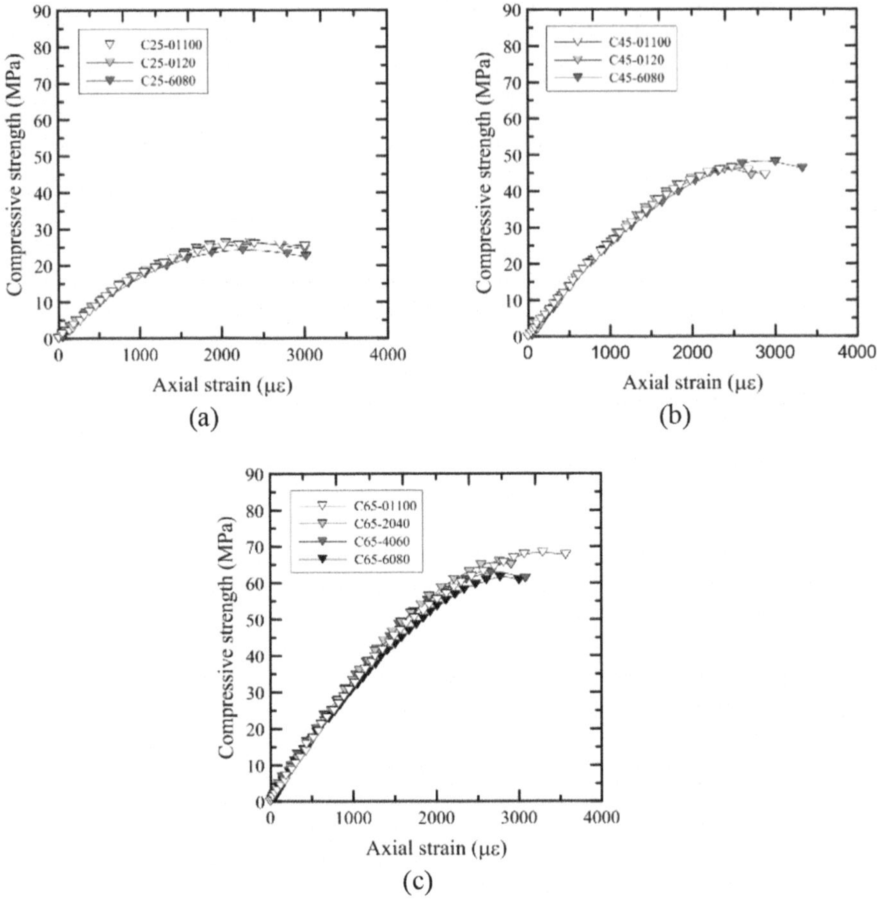

Fig. 1 Compressive stress–strain curves at 28 days: C25 **a**, C45 **b** and C65 **c**

mixing) the 50% of their water absorption capacity; the mechanical properties were not affected by increasing the RCA content in each mixture regardless of strength class. In addition, the experimental results of compressive strength showed a high correlation with the values predicted by the CPM: this demonstrates that this method can be successfully applied for the mixture proportioning of RAC for both normal and high-strength classes; the overall durability indexes were not influenced by the presence of RCA especially for high-strength class concretes.

Table 21 Physical properties and durability indexes for the selected mixtures designed with CPM (Amario et al. 2017)

Concrete mixture	Density (kg/m^3)	Water absorption by immersion (%)	Sorptivity (h$^{1/2}$)	Nitrogen permeability (m^2)
C25-01,100	2166	2.18	8.95	3.39×10^{-16}
C25-0120	2186	1.86	9.24	3.15×10^{-16}
C25-6080	2105	2.58	8.50	3.61×10^{-16}
C45-01,100	2239	1.37	10.13	8.88×10^{-17}
C45-0120	2235	1.35	9.79	8.05×10^{-17}
C45-6080	2227	1.20	9.77	8.42×10^{-17}
C65-01,100	2332	0.60	3.97	3.63×10^{-17}
C65-2040	2323	0.56	4.04	3.56×10^{-17}
C65-4060	2305	0.59	4.43	3.75×10^{-17}
C65-6080	2274	0.58	3.80	6.69×10^{-17}

4 Particle Packing Method of Mix Proportioning for Recycled Aggregate Concrete

The socio-economic expansion is essential to reassure the ever-increasing demand of human society. In this regard, the construction industry has a major role in providing basic infrastructure. The construction industry consumes a huge quantity of non-renewable natural resources, especially in the form of aggregates. Apart from the rapid depletion of natural resources, construction work generates an enormous amount of CDW, which accounts for approximately 30—35% of the global waste (Fischer et al. 2009; Marzouk and Azab 2014). The huge dependency of the concrete industries on non-renewable resources could be reduced if waste concrete is utilized as a source of new raw materials for concrete production. A large portion (about 25–35%) of CDW is waste concrete, crushed to prepare RA (Pradhan et al. 2019). However, the quality of RA is inferior to the conventional aggregate owing to the exposure to aggressive environment, loading history, AM layer and the existence of micro-cracks. The inferior physical and mechanical properties of RA are also translated into the performance of RAC (RAC). To improve the performance of RAC, different techniques such as the incorporation of mineral admixture, using different mix design methods, modifying the mixing process are used.

Direct Weight Replacement (DWR) method, Equivalent Mortar Volume (EMV) method (Fathifazl et al. 2009) and Direct Volume Replacement (DVR) method are three different aggregate replacement methods used to study the resulting performance of RAC. In the DWR method, the weight of coarse aggregate (NCA plus RCA, cement and water content are kept constant for any replacement ratio. In order to produce the same volumetric yields, the fine aggregate amount is reduced. In the EMV method, RCA was considered a two-phase material rather than a single coarse aggregate and the total volume of mortar (residual and fresh mortar) content in RAC was kept the same as the fresh mortar content of target NAC (Fathifazl et al. 2009). The DVR method considers RCA as a single-phase coarse aggregate. In this method, the volume of replacing RCA is same as the volume of NA. The amount of fine aggregate, cement, and water is unaffected as the coarse aggregate volume remains constant. In EMV method workability reduces significantly and a replacement ratio more than 20% needs changes in mix proportions and high dose of water-reducing admixture. The compressive strength is not influenced significantly by any of these mix design methods at any replacement ratio (Knaack and Kurama 2013).

There are some modified mixing approaches, such as Double Mixing Method, Two-Stage Mixing Approach (TSMA) and Triple Mixing Method (Kong et al. 2010) have been employed by different authors to enhance the performance of RAC. (Tam et al. 2005, 2009) have proposed a TSMA and basically, this method divides the mixing process into two parts and water is added in two steps at different times. The TSMA helps develop a stronger ITZ by effectively filling the voids and cracks present in RCA (Tam et al. 2009). The compressive strength of RAC is increased by up to 21% and TSMA is also effective in enhancing durability (Tam et al. 2005). The TM contributes significantly to improving workability, compressive strength, and flexural strength compared to the Double Mixing Method.

The DWR and DVR methods are not very useful in improving the performance of RAC, whereas the performance of RAC using the EMV method of mix design is quite encouraging (Knaack and Kurama 2013). But, the major drawback of EMV method is that 100% replacement of RCA is never achievable. There is no specific code or guidelines available for mix design of RAC. So, most of the researchers follow the mix design procedure used for conventional concrete to produce RAC. Therefore, an attempt is made to use Particle Packing Method (PPM) of mix design (Pradhan et al. 2017) to improve the performance of RAC.

4.1 Particle Packing Method

The primary aim of PPM of mix design is to minimize the voids content (Glavind et al. 1993, De Larrard 1999, De Larrard and Sedran 2002). This is a process of optimization where the smaller particles are selected to fill up the voids between large particles and so on. In PPM the maximum possible packing density can be achieved by using the appropriate proportion of different sized coarse aggregates and fine aggregate. The increase in packing density leads to the minimization of voids and hence the requirement of binder and water will be less.

Traditionally, the void minimization is ensured by using Fuller's curve for the particle size distribution. In PPM the optimal combination of the aggregates is achieved by mixing different sized aggregates in appropriate proportion. The coarse aggregates are segregated in different size zones from coarser to finer, such as CA_1, CA_2, and CA_3, etc. The compacted bulk density and specific gravity of aggregates are determined separately for each size zone. In the hierarchy of aggregate mixture, the mixing of coarse aggregate is carried out in descending order. So, the first mixture is prepared by blending CA_1 and CA_2 in different proportions by mass, such as 90:10, 80:20, 70:30, etc., and for each mixture proportion the bulk density is determined and packing density and voids content (VC) are calculated using equations stated in ASTM C293-00 2001 or BSI 1998.

$$\text{Packing Density} = \sum \frac{\text{Bulk Density} \times \text{Weight Fraction}}{\text{Specific Gravity}} \tag{12}$$

$$\text{Voids Content} = 1 - \sum \frac{\text{Bulk Density}}{\text{Specific Gravity}} \tag{13}$$

The bulk density and PD of the mixture increase, as the smaller-sized aggregate fraction increases up to a certain proportion of the mixture, after which both start decreasing. The mixture proportion for maximum PD obtained is determined for the aggregate mixture of CA_1 and CA_2. the particular weight fraction of CA_1 and CA_2 aggregate combination is kept fixed while preparing mixture and CA_3 aggregate size in different proportions by mass (90:10, 80:20, 70:30, etc.). The bulk density, PD and VC are calculated for each weight fraction of this aggregate mixture. The maximum PD of the mixture and its corresponding weight fraction are determined. Then CA_4 is added in different proportions by mass to the combination of CA_1, CA_2, and CA_3 sized coarse aggregates and the proportion at which maximum PD obtained is determined. Similarly, the smaller-sized coarse aggregates are added to the mixture one after another in descending order and for every mixture the maximum PD and its corresponding weight fraction of each aggregate size are determined. Subsequently, fine aggregate is added to the combination of different sized coarse aggregates in different proportions by mass. Finally, the maximum possible PD for the coarse and fine aggregate combination and the weight fraction of the selected size of coarse and fine aggregates are obtained.

4.2 Determination of Paste Content

The total PD obtained by mixing different sized coarse aggregates and fine aggregate is used to determine the VC of the mixture using Eq. 14.

$$\text{Voids Content (VC)} = 1 - \text{PD} \tag{14}$$

In a concrete mix, the paste (cementitious material and water) fills up the voids present in the aggregate mixture and provides desired workability. Hence, the total paste content in this method is the sum of the voids content and the excess quantity of paste required to lubricate the aggregate particles and also to obtain desirable workability of the concrete. The total quantity of paste is determined by multiple trials to achieve the necessary conditions; i.e. the desired workability (without any segregation and bleeding) as well as the compressive strength at the same time. In order to satisfy the necessary conditions, the different w/c ratio is employed in the trial process. The cement and water content in the PPM design mix are determined by employing the following steps.

(1) Total Voids Content = VC + Excess Paste Content × VC
(2) Volume of aggregates = 1 − Total Voids Content
(3) Total solid volume of aggregates = $\sum \frac{\text{Weight fraction of aggregates}}{\text{Specific Gravity}}$
(4) Weight of the aggregate = $\frac{\text{Volume of aggregates}}{\text{Total solid volume of aggregates}} \times \text{Weight fraction} \times 1000$
(5) If w/c ratio = n; w = n × c

$$\text{Total paste content} = c + w = \frac{c}{\text{Specific Gravity}_{\text{cement}}} + \frac{n \times c}{\text{Specific gravity}_{\text{water}}}$$

(6) Cement content = $\frac{\text{Total voids content}}{\text{Total paste content}} \times 1000$
(7) Water content = w/c ratio × Cement content

4.3 Experimentation

The study by Pradhan et al. (2017) used 20–12.5 mm, 12.5–10 mm, and 10–4.75 mm sized NCA and 20–12.5 mm, 12.5–10 mm 10–6.3 mm and 6.3–4.75 mm sized RCA along with Zone II fine aggregate to optimize the combination of aggregates to achieve maximum packing density. As discussed in the previous section, the compacted bulk density and specific gravity for each size of aggregates were determined. The procedure to determine the aggregate mixture's bulk density and packing density (coarse and fine aggregates) was followed as discussed in the previous section. After conducting the whole exercise, maximum bulk density and packing density for the NCA and fine aggregate mixture were found out to be 1.99 and 0.713, respectively. The proportion of 20–12.5 mm, 12.5–10 mm, 10–4.75 mm sized NCA and fine aggregate to achieve maximum packing density was 38.4:9.6:12:40. Similarly, the maximum bulk density and packing density were found out to be 1.865 and 0.728, respectively for aggregate proportion 31.92:13.68:11.4:3:40 of 20–12.5 mm, 12.5–10 mm, 10–6.3 mm and 6.3–4.75 mm sized RCA and fine aggregate. The bulk density and packing density obtained for different aggregate combinations at different proportions are graphically shown from Figs. 2, 3, 4, 5, 6, 7, 8 and 9.

The total paste content in PPM approach can be determined by employing the steps discussed in the previous section. As mentioned earlier the total paste content is the sum of the voids content and the excess quantity of paste required to bind the aggregate particles and obtain desirable workability of the concrete. Multiple trials were carried out at w/c ratio 0.4, 0.43, 0.45 and 0.47 to achieve target mean strength of M30 grade of concrete with desired workability. In this process, 16% excess paste content was incorporated for each w/c ratio to obtain desired workability as well as strength. The mix proportions of both NAC and RAC obtained by using the conventional method and PPM approach are represented in Table 22.

Fig. 2 Bulk density of NCA mixture

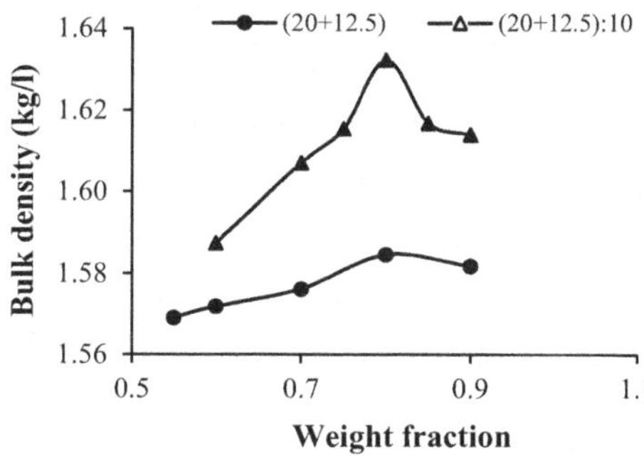

Fig. 3 Bulk density of (NCA + Sand) mixture

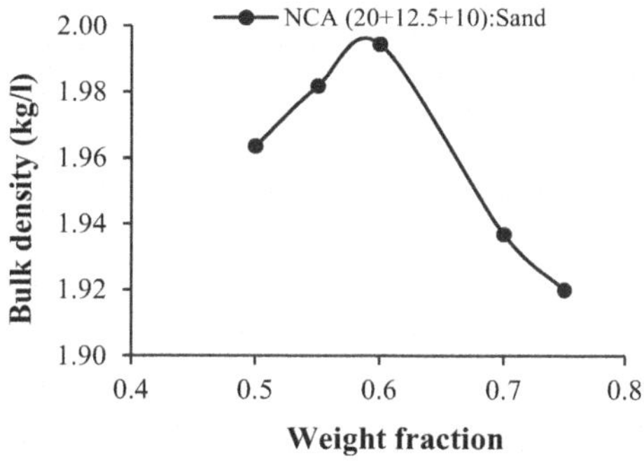

Fig. 4 Density of NCA mixture

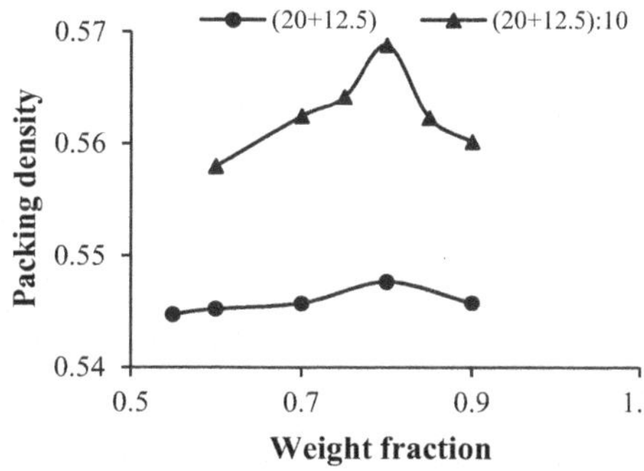

Fig. 5 Packing density of (NCA + Sand) mixture

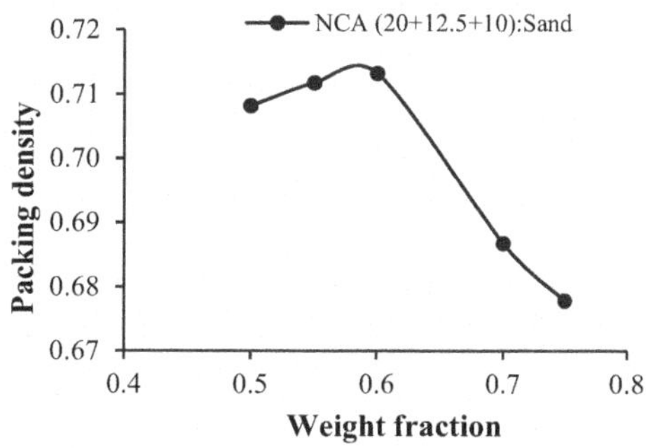

Fig. 6 Bulk density of RCA mixture

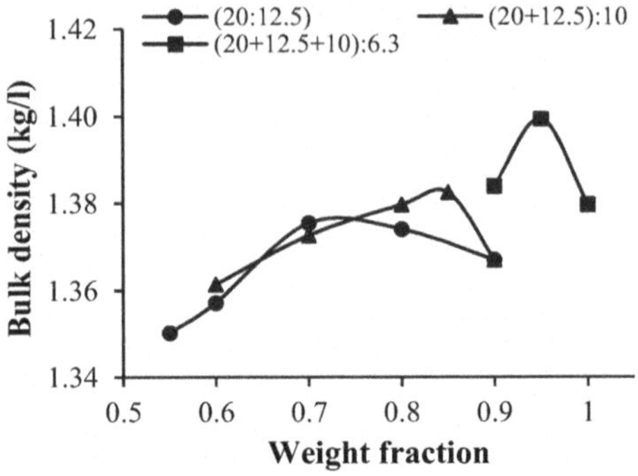

Fig. 7 Bulk density of (RCA + Sand) mixture

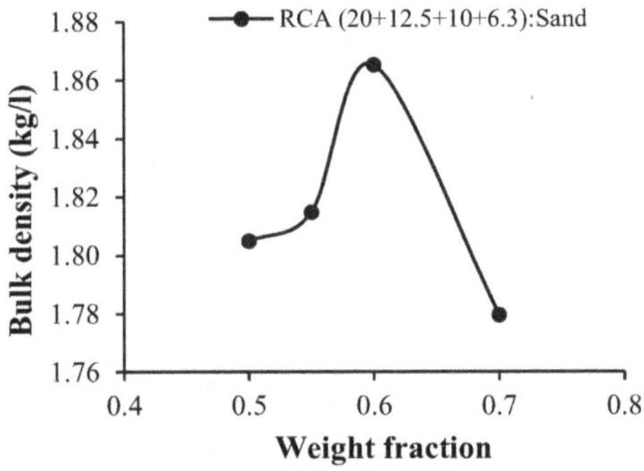

Fig. 8 Packing density of RCA mixture

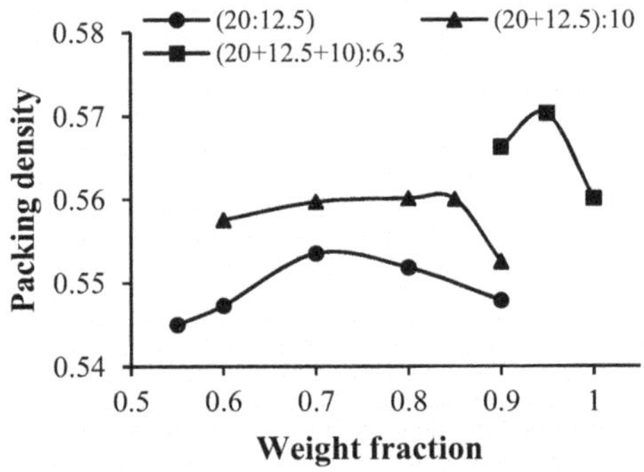

4.4 Mixing Process

The TSMA proposed by Pradhan et al. (2017) is adopted for the study. The established TSMA helps in improving the quality of the concrete at the microstructure level, especially the ITZ and this may be due to the effect of internal curing (Tam et al. 2009). The nomenclature of this mixing method is basically due to the addition of water in two stages and Fig. 10 shows the steps involved in the mixing process.

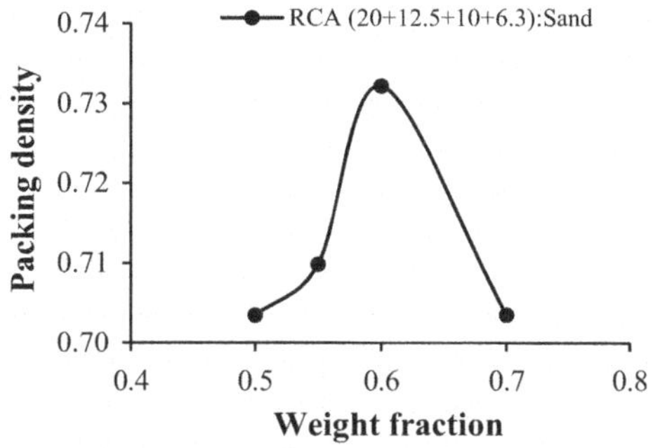

Fig. 9 Packing density of (RCA + Sand) mixture

Table 22 Mix proportions of concrete mixes (Pradhan et al. 2017)

Type of concrete	Mix design method	Water (kg/m^3)	Cement (kg/m^3)	Sand (kg/m^3)	Coarse aggregate (kg/m^3)
NAC IS	IS: 10,262 (2009)	215.9	438.1	610.2	1241.4
NAC PPM	PPM	203.5	410.0	762.1	1143.2
RAC IS	IS: 10,262 (2009)	234.2	422.2	621.2	1088.2
RAC PPM	PPM	225.8	410.0	702.0	1053.1

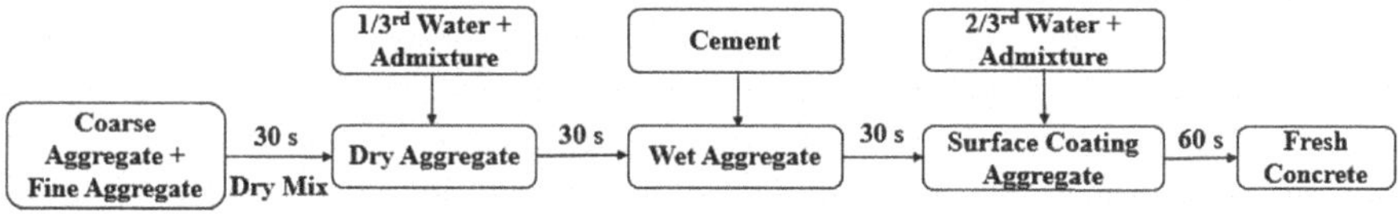

Fig. 10 Schematic diagram of two-stage mixing approach (Pradhan et al. 2017)

4.5 *Results and Discussions*

4.5.1 Compressive Strength

The short-term properties of hardened concrete include compressive strength, tensile strength and modulus of elasticity. The compressive strength of concrete was measured using standard cubes in accordance with IS: 516. The compressive strength of all four types of concrete was evaluated after 7 days, 28 days and 90 days of curing as shown in Table 23. The 7 days compressive strength at w/c ratio 0.45 exhibited negligible difference between NAC and RAC using the PPM mix design. However, for w/c ratio 0.40, 0.43 and 0.47, the 7 days compressive strength of RAC (PPM) were higher than that of NAC (PPM). In PPM mix proportioning except w/c ratio

Table 23 Compressive strength of different types of concrete (Pradhan et al. 2017)

Type of concrete	Mix design method	w/c ratio	7 days strength (MPa)	28 days strength (MPa)	90 days strength (MPa)
NAC (PPM)	PPM	0.40	30.87	51.27	65.49
		0.43	28.74	44.29	58.87
		0.45	25.35	42.75	56.15
		0.47	23.69	38.18	52.52
NAC (IS)	IS: 10,262 (2009)	0.45	29.69	47.68	64.85
RAC (PPM)	PPM	0.40	31.71	49.33	56.34
		0.43	29.48	48.09	53.72
		0.45	25.14	42.82	47.69
		0.47	24.48	42.48	47.59
RAC (IS)	IS: 10,262 (2009)	0.45	35.99	45.23	54.46

0.40, the 28 days compressive strength of RAC (PPM) were higher in comparison to NAC (PPM) and the 28 days compressive strength obtained at w/c ratio 0.40 for NAC (PPM) was 3.9% higher than that of RAC (PPM). So, the 7 days and 28 days compressive strength of RAC (PPM) were slightly on the higher side than NAC (PPM). However, after 90 days of curing the compressive strength of RAC (PPM) at different w/c ratios were 9.6–17.7% lower than NAC (PPM). An increase of 27–37% was observed in compressive strength at different w/c ratio of NAC (PPM) from 28 to 90 days, whereas for RAC (PPM), only 11–14% increment was observed for the same period of curing. The compressive strength of both NAC and RAC prepared using IS: 10,262–2009 were higher than that was observed in the PPM mix designed concrete for all curing ages. It was noticed that the gain in strength from 28 to 90 days of curing is less for RAC irrespective of the mix design method.

4.5.2 Split Tensile Strength

The split tensile strength was measured using a cylindrical specimen after 28 days of curing. Table 24 shows tensile strength results obtained for NAC and RAC. The split tensile strength of RAC was found to be lower than NAC. NAC's split tensile strength values were 21.5% and 25.8% higher than RAC in PPM and IS code method of mix design, respectively. However, for both NAC and RAC split tensile strength observed in the PPM mix design approach was higher than the IS code method of mix design and the increment was found to be 2.6% and 6.3% in NAC and RAC, respectively. This indicates that the concrete mix using PPM has an advantage over IS code method in improving the split tensile strength of RAC. It is observed that, even the compressive strength of both NAC and RAC are higher for mixes using IS:

Table 24 Tensile strength and modulus of elasticity of different types of concrete (Pradhan et al. 2017)

Type of concrete	w/c ratio	Tensile strength (MPa)	
		Split tensile strength	Flexural tensile strength
NAC (PPM)	0.45	3.5	4.75
NAC (IS)	0.45	3.41	4.39
RAC (PPM)	0.45	2.88	4.43
RAC (IS)	0.45	2.71	3.56

10,262 (2009) than PPM, but the split tensile strength value are higher in concrete mixes using PPM.

4.5.3 Flexural Tensile Strength

The flexural tensile strength of both NAC and RAC are shown in Table 24. The behavior of flexural strength is similar to split tensile strength for both NAC and RAC. NAC exhibited 7.2% and 23.3% higher flexural tensile strength than RAC in PPM and IS code method of mix design, respectively. The flexural tensile strength of PPM mix designed concrete mixes exhibited 8.2% and 24.4% higher values than IS code method for NAC and RAC, respectively. This suggests that, the PPM mix design approach is more efficient in improving the flexural tensile strength of RAC than NAC. The concrete proportioned using IS code method exhibited lower flexural tensile strength in comparison to PPM design mix, even the compressive strength obtained in the preceding method is on higher side.

The results of the analysis for compressive strength, split tensile strength, flexural tensile strength and modulus of elasticity are shown in Table 25. The p-values of compressive strength indicate that the PPM design mix has a significant effect on NAC, as the p-value of NAC (PPM) versus NAC (IS) is 0.013, which is less than the acceptable error level of 0.05. But, the higher p-value of 0.52 for RAC (PPM) versus RAC (IS) indicates that the mix proportioning method has no significant effect on the compressive strength parameter for RAC. The p-values of split tensile strength and flexural tensile strength (0.0004 and 0.022, respectively) suggest that the PPM design mix is advantageous to improve the tensile strength parameter of RAC compared to the IS code method design mix. But, the PPM design mix has no significant advantage over the IS code method of mix proportioning for the tensile strength properties of NAC as the obtained p-values (0.602 and 0.873) are much higher than the acceptable error level 0.05.

Table 25 Single factor ANOVA test result (Pradhan et al. 2017)

Types of concrete	p-value		
	Compressive strength	Split tensile strength	Flexural tensile strength
NAC (PPM) versus RAC (PPM)	0.185	0.003	0.395
NAC (IS) versus RAC (IS)	0.156	0.014	0.276
NAC (PPM) versus NAC (IS)	0.013	0.602	0.873
RAC (PPM) versus RAC (IS)	0.520	0.0004	0.022

4.6 Conclusions

The proposed PPM mix design approach requires a lesser quantity of cement, but more fine aggregate than the conventional method of mix proportioning. The comparatively lesser requirement of cement in the PPM mix design approach indicates the cost-effectiveness of the method. The 28 days compressive strength of RAC are found to be very close to that of NAC in PPM mix design approach at w/c ratio 0.45. But, the gain in strength from 28 to 90 days is less for RAC than NAC. The single factor ANOVA analysis indicates that PPM mix proportioning has a significant effect on the compressive strength of NAC than the conventional approach. The PPM mix proportioning was found to be effective in improving the tensile strength (split tensile strength and flexural tensile strength) of both NAC and RAC. The single factor ANOVA analysis suggests that the PPM mix design approach has more influence on the tensile strength of RAC. the material inferiority of RCA can be minimized by using PPM mix design approach. In addition to this, RCA can be effectively used at the 100% replacement level using the PPM mix design method. The short term mechanical properties, such as compressive strength, tensile strength and modulus of elasticity of RAC are very close to that of NAC. However, the durability aspects of RAC require further study.

5 A Rational Mix Design Method for Recycled Aggregate Concrete

5.1 Introduction

Pepe (2015) recently proposed a comprehensive physically-based conceptual approach for designing RAC (i.e., RAC-C) mixtures for a given target compressive strength.

The model results in a conceptual formulation capable to predict and control the key mechanical properties of structural concrete produced with coarse RCAs. As a matter of principle, due to the presence of AM, RCAs are significantly more porous than NAs of the same size fraction, and, consequently, affects the mixture proportioning rules commonly adopted for ordinary concretes. As a matter of the fact, the common rules cannot be applied for cement-based mixtures produced with RCAs to create RAC. However, the definition of accurate relationships capable of predicting the relevant properties of concrete made out with these constituents is still considered an open issue Brito and Saikia 2013. The main reason of this is the lack of knowledge on the actual composition and the resulting engineering properties of RCAs (De Juan and Gutiérrez 2009). In fact, RCAs can be intended as a two-phase composite consisting of the "original" NCAs and the AM, made of sand, hydration products and fractions of un-hydrated cement. The attached mortar content is also the part that is responsible for the increased adsorption capacity of RCAs due to its relatively higher porosity in comparison to NAs (Belin et al. 2014).

The rational mix design approach proposed in this study for designing, controlling and predicting the compressive strength of RAC is mainly based on identification of a key parameter characterizing the RCA properties and, consequently, the resulting concrete performance. As a matter of fact, porosity can be recognized as the main parameter controlling RCAs' physical and mechanical properties and the water transportation phenomena occurring between cement and aggregates during the setting and hardened phases of RAC (Pepe et al. 2016). In fact, the role of free water inside the RAC mixture becomes fundamental as it determines the time evolution of the cement paste strength and the bond strength between aggregates and cement (ITZs). In addition, the presence of a porous medium (i.e. RCAs) in RAC results in a possible weak point for the resulting concrete composite.

5.2 *Physical Properties of RCAs: The Role of Attached Mortar*

Based upon the experimental results achieved for RCAs (Pepe et al. 2016), a possible linear correlation is proposed for describing the variation of the porosity p as a function of the AM content:

$$p = \left[p_{NA} \cdot (1 - AM) + P_{AM}(AM)\right] \cdot \frac{d_0}{d} \tag{15}$$

where d is the diameter of the aggregate expressed millimeters and d_0 is a reference diameter of 20 mm. This formulation can be interpreted as a liner combination of the two phases NA and AM. In fact, p_{NA} represents the porosity of a NA inside RCA, while p_{AM} represents the porosity of the attached mortar residing to it. The calibration of Eq. 16 led to approximate values for p_{NA} and p_{AM} equal to 0.45% and 15%, respectively.

Moreover, the particle density represents another fundamental parameter to be determined for the mix design of RAC mixtures. The results reported in the previous section highlight the particle density and complies with the general trend that its value tends to decrease with an enhancing open porosity. As a first order approximation, the following correlation can be proposed for the relationship between the particle density and the open porosity:

$$\gamma = \gamma_0(1 - \beta \cdot p) \tag{16}$$

where γ_0, representing the particle density of a fictive aggregate in which the open porosity is zero, is equal to 2700 kg/m^3 and β is a constant parameter that is taken equal to 2.

Finally, an autogenous cleaning process was conceived and performed on some employed particles (Pepe et al. 2014). In particular, the autogenous cleaning implemented at the laboratory scale led to interesting results in terms of enhancement of properties of crushed concrete particles, especially in terms of reduction of the attached mortar content on RCA surfaces and, consequently, on their water absorption capacity (Pepe et al. 2014). Consequently, an analytical expression is proposed for evaluating the actual open porosity, p, as a function of the "autogenous cleaning time" t, expressed in minutes:

$$p(t) = \left(1 - \frac{a \cdot t}{b + t}\right) \cdot p_{t=0} \tag{17}$$

where $p_{t=0}$ is the initial open porosity of the aggregates without autogenous cleaning, a is a constant equal to 0.6 for class 1 and 0.2 for the class 2 aggregates, while $b = 2$ in both cases.

5.3 Mechanical Properties of RAC

Due to their high porosity and water absorption capacity, RCAs significantly affect the hydration process of cementitious composites, as they modify the actual amount of water, which is available for the hydration reaction.

Based on the above considerations, (Martinelli et al. 2013) recently proposed an inverse identification procedure capable to predict the time evolution of the degree of hydration based on temperature measurements performed on concrete cubic samples cured under semi-adiabatic boundary conditions. Therefore, the proposed model enables calculation of the time evolution of the degree of hydration based on semi-adiabatic temperature measurements and, based on this, simulate hydration reactions for different thermal boundary conditions. This procedure, already validated and applied to RAC (Pepe et al. 2014), enables determining the value of the degree of hydration supposedly achieved in the concrete specimens subjected to compression tests.

Once the time evolution of the degree of hydration has been calculated, the experimental results obtained in compressive tests at a certain curing age and the corresponding values of the degree of hydration were correlated. In fact, the following conceptual formula is available in the literature for correlating degree of hydration α and compressive strength R_c (Lokhorst 2001):

$$R_c = R_{c,\max} \cdot \frac{\alpha - \alpha_0}{1 - \alpha_0} \tag{18}$$

where α_0 is the minimum degree of hydration representing the moment that the strength in the mixture is starting to build up (i.e. end of setting phase), and $R_{c,max}$ the maximum cubic compressive strength, ideally corresponding to a degree of hydration equal to 1. Then, the correlation between the mechanical properties and the degree of hydration may show the impact of RCAs on the resulting RAC properties. Particularly, it may express the influence of the residual anhydrous cement fractions, the porosity and the moisture conditions of the attached mortar on the overall hydration process that controls the formation of the concrete microstructure and associated strength.

The experimental activities carried out on RAC by (Pepe et al. 2016) were aimed at understanding the role played by RA in affecting the relevant properties of structural concrete. Several experimental tests were performed with the aim of investigating the influence of relevant parameters, such as processing procedures for RCAs, initial moisture condition of aggregates, aggregate replacement ratio and nominal water-to-cement ratio, on the resulting concrete performance. Based on the above considerations and the complete analysis of the results (Pepe et al. 2016), it is clear that the initial moisture condition and the aggregate porosity tend to modify the water-to-cement ratio and, for this reason, an effective w/c ratio taking into account both the initial moisture condition and the aggregate porosity can be defined as follows:

$$\left(\frac{w}{c}\right)_{eff} = \frac{w}{c} + \frac{w_{add}}{c} - \delta \cdot \left(\sum_{i=1}^{n} \frac{p_i \cdot P_i}{c}\right) \tag{19}$$

where w/c is the nominal water-to-cement ratio, w_{add} he extra water added to the mix corresponding to (partially or totally) soaking the aggregates, p_i and P_i represent the absorption capacity and the weight in the mixture of the i-th aggregate fraction, and $\underline{\Omega}$ is a parameter taking into account the initial moisture condition of the aggregates and is zero in saturated conditions and 0.5 in dried ones.

Then, the first step in formulating a model capable of predicting the mechanical strength development for RAC is based on taking into account both the concrete mixture composition and the main parameters characterizing RCAs.

As a matter of fact, the first possible correlation can be elaborated between R_c and $R_{c,max}$ and the effective water-to-cement ratio defined by Eq. 20, which takes into account the amount of free water available into the concrete mixture. However, the aggregate replacement ratio also affects the concrete strength, which is not explicitly considered in the definition of $(w/c)_{eff}$.

Therefore, a further parameter is introduced that accounts for the average porosity of the coarse aggregates employed in the mixture, and is defined as follows:

$$A_{MX} = \sum_{i=1}^{n} p_i \cdot V_i \tag{20}$$

where p_i is the open porosity characterizing the i-th fraction of the coarse aggregates and V_i is the corresponding fraction volume.

This parameter can be useful for describing the reduction in compressive strength due to the replacement of NAs by RCAs. Particularly, it can physically explain the decrease of α_0 and $R_{c,max}$ that characterizes the evolution of the compressive strength of RAC with respect to $\alpha_{0,NAT}$ and $R_{c,max,NAT}$, which represents the compressive strength of a reference concrete made of only natural aggregates (and keeping the grain size distribution constant). Therefore, it is possible to define the ratios of r* and α* for expressing the reduction in strength and its rate development in terms of $R_{c,max}$ and α_0, respectively. Moreover, the following analytical formulations are introduced for describing the existing relation between the above described parameters r* and α* and the A_{NAT}-to-A_{MIX} ratio:

$$r^* = \frac{R_{c,\max}}{R_{c,\max,NAT}} = \frac{a_R \cdot \frac{A_{NAT}}{A_{MIX}}}{b_R \cdot \frac{A_{NAT}}{A_{MIX}}} \tag{21}$$

$$\alpha^* = \frac{\alpha_0}{\alpha_{0,NAT}} = \frac{a_\alpha \cdot \frac{A_{NAT}}{A_{MIX}}}{b_\alpha \cdot \frac{A_{NAT}}{A_{MIX}}} \tag{22}$$

where, A_{NAT} is the corresponding A_{MIX} value for an ordinary cete mixture and a_R, b_R, a_α and b_α are constants that should, in principle, be calibrated on experimental results. Based on tested mixtures, the values of a_R and b_R can be approximated to be 1.06 and 0.06, respectively, while a_R and b_R are equal to 1.05 and 0.05.

As a result, if the individual pairs of $R_{c,max}$ and α_0 are divided by the corresponding values of r* and α* and, then, represented with respect to the effective water-to-cement ratio. The analysis of these results (Pepe et al. 2016) shows that all the points (w/c_{eff}-$R_{c,max}$/r*) can be aligned on a curve that can be analytically expressed by the well know Abram´s law and, particularly, it assumes the expression 5–23. Similarly, a possible linear correlation existing between w/c_{eff} and α_0/α^* can be expressed by the analytical formulation 5–24.

$$R_{c,\max} = \frac{A_R}{B_R^{w/c_{eff}}} \cdot r^* \tag{23}$$

$$\alpha_0 = A_\alpha \cdot \left(\frac{w}{c_{eff}} - B_\alpha \right) \cdot \alpha^* \tag{24}$$

In Eq. 23 the values of A_R and B_R can be assumed to be equal to and 9.5, respectively, while, in Eq. 24 A_α is equal to 1.62 and B_α is equal to 0.28.

5.4 Design ABACI and Model Flow Chart

The possible application of the entire formulation presented in their work is schematically highlighted in Fig. 11.

The scheme reported therein encompasses various boxes. Particularly, the "CDW and RCAs" box reported refers to the production protocol and the physical and mechanical characterization of RCAs. Moreover, the "RAC" box deals with the influence of the RCAs in concrete mixtures. Thus, the proposed formulation aimed

Fig. 11 Flow conceptual model for RAC (Pepe 2015)

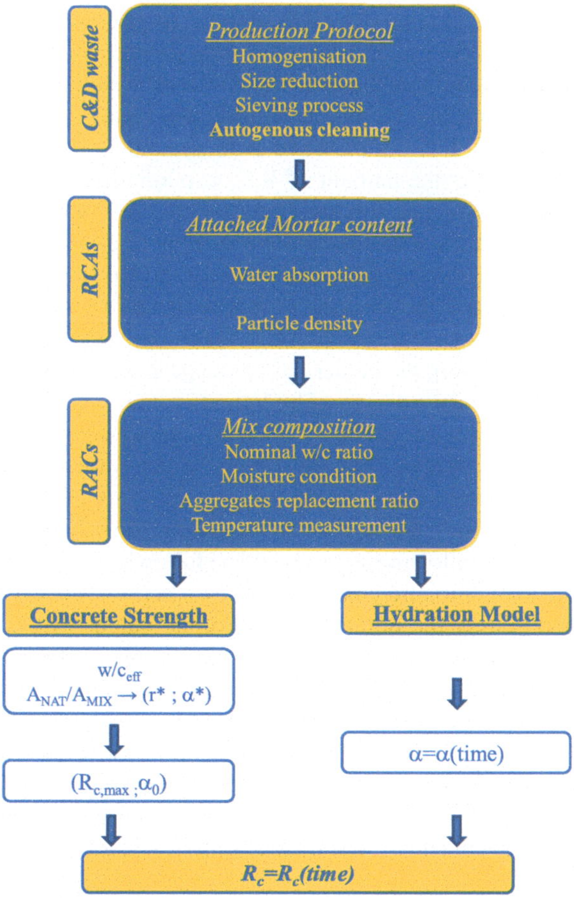

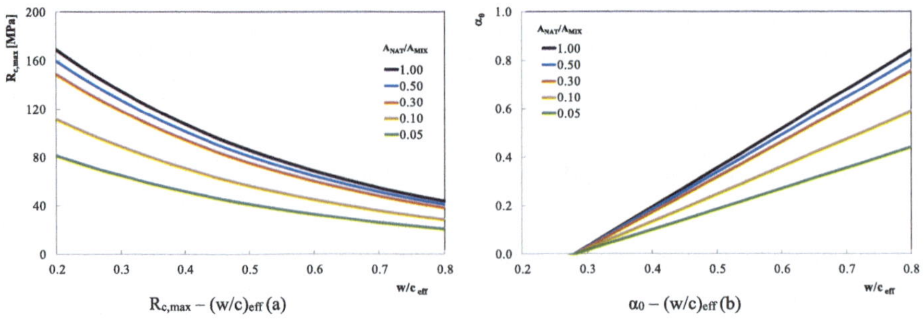

Fig. 12 Influence of A_{NAT}/A_{MIX} on $R_{c,max}$ and α_0

at simulating the hydration processes, is reported in the "Hydration model" box. Finally, the RAC strength is represented in the "Concrete strength" box.

The novel approach methodology is based on defining the parameter A_{MIX} that represents the average porosity of the concrete mixture made with RA. Once defined the composition of the concrete mixture the parameter A_{MIX} is evaluable by applying the formula and, consequently, the ratio A_{NAT}/A_{MIX} is defined.

Figure 12 shows the trend of $R_{c,max}$ and α_0 values related to both the A_{NAT}/A_{MIX} parameter (r the amount of RA or, particularly, the amount of the AM inside the resulting concrete mixture) and the effective water-to-cement ratio (that takes into account of the initial moisture condition of coarse RCAs).

The curves reported in Fig. 12, described by the proposed formulas in Eqs. (23 and 24), have the meaning of mix design abaci for R, once defined the $R_{c,max}$ and α_0 parameters, the time evolution of the compressive strength could be easily determined by considering the numerical model proposed by the authors for estimating the time evolution of the degree of hydration.

Moreover, the model flow charts can be summarized by the following steps: the water absorption capacity should be determined for each aggregate fraction; the composition of a reference mixture in terms of volume fractions of NAs and the corresponding A_{NAT} value characterizing the bulk of its aggregates should be defined. Then, the following steps should be followed for predicting the compressive strength of a RAC mixture characterized by a certain replacement ratio of NAs with RAC: after setting the replacement ratio, the value A_{MIX} for the aggregate mixture can be determined according to Eq. 20; the parameters r* and α * can be determined through Eqs. 21 and 22; the values of $R_{c,max}$ and α_0 can be obtained by applying Eqs. (23) and (24): the value of compressive strength at 28 days can be determined by means of Eq. 18, once the corresponding value of α is determined via a numerical simulation of the hydration process carried out by means of models available in the literature (Martinelli et al. 2013); in the absence of more accurate determinations, the following analytical relationship can be adopted:

$$\alpha_{\max} = \frac{1.031 \cdot \frac{w}{c_{eff}}}{0.194 + \frac{w}{c_{eff}}} \tag{25}$$

Finally, it is worth highlighting that the above procedure may also be inverted. Once the target concrete strength R_c is selected, the corresponding values of A_{MIX} can be determined and, hence, the mixture composition: the limitation in terms of aggregate replacement and the need for a cleaning process possibly needed to reduce the aggregate porosity may result from this process.

References

ACI Committee 311: ACI manual of concrete inspection Detroit. Michigan, American Concrete Institute (1992)

Amario, M., Rangel, C.S., Pepe, M., Toledo Filho, R.D.: Optimization of normal and high strength recycled aggregate concrete mixtures by using packing model. Cement Concr. Compos. **84**, 83–92 (2017)

AS 1012.9: Methods of testing concrete–determination of the compressive strength of concrete specimens. Australian Standards, Australian Government (2014)

AS 1141.14: Methods for sampling and testing aggregates–particle shape, by proportional calliper. Australian Standards, Australian Government (2014)

AS 1141.15: Methods for sampling and testing aggregates–flakiness index. Australian Standards, Australian Government (2014)

AS 1141.20: Methods for sampling and testing aggregates–chloride and sulfate content. Australian Standards, Australian Government (2014)

AS 1141.22: Methods for sampling and testing aggregates–wet/dry strength variation. Australian Standards, Australian Government (2014)

AS 1141.32: Methods for sampling and testing aggregates–weak particle. Australian Standards, Australian Government (2014)

AS 1141.4: Methods for sampling and testing aggregates–bulk density of aggregates. Australian Standards, Australian Government (2014)

AS 1141.6.1: Methods for sampling and testing aggregates–particle density and water absorption of coarse aggregate–weighing-in-water method. Australian Standards, Australian Government (2014)

AS 1391: Metallic materials—Tensile testing at ambient temperature. Australian Standard, Australian Government (2007)

AS 3600: Concrete structure. New South Wales, Standards Australia (2001)

AS1012.: Methods of testing concrete. Australian Standard (1993)

ASTM C293-00: Standard test method for flexural strength of concrete (using simple beam with center-point loading). American Society for Testing and Materials, United States (2001)

Belin, P., Habert, G., Thiery, M., Roussel, N.: Cement paste content and water absorption of recycled concrete coarse aggregates. Mater. Struct. **47**(9), 1451–1465 (2014)

Brito, J.D., Saikia, N.: Recycled Aggregate in Concrete (2013)

BSI: BS EN 1097-3: 1998: Tests for mechanical and physical properties of aggregates. Determination of loose bulk density and voids, BSI London, UK (1998)

Carneiro, A. P., et al.: Construction waste characterization for production of recycled aggregate - Salvador/Brazil. Waste Materials in Construction: WASCON 2000: Proceeding of the International Conference on the Science and Engineering of Recycling for Environmental Protection, Harrogate, England. (2000)

De Juan, M.S., Gutiérrez, P.A.: Study on the influence of attached mortar content on the properties of recycled concrete aggregate. Constr. Build. Mater. **23**(2), 872–877 (2009)

De Larrard, F., Sedran, T.: Mixture-proportioning of high-performance concrete. Cem. Concr. Res. **32**(11), 1699–1704 (2002)

Dong, J.F., Wang, Q.Y., Guan, Z.W.: Structural behavior of recycled aggregate concrete filled steel tube columns strengthened by CFRP. Eng. Struct. **48**(3), 532–542 (2013)

Faella, C., Lima, C., Martinelli, E., Pepe, M., Realfonzo, R.: Mechanical and durability performance of sustainable structural concretes: an experimental study. Cement Concr. Compos. **71**, 85–96 (2016)

Fischer, C., Werge, M., Reichel, A.: EU as a Recycling Society. European Topic Centre on Resource Waste Management. Working Paper 2 (2009)

Fathifazl, G., Abbas, A., Razaqpur, A.G., Isgor, O.B., Fournier, B., Foo, S.: New mixture proportioning method for concrete made with coarse recycled concrete aggregate. J. Mater. Civ. Eng. **21**(10), 601–611 (2009)

Engineer Forum. (2011). https://civil4m.com/showthread.php/55-Methods-of-proportioning-concrete

Glavind, M., Olsen, G.S. Munch-petersen, C.: Packing calculations and concrete mix design. Nord. Concr. Res. **13**(2) (1993)

IS: 10262: Concrete mix proportioning—Guidelines (2019)

Knaack, A.M., Kurama, Y.C.: Design of concrete mixtures with recycled concrete aggregates. ACI Mater. J. **110**(5), 483 (2013)

Kong, D., Lei, T., Zheng, J., Ma, C., Jiang, J., Jiang, J.: Effect and mechanism of surface-coating pozzalanics materials around aggregate on properties and ITZ microstructure of recycled aggregate concrete. Constr. Build. Mater. **24**(5), 701–708 (2010)

Levy, A.B., Rockafellar, R.T.: Sensitivity analysis of solutions to generalized equations. Trans. Am. Math. Soc. **345**(2), 661–671 (1994)

Lokhorst, S.J.: Deformational behavior of concrete influenced by hydration related changes of the microstructure. Delft University of Technology (2001)

Martinelli, E., Koenders, E.A., Caggiano, A.: A numerical recipe for modelling hydration and heat flow in hardening concrete. Cement Concr. Compos. **40**, 48–58 (2013)

Marzouk, M., Azab, S.: Environmental and economic impact assessment of construction and demolition waste disposal using system dynamics. Resour. Conserv. Recycl. **82**, 41–49 (2014)

Morledge, R., Jackson, F.: Reducing environmental pollution caused by construction plant. Environ. Manag. Health **12**(2), 191–206 (2001)

Noguchi, T., Tamura, M.: Concrete design towards complete recycling. Struct. Concr. **3**(2), 155–167 (2001)

Pepe, M., Toledo Filho, R.D., Koenders, E.A., Martinelli, E.: Alternative processing procedures for recycled aggregates in structural concrete. Constr. Build. Mater. **69**, 124–132 (2014)

Pepe, M., Toledo Filho, R.D., Koenders, E.A.B., Martinelli, E.: A novel mix design methodology for recycled aggregate concrete. Constr. Build. Mater. **122**, 362–372 (2016)

Pepe: A Conceptual Model for Designing Recycled Aggregate Concrete for Structural Applications (2015)

Polster, B., Peuportier, B., Sommereux, I.B., Pedregal, P.D., Gobin, C., Durand, E.: Evaluation of the environmental quality of buildings towards a more environmentally conscious design. Sol. Energy **57**(3), 219–230 (1996)

Pradhan, S., Kumar, S., Barai, S.V.: Recycled aggregate concrete: particle packing method (PPM) of mix design approach. Constr. Build. Mater. **152**, 269–284 (2017)

Pradhan, S., Tiwari, B., Kumar, S., Barai, S.V.: Comparative LCA of recycled and natural aggregate concrete using particle packing method and conventional method of design mix. J. Clean. Prod. **228**, 679–691 (2019)

Saltelli, A., Ratto, M., Andres, T., Campolongo, F., Cariboni, J., Gatelli, D., Saisana, M., Tarantola, S.: Global sensitivity analysis, the primer. John Wiley & Sons (2008)

Sedran, T., De Larrard, F.: Optimization of self-compacting concrete thanks to packing model. In: Proceedings 1st SCC Symp, CBI Sweden, RILEM PRO7 (1999)

Tam, V.W., Gao, X., Tam, C.M.: Microstructural analysis of recycled aggregate concrete produced from two-stage mixing approach. Cem. Concr. Res. **35**(6), 1195–1203 (2005)

Tam, V.W., Gao, X., Tam, C.M., Chan, C.: New approach in measuring water absorption of recycled aggregates. Constr. Build. Mater. **22**(3), 364–369 (2008)

Tam, V.W., Gao, X., Tam, C.M., Ng, K.M.: Physio-chemical reactions in recycle aggregate concrete. J. Hazard. Mater. **163**(2–3), 823–828 (2009)

Tam, W.Y.V., Tam, L., Le, K.N.: Cross-cultural comparison of concrete recycling decision-making and implementation in construction industry. Waste Manag. **30**(2), 291–297 (2010)

Tam, V.W., Le, K.N., Zeng, S.: Review on waste management systems in the Hong Kong construction industry: use of spectral and bispectral methods. J. Civ. Eng. Manag. **18**(1), 14–23 (2012)

Tam, W.Y.V., Wang Z.B., Tao Z.: Behavior of recycled aggregate concrete filled stainless steel stub columns. Materials and Structures In Press (2014)

Tech Data Sheet: Recycling spent sandblasting grit and similar wastes as aggregate in asphaltic concrete, Naval Facilities Engineering Service Center, Port Hueneme, Falifornai 4370. (1998)

Xiao, J.Z., Xie, H., Yang, Z.: Shear transfer across a crack in recycled aggregate concrete. Cem. Concr. Res. **42**(5), 700–709 (2012)

Yuan, H.P., Shen, L.Y.: Trend of the research on construction and demolition waste management. Waste Manag. **31**(4), 670–679 (2011)

Constitutive Relationships of Recycled Aggregate Concrete

Belen González-fonteboa, Cesar Medina, Fernando Abella Martinez, Sindy Seara-paz, and Takafumi Noguchi

1 Introduction

To predict constitutive relationship of structural RAC, it is necessary to determine the main parameters. Based on literature review, it can be concluded that these main parameters are the used quantity, the quality of RA, and also the production method (Etxeberria et al. 2007; Paine and Dhir 2010a; Ferreira et al. 2011; García-González et al. 2016).

Therefore, the final objective is to develop state-of-the-art in structural RAC and provide recommendations that allow the engineers to use RAC with similar guarantees as with conventional ones. In some cases, modified expressions are proposed to predict RAC properties considering (if possible) the three primary parameters above: the recycled quantity, the quality of RA, and also the production method.

B. González-fonteboa · F. A. Martinez
Civil Engineering School, Centro de Innovación Tecnolóxica en Edificación E Enxeñaría Civil (CITEEC), Universidade da Coruña (UDC), A Coruña, Spain

C. Medina
Departamento de Construcción, Instituto de Investigación en Desarrollo Territorial Sostenible (INTERRA), Universidad de Extremadura, Grupo de Investigación MATERIA, Avda. de La Universidad, Cáceres, Spain

S. Seara-paz
Department of Civil Engineering, Universidade da Coruña, Coruña, Spain

T. Noguchi (✉)
Building Materials Engineering Laboratory, Department of Architecture, Graduate School of Engineering, The University of Tokyo, Tokyo, Japan
e-mail: takafuminoguchi@g.ecc.u-tokyo.ac.jp

J. Xiao et al. (eds.), *Structural Behaviour and Innovation of Recycled Aggregate Concrete*, RILEM State-of-the-Art Reports 41,
https://doi.org/10.1007/978-3-031-84136-1_4

1.1 Recycled Aggregate Quality

1.1.1 Factor Affecting the Quality

The RA quality is the first parameter that has to be considered to produce RAC. The most significant difference between RA and conventional aggregate is the AM from the original concrete and the number of impurities it presents. This chapter deals with RA from concrete waste, which are the most suitable aggregates to produce structural RAC. In this case, impurities' presence is low, being the AM, the main parameter that controls RA quality.

Moreover, it should be noted that there is a significant difference between the properties of RCA and recycled fine aggregate, which always contain a much higher proportion of AM. Therefore, to produce structural RAC, only RCA is recommended. Focusing on RCA from concrete waste, it can be concluded that its quality is linked to the amount and characteristics of the AM that it contains. This amount is related to the original concrete. Low-grade original concrete leads to low AM and, consequently, low-grade RA. Secondly, the recycling process is also essential because it influences the amount (quantity) of the attached mortar, decreasing as the stages in the crushing process increase (Nagataki et al. 2004). One of the best procedures to define RA quality is its physical properties, especially through its water absorption capacity. A high water absorption capacity indicates a high amount or a low-grade AM. In contrast, a low water absorption capacity means a high grade or a low AM. In this regard, in general, RCA from concrete waste presents a water absorption capacity of 8.5%.

The RA quality influences concrete properties in different ways. There will be a higher strength loss for a given water-to-cement ratio as the quality of RA worsens. However, this effect is more noticeable for lower water-to-cement ratios since the ultimate compressive strength becomes more dependent on RA's strength instead of that of the cement matrix (Silva et al. 2015a, b, c, d, e, f). It was also found that RAC produced with rounded NA presents an entirely different strength behavior from RAC produced with crushed NA. The former may not show a decrease in the compressive strength, and it may even produce an increase of up to 15% (Laserna and Montero 2016).

1.1.2 The Interfacial Transition Zone in RAC

The ITZ spreads radially outward from the aggregate's surface into the cement paste. The microstructure of this area, which differs from that of the cement matrix (cement paste + sand = mortar), is characterized by the presence of $Ca(OH)_2$ platelets positioned perpendicularly to the surface of the aggregate, and by higher ettringite content, porosity and mean pore size. Those features make it the weakest area in concrete, with properties that distinguish it from the cement matrix. Its elastic modulus is just 80% of the matrix value (Wang et al. 2016). Its water permeability is 100-fold, and its diffusion coefficient 6- to 12-fold more significant than in the cement paste (Zhang et al. 2019b, a). As a result, the intrinsic properties of the ITZ (primarily thickness and elastic modulus) directly impact mechanical behavior, modulus of elasticity, drying shrinkage, creep, and durability. RAC has two ITZ, one between the original aggregate and the old AM (old ITZ) and the other between RA (AM or original aggregate) and the new mortar matrix (new ITZ) (Otsuki et al. 2003) (Figs. 1 and 2). The tiny pores and cracks in the new ITZ is the weakest part of RAC, have an adverse effect on material performance (Zhang et al. 2019b, a). The RA/mortar ITZ is characterized by porosity 1.6–1.9 times greater than in the natural aggregate/paste ITZ (Djerbi 2018); a thickness of 55–65μm, compared to 10μm to 50μm at the natural aggregate/paste interface; and less favorable elastic modulus and microhardness values than in the natural aggregate/paste ITZ (Sáez del Bosque et al. 2017). So, RAC characteristics depend on the new ITZ's quality when the old ITZ's quality is better than that of the new ITZ. Whereas RAC characteristics depend on the old ITZ's quality, the new ITZ's quality is better than that of the old ITZ (Ryu 2002). In this way, when the w/c ratio is high, the influence on concrete properties of the presence of RA is not so significant as when it is low. Moreover, in these cases, the effect of the new cement paste's low-quality is more significant than the presence of the RA.

Understanding ITZ properties is highly instrumental in modeling the microstructure of RAC for the assessment and prediction of their complex, non-elastic behavior across their service life. One of the most critical stages in this process is simulating each component's micro-properties (in the ITZ, thickness, and elastic modulus). When modeling new RAC, five phases (new mortar, new ITZ, old mortar, old ITZ, and aggregate) must be envisaged instead of the three needed for conventional concretes (mortar, ITZ, and aggregate) (Medina et al. 2019).

Of the various instrumental techniques currently in place to study this region, the research community most favored includes scanning or atomic force microscopy, nanoindentation, and scanning and backscattered electron microscopy. With recent

Fig. 1 The RAC interfacial transition zones (Otsuki et al. 2003)

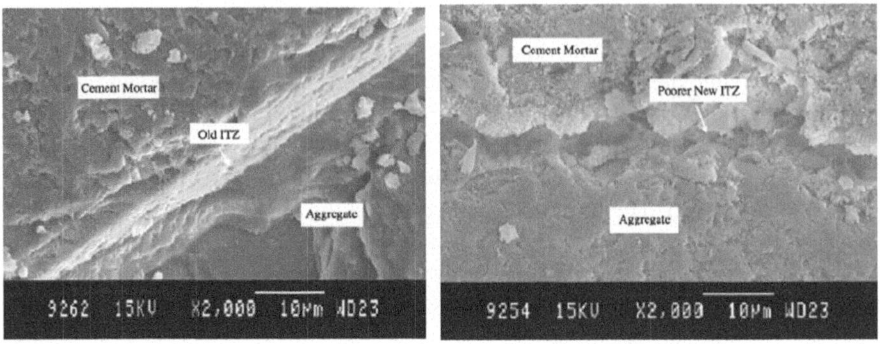

Fig. 2 The microstructure characteristics of ITZs in RAC (Marí et al. 2006)

developments that have improved the application of nanoindentation to these materials, the modulus of elasticity and thickness of the ITZ, which cannot be determined with traditional techniques, can now be measured empirically (Xiao et al. 2013a, b).

1.2 Recycled Aggregate Quantity

The second parameter that has to be considered is the quantity of RA incorporated in the RAC mix. The amount of RA has a remarkable influence on all RAC properties. Replacement percentages from 20 to 100% by volume of natural aggregate with RA are typically considered.

1.3 Production Method

Finally, it is well known that the production method (i.e., the mixing procedure) chosen to compensate the RA high water absorption modifies the ITZ quality between the RCA and the new paste and, therefore, the RAC properties (Corinaldesi 2010; Belén et al. 2011; de Brito et al. 2011; Seara-Paz et al. 2014; Silva et al. 2015a, b, c, d, e, f).

Three alternative methods have been essentially proposed to compensate the high water absorption of the RA (Hansen 1986; Sagoe-Crentsil et al. 2001; Poon et al. 2004; Etxeberria et al. 2007; Corinaldesi 2010; González-Fonteboa et al. 2012; Pedro al. 2014; Seara-Paz et al. 2016). In the first one, the aggregate is added after being immersed in water for a pre-established time (pre-soaking), generally 10 min (Nealen 1998). In the second one, the aggregate is added with its natural moisture, compensating the aggregate absorption by adding extra water (extra water), generally the necessary water to get the 80% of saturation. Lastly, other authors produce concrete without considering the high absorption of RA (dry), controlling workability with supplementary superplasticizer quantities. This last method is almost given up because very high superplasticizer quantities are required to maintain workability, and, in some cases, it may not be possible to get it. So, in this chapter, only the pre-soaking and extra water methods are going to be considered.

In general terms, authors conclude that the pre-saturated RA cause damage in the ITZ between the RCA and the new cement paste (Etxeberria et al. 2007), which supposes a negative effect on the concrete compressive strength (Etxeberria et al. 2007; Belén et al. 2011). Using RA in the saturated surface-dry state, the high water content inside the aggregate particles may result in bleeding. The water inside RA particles can move to cement paste, creating a relatively high water-to-cement ratio in particles' proximity. This process may weaken the link between the RA and the cement matrix. Consequently, a high degree of bleeding leads to lower RAC strength (Poon et al. 2007).

The current tendency suggests that the extra water method avoids the bleeding effect and leads to a better behavior than the pre-soaking method (Poon et al. 2004).

1.4 Methodology

Each section of this chapter is divided into two stages. The first one aims to know how much one property can be affected when RAC coarse aggregates are used, analyzing different incorporation ratios and different mixing procedures.

With the literature results and using linear regression, a coefficient can be obtained to estimate how much the RAC property is modified compared to conventional concrete one (Eq. 1).

$$\mathrm{Pro}_{\mathrm{c,RC}} = \mathrm{X}(\%\ \mathrm{RCA},\ \mathrm{production\ method})\cdot Pro_{\mathrm{c,CC}} \tag{1}$$

$\mathrm{Pro}_{\mathrm{c,RC}}$: RAC property; RCA: recycled coarse aggregate; $\mathrm{Pro}_{\mathrm{c,CC}}$: conventional concrete property, X: adjusted coefficient.

The second stage aims to correct the code expressions (Eurocode) to predict the RAC properties with the same approximation degree as in conventional concretes. To do so, the theoretical value of any property is calculated according to code proposals and taking into account the experimental value of the compressive strength. Using this result, the ratios "experimental property/calculated property" are obtained. Suppose the ratios of RAC are similar to those obtained with conventional ones. In that case, it can be considered that code expressions provide the same approximation degree in RAC than in conventional ones to calculate the analyzed property, and, therefore, it is not necessary to modify these expressions, however, if these ratios decrease when the content of RCA increases, code expressions should be corrected using a coefficient that, in this case, will be adjusted by multivariable regression.

$$\mathrm{Ratio\ conventional\ concrete} = \mathrm{Ratio\ CC} = \frac{\mathrm{Pro}_{\mathrm{CC,exp}}}{\mathrm{Pro}_{\mathrm{CC,cal}}}$$

$$\text{Ratio RAC} = \text{Ratio RC} = \frac{\text{Pro}_{\text{RC,exp}}}{\text{Pro}_{\text{RC,cal}}}$$

When $RatioCC \neq RatioRC \Rightarrow \frac{\text{Pr}\,o_{CC,\text{exp}}}{\text{Pr}\,o_{CC,cal}} \neq \frac{\text{Pr}\,o_{RC,\text{exp}}}{\text{Pr}\,o_{RC,cal}}$ a correction coefficient is necessary:

$$\frac{\text{Pro}_{\text{CC,exp}}}{\text{Pro}_{\text{CC,cal}}} = \frac{\text{Pro}_{\text{RC,exp}}}{\text{Pro}_{\text{RC,cal}} \cdot \text{coefficient}\left(f_{c,RC}, \%\ \text{RCA}, \text{WA}, \text{production method}\right)} \tag{2}$$

$$\text{Coefficient}\left(f_{c,RC}, \%\ \text{RCA}, \text{WA}, \text{production method}\right) = \text{CF} = \frac{\text{Ratio RC}}{\text{Ratio CC}} \tag{3}$$

where, $\text{Pro}_{\text{CC, cal}}$: calculated property of conventional concrete using code expressions; $\text{Pro}_{\text{CC, exp}}$: experimental property of conventional concrete; $\text{Pro}_{\text{RC, cal}}$: calculated property of RAC using code expressions; $\text{Pro}_{\text{RC,exp}}$: experimental property of RAC.

Using this method, the proposed prediction expressions will be:

$$\text{Pro}_{\text{RC,cal}} = \text{CF}\left(f_{c,RC}, \%\ \text{RCA}, \text{WA}, \text{production method}\right) \cdot \text{EUROCODE}_{\text{expression}} \tag{4}$$

where, $\text{Pro}_{\text{RC,cal}}$: property of RAC; CF: correction coefficient for the Eurocode expression, $f_{c,RC}$: RAC compressive strength; %RCA: replacement ratio (RAC coarse aggregate percentage); WA: recycled coarse aggregate water absorption.

The mixing procedure and the RAC coarse aggregate quality (considering it through its water absorption) have only been considered in predicting some properties (those that have been more widely studied). However, due to the lack of enough data, the other properties have been analyzed only considering the replacement ratio and regardless of the mixing procedure or the aggregate quality.

2 Compressive Strength

2.1 Introduction

In general, RAC compressive strength is lower than the conventional concrete one, decreasing when the coarse aggregate substitution percentage increases (Poon et al. (2007); Kovler and Roussel (2011); Xiao et al. (2012); Silva et al. (2015a, b, c, d, e, f); Seara-Paz et al. (2016)). However, the extent of this decrease mainly depends on the RA percentage, concrete production method (González-Taboada et al. 2016), and RA' type, size, and origin (Silva et al. 2015a, b, c, d, e, f).

2.2 *Compressive Strength of RAC*

The compressive strength of RAC is lower compared with the strength of the control mixes of conventional concrete. Using a database (González-Taboada et al. 2016), the relative compressive strength, defined as the ratio of the compressive strength of RAC to that of the conventional concrete, can be known (Eqs. 4 and 5). This ratio allows the engineer to estimate the expected compressive strength decrease in a structural RAC as a function of the RCA content and the production method (Table 1)

$$C = X \cdot f_{c,CC} \tag{5}$$

Figures 3, 4 and 5 shows the relationship "RAC compressive strength vs. conventional concrete compressive strength" of the different recycled percentages obtained using each production method (González-Taboada et al. 2016).

It can be seen that the lower the RA percentage is, the lower the strength loss will be. The pre-soaking method presents the highest reduction of the compressive strength due to the bleeding effect. With the "adding extra water" method, the reductions in compressive strength are lower. Also, it can be seen that low replacement percentages (20%) hardly influence the compressive strength of concrete (Rao et al. 2011).

2.3 *Statistical Characteristics for Compressive Strength*

Xiao et al. (2005a, b) investigated the statistical characteristics of the compressive strength of RAC with relatively uniform quality. It is found that a standard distribution model can be applied to fit the compressive strength results of the RAC at a 95% confidence level with different RCA contents Fig. 6. Researchers from other countries also prove that the coefficient of variation for compressive strength of RAC in the

Table 1 Correction coefficients (X) to estimate $f_{c,RC}$ (González-Taboada et al. 2016)

%RCA		ADwEW	PS
100	Data	140	89
	X	0.888	0.828
	R2	0.960	0.908
50	Data	49	31
	X	0.937	0.861
	R2	0.911	0.874
20	Data	38	37
	X	0.969	0.936
	R2	0.954	0.958

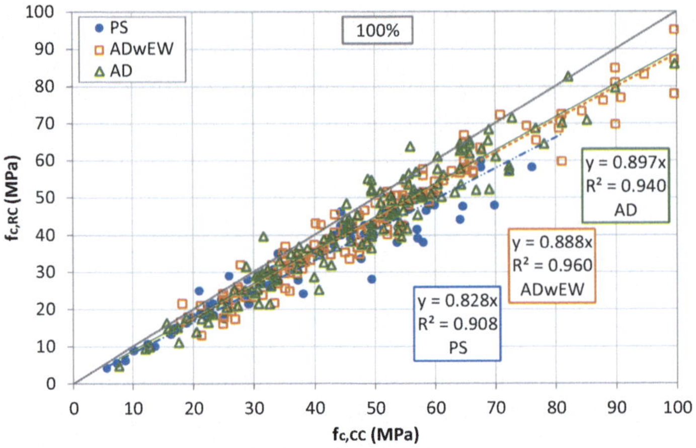

Fig. 3 $f_{c,RC}$ versus $f_{c,CC}$ (100%)

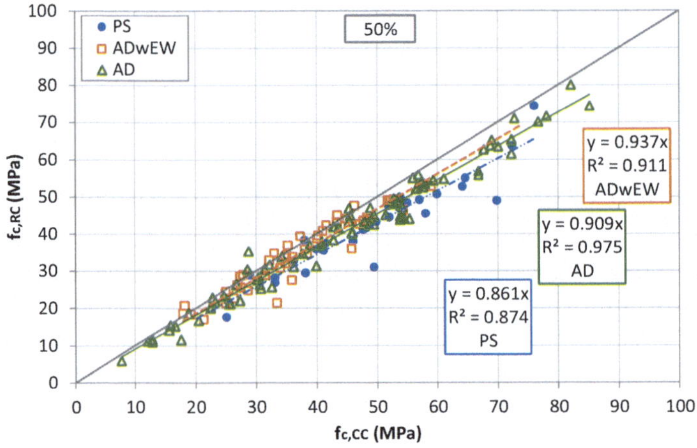

Fig. 4 $f_{c,RC}$ versus $f_{c,CC}$ (50%)

laboratory is not much different from that of conventional concrete when the same RCA was used throughout the production.

Xiao et al. (2005a, b) conclude that further research is required in terms of statistical characteristics for the compressive strength of concrete with RA from different resources. The test result reveals that the RA with non-uniform quality increases the discrete for the distribution of the compressive strength of RAC. The normal model can still be used to describe the probability distribution for the compressive strength of RAC. However, the coefficient of variation for compressive strength is slightly larger.

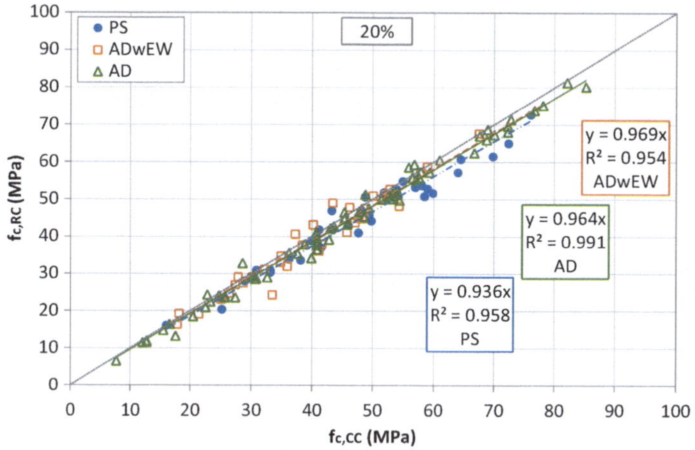

Fig. 5 $f_{c,RC}$ versus $f_{c,CC}$ (20%)

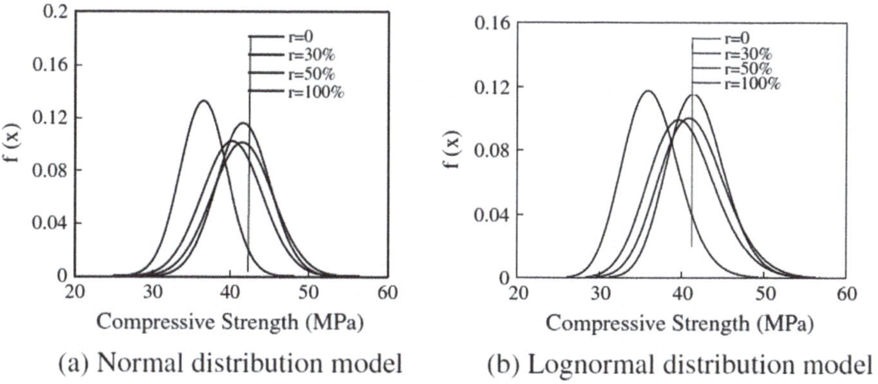

Fig. 6 Distribution model for compressive strength of RAC

This conclusion has also been achieved by Rahal (2007), who found similar coefficients of variation in recycled and conventional concretes and explained that this difference is insignificant, possibly due to the limited number of RA sources.

2.4 The Relationship for Cubic-Cylindrical Compressive Strength

To obtain the relationship between the cube compressive strength f_{cu} and the cylinder compressive strength f_{cy}, Li (2008) carried out an extensive experimental study with only concrete with 100% RA. The cylinder compressive strength was determined

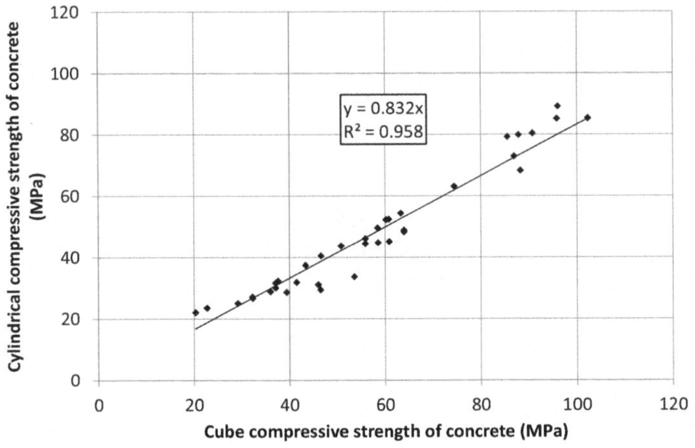

Fig. 7 $f_{c,cub}$ versus $f_{c,cyl}$ (0–100% RCA)

on cylindrical specimens of Ø 150mm × 300mm. The test results indicate a linear relationship between the cylinder and the cube compressive strength, which can be described as follows. It should be noted that Eq. (6) is only valid up to cube compressive strength of about 45MPa.

$$f_{cy} = 0.82 f_{cu} \tag{6}$$

In this way, to confirm this result, the ratio $f_{c,cyl}$ to $f_{c,cub}$ was calculated using the researcher works that provide both results. Using regression analysis and with a coefficient of determination $R^2 = 0.96$, the ratio was found to be 0.832 regardless of the RCA percentage used Fig. 7. Analyzing the results considering different replacement percentages, 0–20% and 50–100%, the ratio was found to be 0.810 and 0.861, respectively (Fig. 8).

Zhou and Chen (2017) also analyze the relationship between cube compressive strength and axial compressive strength of RAC (Fig. 9). They conclude that equations adopted by codes for conventional concrete do not apply to RAC. As a result, they suggest regression equations based on the experimental results to obtain the compressive strength of RAC (f_c) through the cube compressive strength (f_{cu}) and taking into account the replacement percentage of RCA (r), Eqs. 4–7:

$$f_c = \left(ar^2 + br + c\right) 0.89 f_{cu} \tag{7}$$

For RA obtained from crushed waste concrete with crushed rock, a is 0.0227, b is 0.1030, and c is 0.9620. For RA from crushed waste concrete with pebbles, a is –0.1037, b is 0.1030, and c is 0.9620.

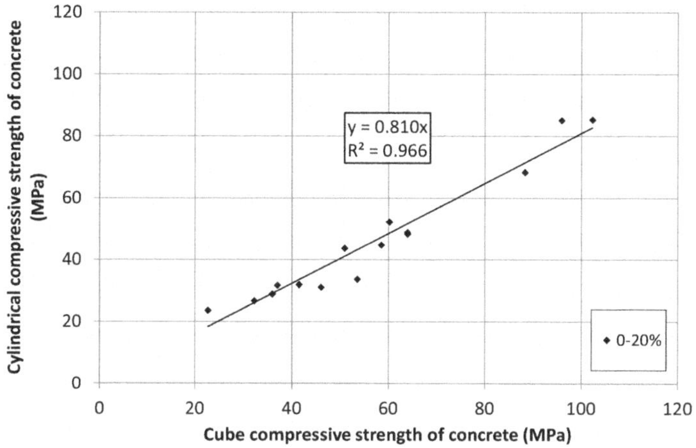

Fig. 8 $f_{c,cub}$ versus $f_{c,cyl}$ (0–20% RCA)

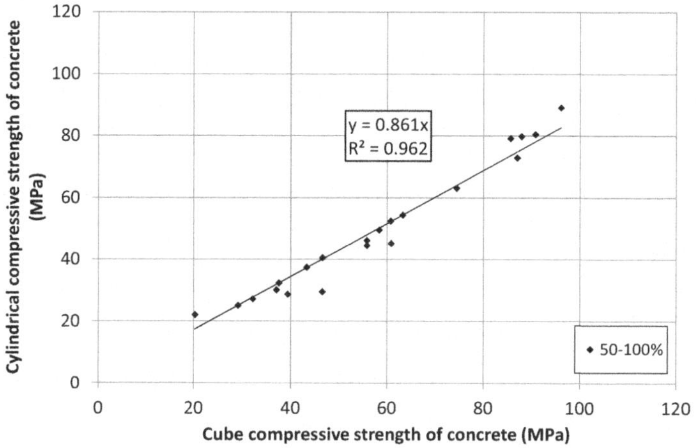

Fig. 9 $f_{c,cub}$ versus $f_{c,cyl}$ (50–100% RCA)

2.5 Time-Dependent Compressive Strength

2.5.1 Time-Dependent Compressive Strength of RAC

Some authors observed that, over a long period, RAC might demonstrate more significant strength development than the corresponding conventional concrete. This was explained by the mortar's latent cementitious properties adhered to old aggregates (Silva et al. 2015a, b, c, d, e, f; Eiras-López et al. 2017). Eiras-López et al. (2017) studied the compressive strength at 3, 7, 90, and 365 days. The relative strengths

Fig. 10 fc(t)/fc(28d) at 3, 7, 28, 90 and 365 days (Eiras-López et al. 2017)

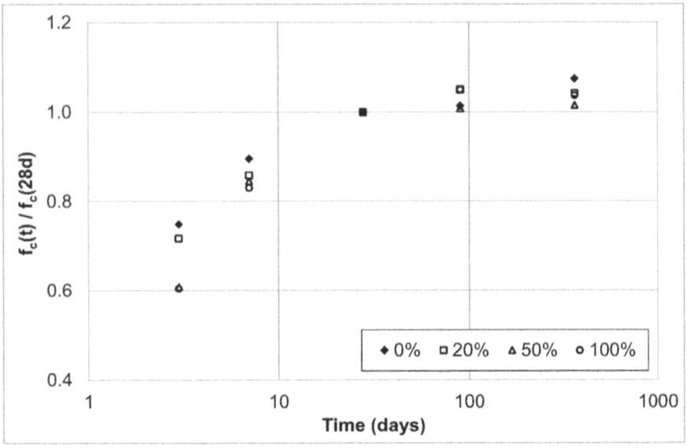

were calculated using the compressive strength at different ages obtained at 28 days (Fig. 10).

These curves confirm that RAC presents a further development of compressive strength than conventional ones. According to these results, a delay in the time-dependent development of compressive strengths of RAC compared with conventional concrete can be seen, especially significant at 3 and 7 days. While RAC reaches 3 days 64% of 28 days compressive strength, the corresponding conventional concrete has already developed 74%. Although, to a less extent, these differences are also detected at 7 days. At this age, the RAC has reached 85% of their compressive strength at 28 days, and the conventional concrete, 90%. Lastly, at 90 and 365 days, there are no significant differences in time-dependent development of compressive strength between recycled and conventional concretes.

These differences are attributed to RA's use under pre-saturation conditions, which produces a curing effect in RAC at early ages. For this reason, it has been named by other authors as "internal curing" (Corinaldesi 2010; Pedro et al. 2015). The extra water inside the pores of pre-soaked RA bleeds to cement paste during the hydration process resulting in a weaker ITZ. This results in further time-dependent development of compressive strength between recycled and conventional concretes, especially significant at an early age (Corinaldesi 2010; Manzi et al. 2013; Pedro et al. 2015).

2.5.2 Prediction of Time-Dependent Compressive Strength

To adjust the expressions proposed by different codes (EN 1992, del Hormigón Estructural 2008; Taerwe and Matthys 2013) to predict the compressive strength of RAC at different ages, Eiras-López et al. (2017) calculated the predicted values of compressive strength, taking into account the experimentally obtained fc (28d) of each concrete.

Table 2 shows the ratios "experimental fc/ calculated fc" obtained at 3, 7, 28, 90, and 365 days. It was thus noted that RAC shows lower ratios than conventional concretes, which decreases as the replacement percentage of RCA increases, especially at an early age. Consequently, a coefficient was proposed to modify code expressions taking into account the RA content.

Therefore, an exponential coefficient was adjusted by multivariable regression, taking into account the replacement percentage used (RCA = %RCA/100) and the concrete age (t, days), Eqs. (4–8).

$$f_c(t) = f_{c,code}(t) \cdot e^{\left[(0.032 \cdot RCA)\left(1-\left(\frac{28}{t}\right)^{(1.28-0.34 \cdot RCA)}\right)\right]} \tag{8}$$

Being, fc (t) = compressive strength of concrete at different ages; fc, code (t) = compressive strength of concrete according to code expressions at different ages calculated with the 28 days compressive strength of each concrete (conventional or recycled).

The proposal was evaluated using three statistical indexes and validated with the compressive strength results at different ages obtained from other researchers (Gomez-Soberon 2002a, b; Rahal 2007; Corinaldesi et al. 2011; De Brito et al. 2011) (Fig. 11). In this regard, it was noted that the modified expression predicts the time-dependent compressive strength of RAC with the same approximation degree as that for conventional concrete.

Table 2 Ratios "experimental f_c/ calculated f_c"

Age	H50–0	H50–20	H50–50	H50–100	H65–0	H65–20	H65–50	H65–100
3 days	1.19	1.21	0.94	0.96	1.07	0.96	0.89	0.87
7 days	1.11	1.15	1.05	1.05	1.08	0.95	1.01	0.98
28 days	1.00	1.00	1.00	1.00	1.00	1.00	1.00	1.00
90 days	0.92	0.97	0.95	0.99	0.93	0.95	0.89	0.93
365 days	0.94	0.93	0.90	0.94	0.92	0.87	0.85	0.86

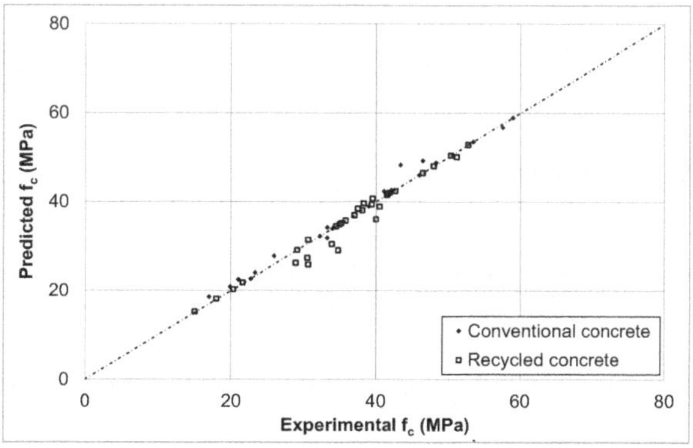

Fig. 11 Compressive strength using the proposed expression vs. experimental results. Data from (Gomez-Soberon 2002a, b; Rahal 2007; Corinaldesi et al. 2011; De Brito et al. 2011)

It should be noted that the proposal is only valid when the RA is added after a pre-soaking period. More test data of the "adding extra water" method are needed to extend the proposal or modify that coefficient.

3 Modulus of Elasticity (EC)

It is well known that concrete elasticity modulus is related to the aggregates and cement paste rigidity. The cement paste modulus is lower than the aggregate one and lower than the concrete one (López Gayarre 2008). Moreover, RA (with AM) presents a modulus of elasticity lower than NA and, therefore, RAC modulus will be lower than a conventional concrete one. This is also due to the new interface's weakness (new cement mortar aggregate, old cement mortar aggregate, and old cement mortar-new cement mortar) that can lead to progressive cracks, which will affect the concrete deformability (Bairagi et al. 1993).

Most of the codes and standards suggest equations to predict elasticity's modulus as a compressive strength function. It is well known that these expressions are not suitable when RAC is used. In this way, many authors have proposed different expressions to predict it (Table 3). However, these expressions have been adjusted with the authors' particular results and, so, it has been revealed that they present some scatterings when other researchers' results are analyzed.

Noguchi et al. (2015), with more than 1300 experimental data obtained from studies performed between 1980 and 2012 in Japan and Korea and statistical analyses, proposed a relationship between compressive strength and modulus of elasticity RAC that considers the ratio and types of RCAs.

Duan et al. (2013) have explored the applicability of artificial neural networks (ANN) in modeling the RAC modulus. Their results show that the constructed ANN models can accurately predict the elastic modulus of concrete made with RA derived from different sources. However, these techniques do not provide the structural designer with a precise mathematical expression (Duan et al. 2013; Behnood et al. 2015).

3.1 RAC Modulus

The RAC modulus is lower compared with a conventional concrete one. Using a database (González-Taboada et al. 2016), the relative modulus, defined as the ratio of the RAC modulus to that of the conventional concrete can be known (Eq. 4–9). This ratio takes into account the recycled content and the mixing procedure used. The adjustment goodness has been evaluated with the coefficient of determination, R2 (Table 4).

$$E_{c,RC} = X \cdot E_{c,CC} \tag{9}$$

Table 3 Equations for $E_{c,RC}$

References	Equations
Ravindrarajah and Tam (1987)	$E_c = 7770.f_c^{0.33}$ $E_c = 4630.f_c^{0.5}$ $E_c = 3480.f_c^{0.5} + 13050$
Kazikaki et al. (1988)	$E_c = 1.9 \cdot 10^5 \cdot \left(\frac{\rho}{2300}\right)^{1.5} \cdot \left(\frac{f_c}{2000}\right)^{0.5}$
Bairagi (1993)	$E_c = \alpha \cdot f_c^{0.5}; \alpha = \alpha(r) = 5780 - 1340 \cdot r \ (0 \leq r \leq 1)$
Dillmann (1998)	$E_c = 634.43 \cdot f_c + 3057.6$
Dhir et al. (1999)	$E_c = 370.f_c + 13100$
Mellmann (1999)	$E_c = 378 \cdot f_c + 8242$
Zilch and Roos (2001)	$E_c = 9100 \cdot (f_c + 8)^{1/3} \cdot \left(\frac{\rho}{2400}\right)^2$
Katz (2003)	$E_c = 0.043 \cdot \rho^{1.5} \cdot f_c^{0.5}$
Xiao (2006)	$E_c = \frac{10^5}{2.8 + \frac{40.1}{f_c}}$
Li (2008)	$E_c = 5.5 \cdot 10^3 \cdot f_c^{0.5} \cdot \left(\frac{\rho}{2400}\right) \cdot \left(1 - \frac{r}{\alpha}\right);$ $\alpha = 2.2876 \cdot r + 0.1288 \ (0 \leq r \leq 1)$
Tang (2010)	$E_c = 4450 \cdot f_c^{0.5}$
Corinaldesi (2010)	$E_c = 18800 \cdot \sqrt[3]{\frac{0.83 \cdot f_{cu}}{10}}; E_c = 909 \cdot f_{cu} + 8738$
González-Fonteboa et al. (2011)	$E_c = (1 - 0.0020 \cdot r) \cdot 22000 \cdot \left(\frac{f_{cm}}{10}\right)^{0.3}$

where, E_c = Modulus of elasticity (MPa); f_c or f_{cu} = Compressive strength (MPa); ρ = Density (kg/m3); r = Replacement percentage

Figures 12 and 13 the relationship "RAC elasticity modulus vs. conventional concrete elasticity modulus" for the different replacement percentages and the different mixing procedures. As in the compressive strength analysis, the PS method presents the highest reduction of the elasticity modulus, which is, once again, due to the bleeding effect (Figs. 12 and 13).

Table 4 Correction coefficients (X) to estimate Ec, RC

Production method		< 50%	50–100%
ADwEW	Data	19	73
	X	0.882	0.856
	R2	0.931	0.763
PS	Data	39	68
	X	0.886	0.792
	R2	0.437	0.579

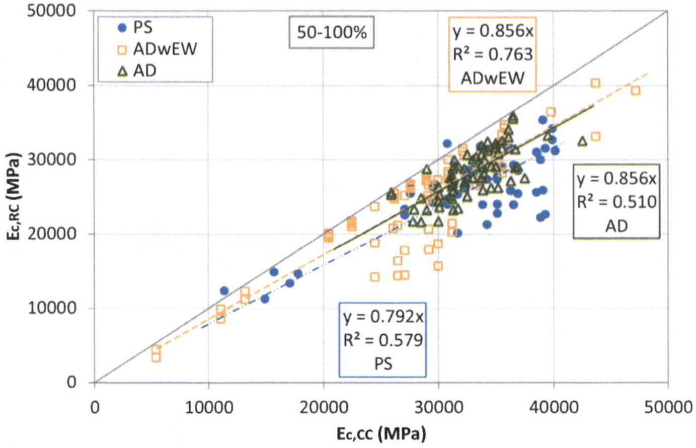

Fig. 12 $E_{c,RC}$ versus $E_{c,CC}$ (50–100%)

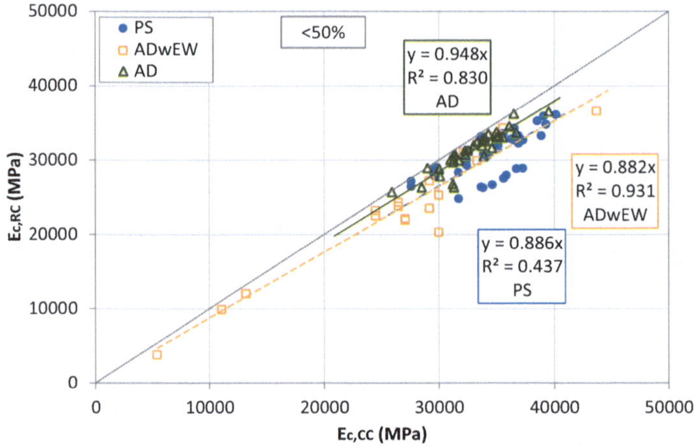

Fig. 13 $E_{c,RC}$ versus $E_{c,CC}$ (<50%)

3.2 Prediction of RAC Modulus

Taboada, Fonteboa et al. (2016) performed the experiments, and the ratios "experimental modulus/calculated modulus" of recycled vs. conventional concretes are analyzed for each mixing procedure (ADwEW, PS). It was observed that the ratios of conventional concretes were higher than those obtained with recycled ones. This means that the code expression will not provide the same approximation degree in RAC as in conventional ones to calculate the modulus. Therefore, it is necessary to modify it by introducing a correction coefficient.

Therefore, trying to get the same approximation degree as with the Eurocode expression, that is trying to obtain the same ratio "experimental modulus/calculated modulus" in conventional and RAC using that expression, a correction coefficient (CF) was adjusted by multivariable regression (González-Taboada et al. 2016). The variables considered were the replacement ratio and the RAC coarse aggregate quality (considering it through its water absorption). A different coefficient was adjusted for each of the different mixing procedures considered.

The expression is as follows:

$$E_{c,RC} = CF\left(f_{c,RC}, \%\ RCA, WA\right) \cdot EUROCODE_{expresssion} \tag{10}$$

$$\text{Where, } EUROCODE_{expresssion} = Ec_{cal} = 22000 \cdot \left(\frac{fc}{10}\right)^{0.3} \tag{11}$$

The following general expression was selected to adjust the coefficient CF:

$$CF = \frac{\left(1 - b \cdot \%\ RCA/_{100} \cdot WA/_{5}\right)}{\left(\frac{f_{c,RC}}{10}\right)^{\left(0.3 \cdot a^{\%\ RCA}/_{100}\right)} \cdot \left(1 - d \cdot \%\ RCA/_{100}\right)} \tag{12}$$

The parameters a, b, and d were adjusted (Table 5). The adjustment performance was analyzed using the Pearson correlation coefficient (r) and the mean squared error (MSE).

Figures 14 and 15 show the ratios "experimental modulus/calculated modulus" of RAC vs conventional ones for each mixing procedure with and without the correction coefficient. As can be observed, RAC ratios are similar to those obtained with conventional concretes when the correction coefficient is used.

Table 5 Adjustment of the correction coefficient to calculate the RC modulus (González-Taboada et al. 2016)

Production method	Data		Adjustment			R		MSE	
	Training	Testing	a	b	d	Training	Testing	Training	Testing
ADwEW	164	42	–0.0697	0.0677	–0.1044	0.4019	0.4936	0.003843	0.003038
PS	73	17	–0.3433	0.1842	–0.1120	0.5106	0.7252	0.007868	0.004857

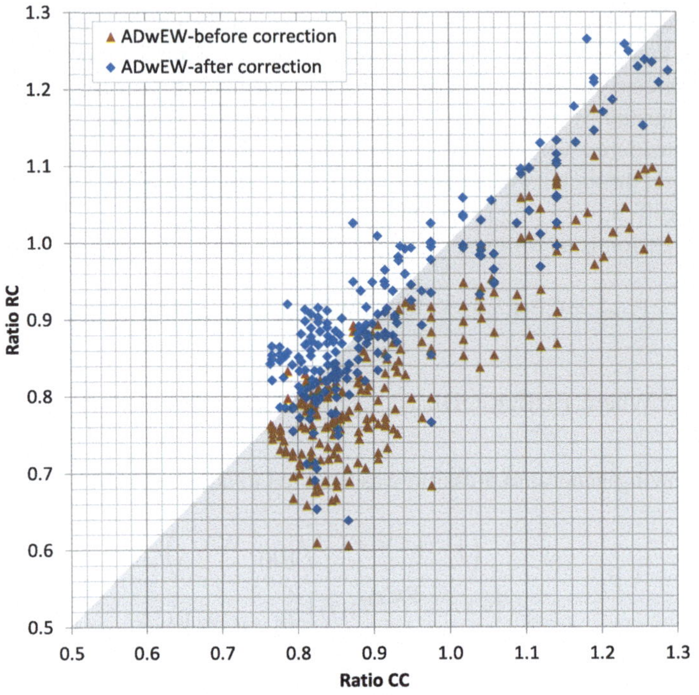

Fig. 14 Ratio RC versus Ratio CC (pre-soaking method)

4 Splitting Tensile Strength

As with the compressive strength and modulus Ec, there are several exciting works (Brito and Robles 2010; Silva et al. 2015) that have tried to predict the splitting tensile strength of RACs. These authors have also stated that splitting tensile strength graphs obtained to confirm the scatter of the test results for this property in different campaigns.

Recently, Silva et al. (2015) have suggested that the tensile strength is also affected, as the compressive strength (Silva et al. 2015), by the RA content, size, type, and quality. They have found that the relationship between tensile and compressive strengths appears to have been unaffected by RA's use. They conclude that RA's inclusion causes a proportional decrease in tensile and compressive strengths at a rate that follows the same relationship observed in conventional concrete.

However, other authors suggest that the equations proposed by codes and standards to predict the splitting tensile strength as a function of the compressive strength are not suitable when RAC is used. In this way, these authors (Katz 2003; Xiao et al. 2006; Li 2008), based on their experimental campaigns, have proposed specific expressions adjusted to predict the RAC splitting tensile strength (Table 6).

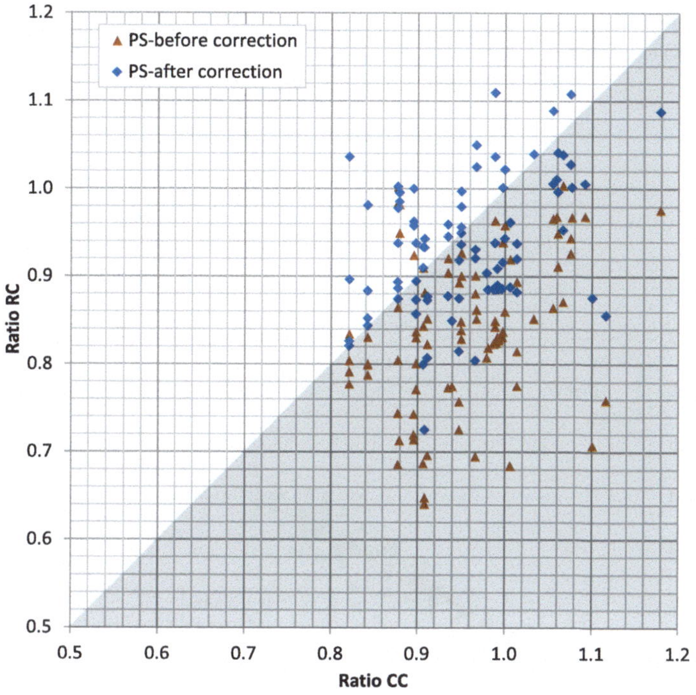

Fig. 15 Ratio RC versus Ratio CC (pre-soaking method)

Table 6 Authors' equations–$f_{sp,RC}$

References	Equations
Xiao et al. (2006)	$f_{sp} = 0.24 \cdot f_{cu}^{0.65}$
Katz (2003)	$f_{sp} = 0.59 \cdot \sqrt{f_c}$
Li (2008)	$f_t = (0.24 - 0.06 \cdot r) \cdot f_{cu}^{2/3}$

4.1 Splitting Tensile Strength of RAC

The same analysis as with the modulus of elasticity was made in González-Taboada et al. (2016) with the splitting tensile strength. The results obtained with this parameter show higher scatterings than the ones obtained with the modulus.

Firstly, with a database and using linear regression (Figs. 14 and 15), a coefficient to estimate RAC splitting tensile strength as a function of conventional concrete one has been adjusted (Eq. 4–13). This coefficient takes into account the recycled content and the mixing procedure used. The adjustment goodness has been evaluated with the coefficient of determination, R2 (Table 7).

$$f_{sp,RC} = X \cdot f_{sp,CC} \tag{13}$$

Table 7 Correction coefficients (X) to estimate $f_{sp,RC}$

Production method		<50%	50–100%
ADwEW	Data	19	73
	X	0.882	0.856
	R^2	0.931	0.763
PS	Data	39	68
	X	0.886	0.792
	R^2	0.437	0.579

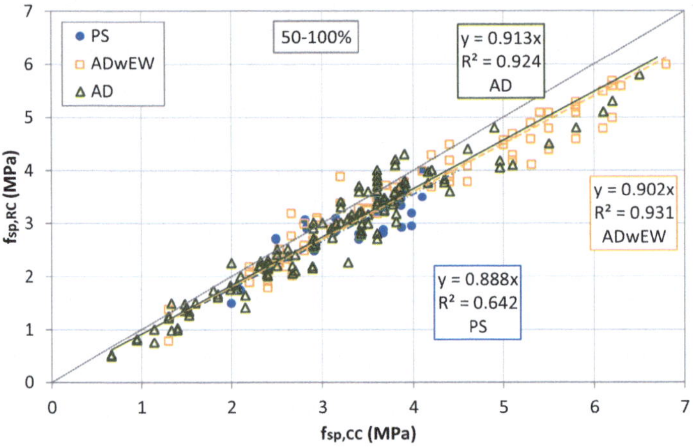

Fig. 16 $f_{sp,RC}$ versus $f_{sp,CC}$ (50–100%)

As in the compressive strength analysis, the PS method presents the highest reduction of the splitting tensile strength, higher than the ADwEW method (Figs. 16 and 17). This can be explained, again, by the bleeding effect and, according to Silva et al. (2015a, b, c, d, e, f) and de Brito et al. (2011), by the "nailing effect", which is caused by the cement paste filling the surface pores of aggregate particles, achieving an improved bond strength.

4.2 Prediction of Splitting Tensile Strength in RAC

As in the elasticity modulus section, the second step developed in González-Taboada et al. (2016) was to analyze if necessary adapt the proposed Eurocode expression for the splitting tensile strength when RAC is used. The ratio "experimental splitting tensile strength/calculated splitting tensile strength" of RAC and conventional ones were calculated for each mixing procedure (ADwEW, PS).

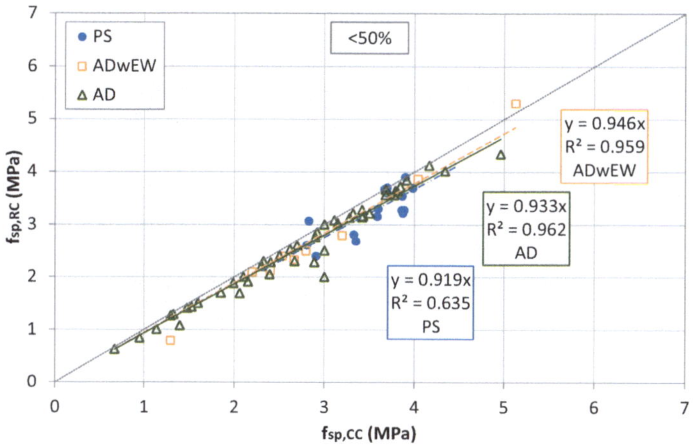

Fig. 17 $f_{sp,RC}$ versus $f_{sp,CC}$ (<50%)

It was observed (Figs. 18 and 19) that the points (ratio CC versus ratio RC) were noticeably closed and around (up and down) the midline in the case of the ADwEW method. This means that the code expression provides the same approximation degree in RAC as in conventional ones to calculate the splitting tensile strength. Therefore, it is not necessary to modify the Eurocode proposed expression with any correction coefficient. In these cases, the use of the experimental compressive strength for predicting the splitting tensile strength is enough to get the same approximation degree in RAC as in conventional ones.

High scatterings were observed in the PS method (Fig. 19) (points are not as close as in the ADwEW method). However, they do not show a clear trend in their position, being around the midline. In this way, it can be considered, again, that a correction coefficient is not necessary.

5 Stress–strain Relationship

For the nonlinear analysis of structural concrete members, it is necessary to determine their real stress–strain curves. The curves proposed in different studies for different conventional concretes are based on experimental results.

The actual stress distribution in the concrete members' compression zone is complicated to measure and model. The most common parameters with physical significance used to define the stress–strain in curves are the following:

- Maximum stress: usually considered to be the compressive strength of concrete
- Modulus of elasticity: the initial tangent modulus and the secant modulus of elasticity valued between stress 0 and 40% of the strength of concrete
- Peak strain: strain corresponding to the maximum stress

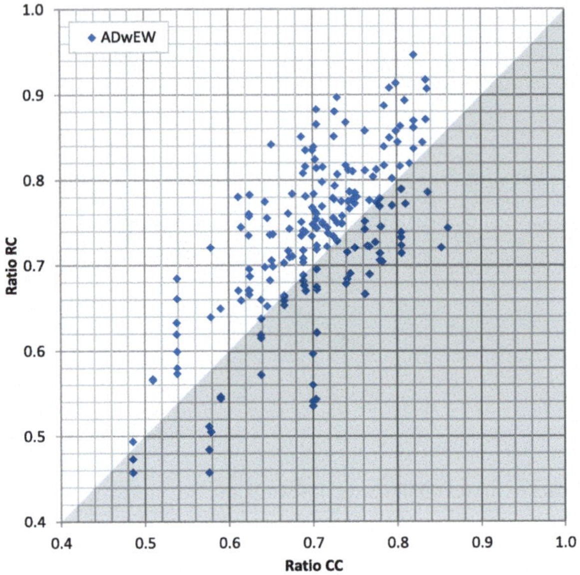

Fig. 18 Ratio RC versus Ratio CC (extra water method)

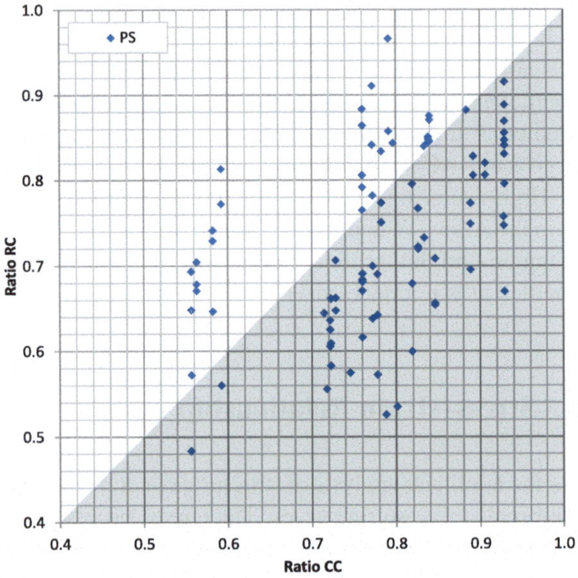

Fig. 19 Ratio RC versus Ratio CC (pre-soaking method)

- Ultimate strain: strain at which failure is defined. According to their analytical expressions, many codes (EN-1992-1-1-1 2008; EHE-08 2010) consider 3.5% as ultimate strain, which corresponds, according to their analytical expressions, to a compressive stress value in the whereabouts of 0.6fc measured in the descending branch of the curve.

The proposed stress–strain curves are either continuous functions for the ascending and descending branches or two separate functions. The equation in EN-1992-1-1-1 (2008) for the ascending and the descending branch up to a strain level of 3.5 ‰ and up to a concrete strength level of 50 MPa is defined as a continuous function.

5.1 Stress–strain in Compression of RAC

5.1.1 Ascending Branch and Peak Strain of RAC

Concerning the shape of the ascending branch of the stress–strain curve, shown in Fig. 20, it can be concluded that all the concretes present very similar stress–strain curves regardless of the RA replacement ratio. The shape is linear up to 0.4 fc, and then it becomes parabolic.

However, a considerable shift in the curves was observed. In this way, with the same unit stress, the RAC develops strains that are higher than the strains achieved

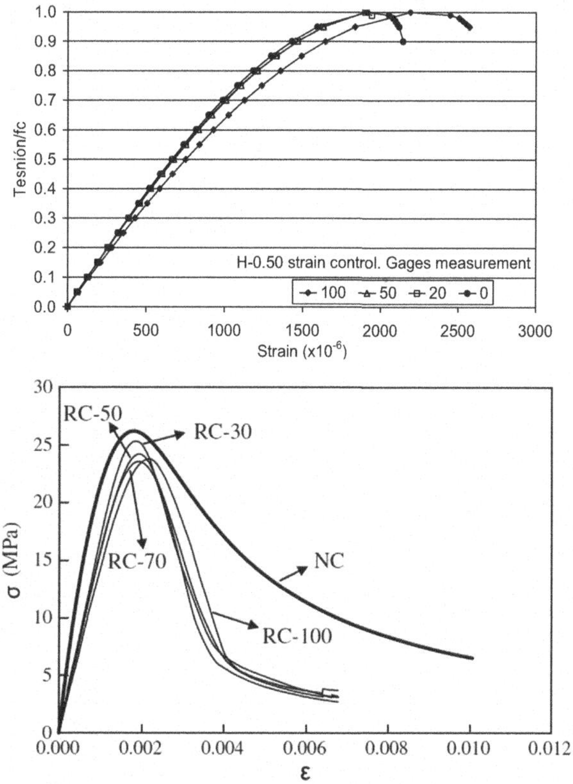

Fig. 20 Stress–strain curve in compression (Belén et al. 2011; Xiao et al. 2012)

with the conventional concretes, and the higher the replacement percentage is, the higher they will be. The presence of mortar adhered to the RA allows the increase of strains in the ascending branch. This affects the elasticity modulus of RAC (reducing it), leading to a more considerable deformation. Consequently, this increment increases the peak strain value, particularly when the RA is used with a replacement ratio of 100%. When RAC with 100% of RCAs is analyzed, this increase reaches values by about 20%. These results had been observed by most of the authors (Xiao et al. 2005a, b; Rahal 2007; Li 2008).

Xiao et al. (2012) studied RAC prisms' peak strain (100 mm × 100 mm × 300 mm) under uniaxial compression with different RCA contents. Du, Wang et al. 2010 also proved that the peak strain of RAC is much higher than that of conventional concrete. Besides, Rahal (2007) found that the strain at the peak compressive stress in RAC was 5.5 larger than conventional concrete. Belén et al. (2011) analyzed the peak strain in cylindrical specimens of RAC with saturated RA and also confirmed this result. Different authors' test results for the RAC's peak strains with different RCA contents are shown in Fig. 21 Zhou and Chen (2017) have achieved the same results.

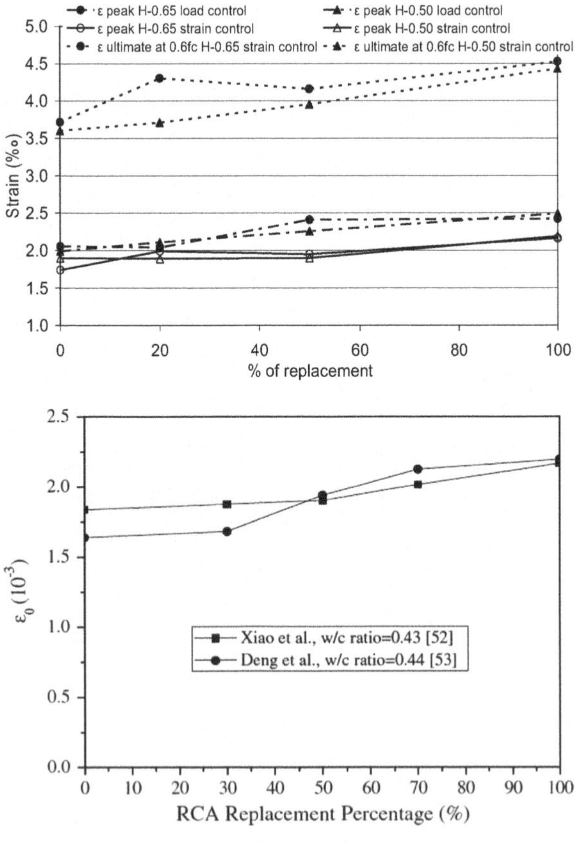

Fig. 21 Peak strain and ultimate strain (Belén et al. 2011; Xiao et al. 2012)

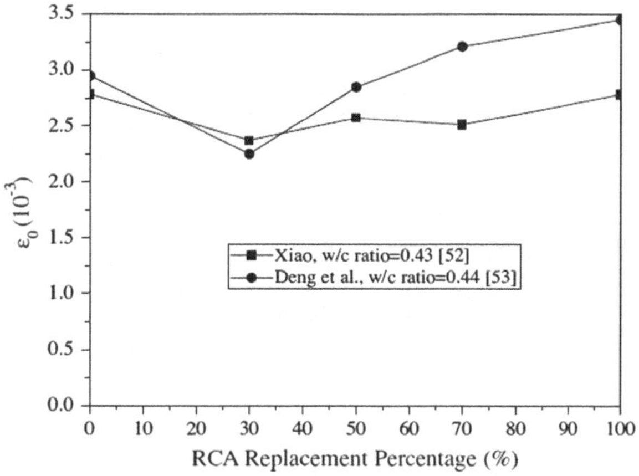

Fig. 22 Ultimate strain as a function of RCA content (Xiao et al. 2012)

5.1.2 Descending Branch and Ultimate Strain of RAC

The stress–strain curve's descending branch shape is similar in recycled and conventional concretes regardless of the RA replacement ratio. However, the conclusions achieved concerning the slope of this descending branch are not conclusive.

Belén et al. (2011) observed a decrease in the slope when the RA percentage is increased in Fig. 21. This reduction implies an increase in the ultimate strain (strain at which failure is defined). As in the peak strain case, this increase is especially significant when the percentage of replacement of conventional coarse aggregate with recycled coarse aggregate reaches 100% (again by about 20%).

Bernier et al. (1978) found that the ultimate strain was also higher in RAC, obtaining values of about 1.75 for concretes made with recycled coarse aggregate. This fact can also be observed in the curves published by Ruhl and Atkinson (1999). However, these results do not agree with those obtained by Xiao et al. (2012). They indicate that the slope of the RAC curves' descending branch decreases as the RAC content increases. This leads to different trends in terms of the ultimate strain. The ultimate strain decreases for a low RA content, while the opposite may be the case for a higher content of the replacement percentage (Fig. 22).

It should be taken into account that the ultimate strain has been measured by Belén et al. (2011) in the descending branch of the curve in a compressive stress value in the whereabouts of 0.6fc, while by Xiao et al. (2012) in a compressive stress value of 0.85 fc.

5.1.3 Ductility

The influence of the replacement percentage on the $\varepsilon_u/\varepsilon_0$-value of RAC is shown in Fig. 23. Using the results obtained by Xiao et al. (2012), it can be concluded that the

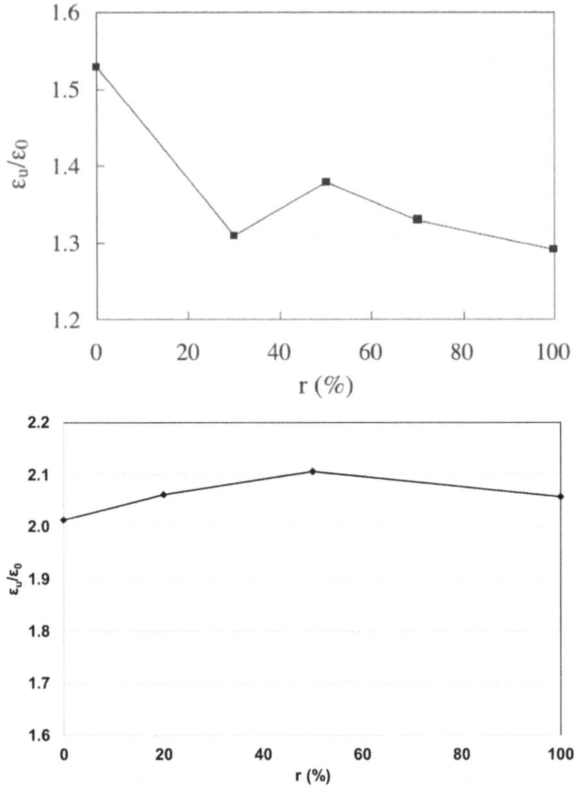

Fig. 23 Ultimate strain as a function of RA content

$\varepsilon_u/\varepsilon_0$-value of RAC is smaller than that of conventional concrete. This means that RAC is more brittle presents low ductility than the conventional one.

However, using the results of Belén et al. (2011) the $\varepsilon_u/\varepsilon_0$-value of RAC is slightly higher than that of the conventional concrete. This means that RAC presents similar ductility to the conventional one.

5.2 Stress–strain in the Tension of RAC

Xiao has studied the uniaxial tensile stress–strain relationship of RAC with different replacement percentages, Li et al. 2012 and Liu et al. (2011), Fig. 24. With an increase in the replacement percentage, the tensile peak strain of RAC slightly increased.

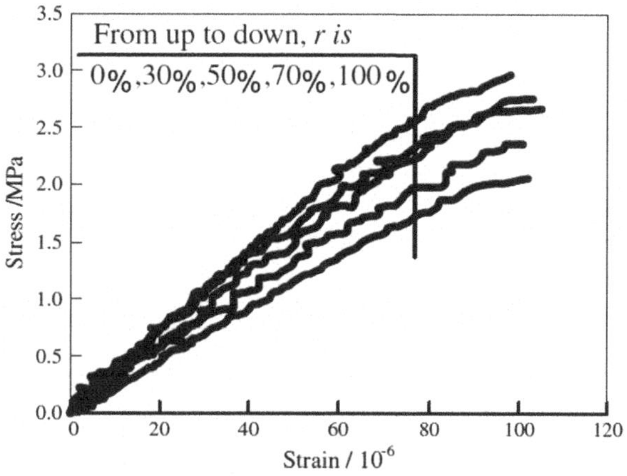

Fig. 24 Stress–strain curve in tension (Xiao et al. 2012)

5.3 Prediction of Stress–strain Relationship in RAC

Different stress–strain relationships for RAC in compression have been proposed (Xiao et al. 2005a, b; Belén et al. 2011; Zhou and Chen 2017). Belén et al. (2011) suggest a modified expression based on the Eurocode proposal because it is one of the most widely used equations. This expression, for the ascending branch and for the descending branch up to stress level 3.5 ‰ (ε_{cu1}) strain, is as follows:

$$\sigma_c/f_{cm} = \left(k\eta - \eta^2\right)/(1 + (k-2)\eta) \tag{14}$$

where, σ_c = compression stress; $\eta = \varepsilon_c/\varepsilon_{c1}$ where ε_{c1} = is the strain at peak stress and ε_c = strain in compression; $k = 1,05E_{cm}|\varepsilon_{c1}|/f_{cm}$; f_{cm} is the maximum compression stress value at 28 days, $f_{cm} = f_{ck} + 8$ and f_{ck} is the characteristic (5%) cylinder strength; E_{cm} is the longitudinal secant modulus of elasticity between stress 0 and stress 0.40 f_{cm}. The current codes consider the values of the longitudinal secant modulus of elasticity, depending on the compression strength. Specifically, the Eurocode provides the following expression for the determination of this modulus $E_{cm} = 22(f_{cm}/10)^{0.3}$.

According to the results obtained by Belén et al. (2011), the correction of stress–strain curves would imply the use of code equations for conventional concrete with three coefficients that take into account the replacement percentage of RA. These coefficients would use to calculate the secant modulus $E_{cm}\left(\varphi^{rec}_{cm}\right)$, another for the peak strain, ε_{c1} and $\varepsilon_{c2}\left(\varphi^{rec}_{cm}\right)$ and the last one for the ultimate strain, ε_{c1} and $\varepsilon_{c2}\left(\beta^{rec}_{cm}\right)$.

These coefficients were adjusted from regression analyses. However, as aforementioned, recent studies developed by the same researchers (González-Taboada et al. 2016) have suggested a more general coefficient to correct the modulus equation. The following expressions depending on the percentage of replacement with RCA (%) were suggested to $\left(\alpha^{rec}_{cm}\right)$ and $\left(\beta^{rec}_{cm}\right)$ coefficients:

$$\begin{aligned} \alpha_c^{rec} &= 0,0021 \times \%RCA + 1 \\ \beta_{cu}^{rec} &= 0,0022 \times \%RCA + 1 \end{aligned} \tag{15}$$

Figure 25 presents the stress–strain curve experimentally obtained and the predicted curves according to model code and modified expression for RAC with 50% and 100% replacement percentage. These curves reveal a better adjustment between fitted and experimental curves than those obtained between model code and experimental curves.

For structural analysis and practical design, Xiao et al. (2005a, b) and Zhou and Chen (2017) proposed a modified expression to calculate the stress–strain curve for RAC based on the proposal adopted by Chinese Code for the Design of Reinforced

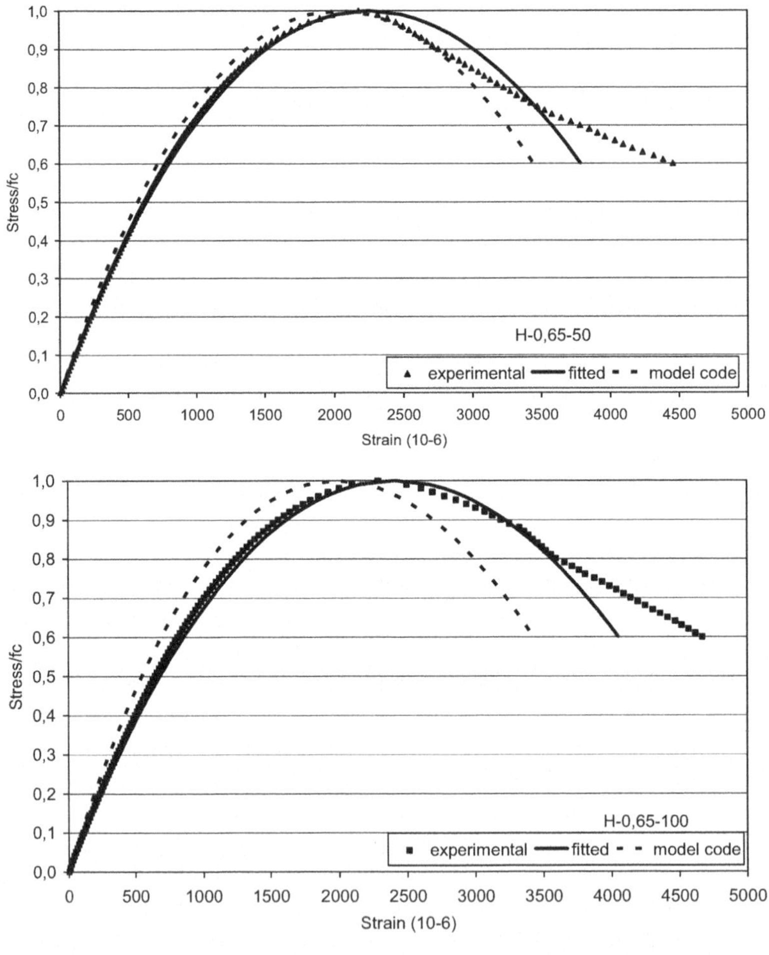

Fig. 25 Experimental and predicated stress–strain curves obtained by González-Fonteboa et al. (Belén et al. 2011)

Concrete Structures, GB50010 2010:

$$y = \begin{cases} ax + \left(3 - 2ax^2 + (a-2), 0 \le x < 1\right) \\ \frac{x}{b(x-1)^2 + x, x \ge 1} \end{cases} \tag{16}$$

where, x is $\varepsilon/\varepsilon_0$ (being ε_0 the strain at peak stress); y is σ/fc; a and b are fitting constants, represented as the initial elastic modulus and related to the shape of the descending branch of the stress–strain curve, respectively.

Based on the experimentally obtained stress–strain curves of RAC, the parameters a and b fitted by Xiao et al. (2005a, b) using a data regression analysis are as follows:

$$\mathrm{a} = 2.2\mathrm{x}\left(0.748\mathrm{r}^2 - 1.231\mathrm{r} + 0.975\right) \tag{17}$$

$$\mathrm{b} = 0.8\mathrm{x}(7.6483\mathrm{r} + 1.142) \tag{18}$$

where r is the replacement percentage of RA.

Figure 26 shows a comparison between the experimentally obtained and the approximated stress–strain curves provided by this proposal. The approximate stress–strain curves agree quite well with those obtained experimentally. Hence, this expression can be used in practical engineering applications of RAC.

Regarding Zhou and Chen (2017) proposal, parameters *a* and *b* obtain the stress–strain curves, depending on the type of RA used. If the concrete with RA obtained from crushed waste concrete with crushed rock is analyzed, these parameters are:

a = 1.15, b = 12.43.

Otherwise, if the concrete is made with RA obtained from crushed waste concrete with natural pebbles, these values are:

a = 1.17, b = 4.58.

Again, the theoretical stress–strain curve of RAC shows a good adjustment, being similar to that experimentally obtained, Fig. 27.

The analytical model of the stress–strain relationship proposed by Xiao et al. (2012) for recycled under uniaxial tension is as follows:

$$\mathrm{y} = \mathrm{gx} - (\mathrm{g} - 1)\mathrm{x}^6 \tag{19}$$

where, x is $\varepsilon/\varepsilon_0$ (being ε_0 the strain at peak stress); y is σ/fc; g is related to the RCA replacement percentage (r) and can be determined by:

$$\mathrm{g} = 0.07\mathrm{r} + 1.19 \tag{20}$$

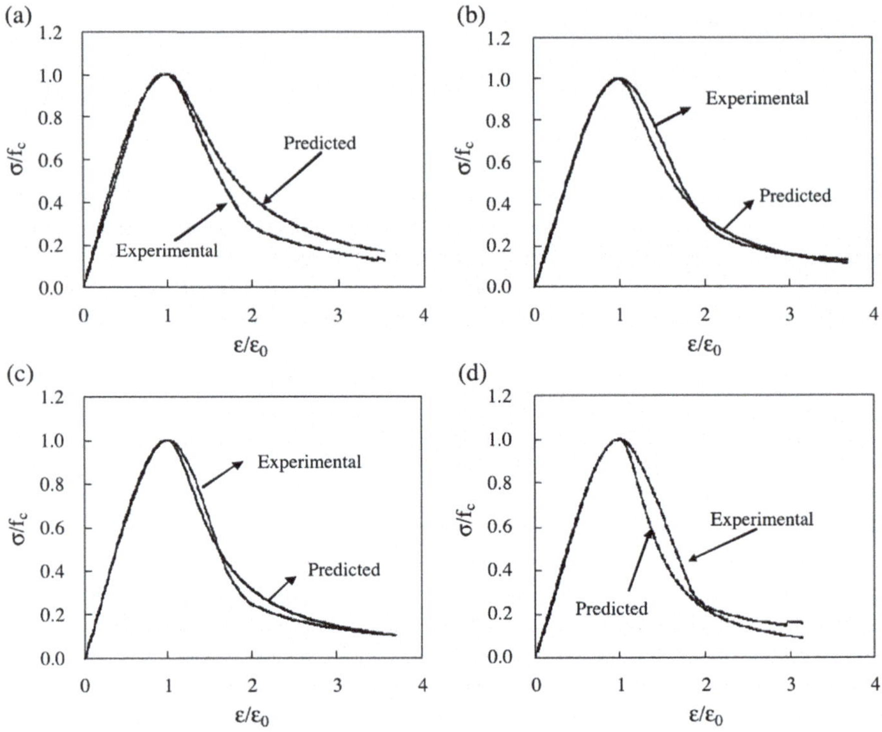

Fig. 26 Experimental and predicated stress–strain curves obtained by Xiao et al. (2005a, b)

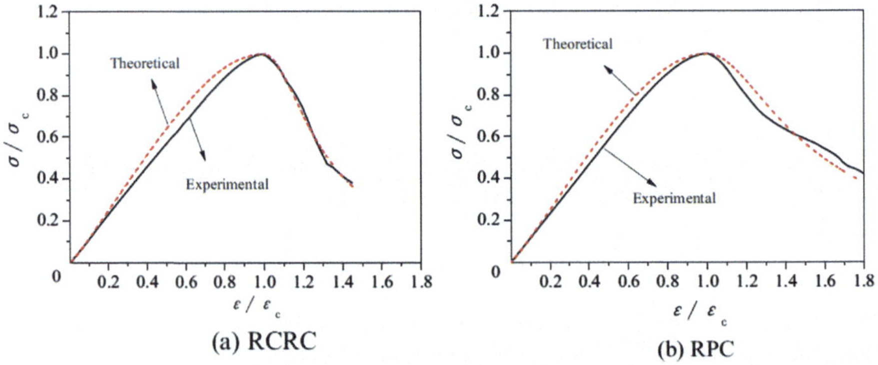

Fig. 27 Experimental and predicated stress–strain curves obtained by Zhou and Chen (2017)

References

Bairagi, N.K., Ravande, K., Pareek, V.K.: Behavior of concrete with different proportions of natural and recycled aggregates. Resour. Conserv. Recycl. **9**(1), 109–126 (1993)

Behnood, A., Olek, J., Glinicki, M.A.: Predicting modulus elasticity of recycled aggregate concrete using M5′ model tree algorithm. Constr. Build. Mater. **94**, 137–147 (2015)

Belén, G.F., Fernando, M.A., Diego, C.L., Sindy, S.P.: Stress–strain relationship in axial compression for concrete using recycled saturated coarse aggregate. Constr. Build. Mater. **25**(5), 2335–2342 (2011)

Bernier, G., Malier, Y., Mazars J.: New material form concrete demolition waste. In: International Conference on the Use of By-products and Waste in Civil Engineering (1978)

Brito, J.D., Robles, R.: Recycled aggregate concrete (RAC) methodology for estimating its long-term properties **17**(6), 449–462 (2010)

Corinaldesi, V.: Mechanical and elastic behavior of concretes made of recycled-concrete coarse aggregates. Constr. Build. Mater. **24**(9), 1616–1620 (2010)

Corinaldesi, V., Letelier, V., Moriconi, G.: Behavior of beam–column joints made of recycled-aggregate concrete under cyclic loading. Constr. Build. Mater. **25**(4), 1877–1882 (2011)

De Brito, J., Barra, M., Ferreira, L.: Influence of the pre-saturation of recycled coarse concrete aggregates on concrete properties. Mag. Concr. Res. **63**(8), 617–627 (2011)

del Hormigón Estructural, I.: EHE-08. Ministerio de Fomento, Secretaría General Técnica, Madrid (2008)

Dhir, R.K., Limbachiya, M.C., Leelawat, T.: Suitability of recycled concrete aggregate for use in BS 5328 designated mixes. Proceedings of the Institution of Civil Engineers - Structures and Buildings **134**(3): 257–274 (1999)

Dillmann, R.: Concrete with recycled concrete aggregates. Proceedings of the International Symposium on Sustainable construction: Use of recycled concrete aggregate. R. K. Dhir, N. A. Henderson and M. C. Limbachiya. London, UK, Thomas Telford: 239–254 (1998)

Djerbi, A.: Effect of recycled coarse aggregate on the new interfacial transition zone concrete. Constr. Build. Mater. **190**, 1023–1033 (2018)

Duan, Z.H., Kou, S.C., Poon, C.S.: Using artificial neural networks for predicting the elastic modulus of recycled aggregate concrete. Constr. Build. Mater. **44**, 524–532 (2013)

EHE-08: Code on structural concrete. Madrid, Spain, Publication Center, General Technical Secretariat, Ministry of Development 556 (2010)

Eiras-López, J., Seara-Paz, S., González-Fonteboa B., Martínez-Abella F.: Bond behavior of recycled concrete: Analysis and prediction of bond stress-slip curve. J. Mater. Civ. Eng. **29**(10) (2017)

EN-1992–1-1: Eurocode 2–design of concrete structures: part 1–1: general rules and rules for buildings. Brussels, Belgium, Comité Européen de Normalisation (CEN) 259 (2008)

Etxeberria, M., Vázquez, E., Marí, A., Barra, M.: Influence of amount of recycled coarse aggregates and production process on properties of recycled aggregate concrete. Cem. Concr. Res. **37**(5), 735–742 (2007)

Ferreira, L., de Brito, J., Barra, M.: Influence of the pre-saturation of recycled coarse concrete aggregates on concrete properties. Mag. Concr. Res. **63**(8), 617–627 (2011)

García-González, J., Barroqueiro, T., Evangelista, L., de Brito, J., De Belie, N., Morán-del Pozo, J., Juan-Valdés, A.: Fracture energy of coarse recycled aggregate concrete using the wedge splitting test method: influence of water-reducing admixtures. Mater. Struct. **50**(2) (2016)

Gomez-Soberon, J.M.V.: Porosity of recycled concrete with substitution of recycled concrete aggregate–an experimental study. Cem. Concr. Res. **32**(8), 1301–1311 (2002a)

Gomez-Soberon, J.M.V.: Creep of concrete with substitution of normal aggregate by recycled concrete aggregate. In: Innovations in Design with Emphasis on Seismic, Wind, and Environmental Loading: Quality Control and Innovations in Materials/Hot-Weather Concreting. V. M. Malhotra. **209**, 461–474 (2002b)

González-Fonteboa, B., Martinez-Abella, F., Eiras-Lopez, J., Seara-Paz, S.: Effect of recycled coarse aggregate on damage of recycled concrete. Materials and structures, 44, 1759–1771 (2011)
González-Fonteboa, B., Martínez-Abella, F., Herrador, M.F., Seara-Paz, S.: Structural recycled concrete: behavior under low loading rate. Constr. Build. Mater. **28**(1), 111–116 (2012)
González-Taboada, I., González-Fonteboa, B., Martínez-Abella, F., Pérez-Ordóñez, J.L.: Prediction of the mechanical properties of structural recycled concrete using multivariable regression and genetic programming. Constr. Build. Mater. **106**, 480–499 (2016)
González-Taboada, I., González-Fonteboa, B., Roussel, N. and Martínez-Abella, F.: Influence of Recycled Concrete Coarse Aggregate on Self-Compacting Concrete Robustness at Short and Long Times. Washington DC, USA. 15–18 May 2016. Edited by Kamal H. Khayat, p. 219 (2016)
Hansen, T.C.: Recycled aggregates and recycled aggregate concrete second state-of-the-art report developments 1945–1985. Mater. Struct. **19**(3), 201–246 (1986)
Katz, A.: Properties of concrete made with recycled aggregate from partially hydrated old concrete. Cem. Concr. Res. **33**(5), 703–711 (2003)
Kazikaki et al. (1988) https://doi.org/10.1299/kikaic.54.1769
Kovler, K., Roussel, N.: Properties of fresh and hardened concrete. Cem. Concr. Res. **41**(7), 775–792 (2011)
Laserna, S., Montero, J.: Influence of natural aggregates typology on recycled concrete strength properties. Constr. Build. Mater. **115**, 78–86 (2016)
Li, X.: Recycling and reuse of waste concrete in China: part I. material behavior of recycled aggregate concrete. Resour. Conserv. Recycl. **53**(1), 36–44 (2008)
Liu, Q., Xiao, J., Sun Z.: Experimental study on the failure mechanism of recycled concrete **41**(10), 1050–1057 (2011)
López Gayarre, F.: Influencia de la variación de los parámetros de dosificación y fabricación de hormigón reciclado estructural sobre sus propiedades físicas y mecánicas (2008)
Manzi, S., Mazzotti, C., Bignozzi, M.C.: Short and long-term behavior of structural concrete with recycled concrete aggregate. Cement Concr. Compos. **37**(37), 312–318 (2013)
Marí, A., Etxeberria, M., Vázquez, E.: Microstructure analysis of hardened recycled aggregate concrete. Mag. Concr. Res. **58**(10), 683–690 (2006)
Medina, C., Sánchez, J., Sáez del Bosque, I.F., Frías, M., Sánchez de Rojas, M.I.: 15–Meso-structural modelling in recycled aggregate concrete. New Trends in Eco-efficient and Recycled Concrete. de Brito, J., Agrela, F.: Woodhead Publishing, pp. 453–476 (2019)
Mellmann, G., Meinhold, U., Maultzsch, M.: Processed concrete rubble for the reuse as aggregates. In Exploiting Wastes in Concrete. Thomas Telford Publishing, pp. 171–178 (1999)
Nagataki, S., Gokce, A., Saeki, T., Hisada, M.: Assessment of recycling process induced damage sensitivity of recycled concrete aggregates. Cem. Concr. Res. **34**(6), 965–971 (2004)
Nealen, A.: The influence of recycled aggregate core moisture on freshly mixed and hardened concrete properties. Darmstadt Concr. (1998)
Noguchi, T., Park, W.-J., Oh, D.-Y., Shin, S.-H.: Modulus of elasticity of recycled aggregate concrete. Mag. Concr. Res. **67**(11), 585–591 (2015)
Otsuki, N., Miyazato, S., Yodsudjai, W.: Influence of recycled aggregate on interfacial transition zone, strength, chloride penetration and carbonation of concrete. J. Mater. Civ. Eng. **15**(5), 443–451 (2003)
Paine, K.A., Dhir, R.: Recycled aggregates in concrete: a performance-related approach. Mag. Concr. Res. **62**(7), 519–530 (2010a)
Pedro, D., Brito, J.D., Evangelista, L.: Influence of the use of recycled concrete aggregates from different sources on structural concrete. Constr. Build. Mater. **71**, 141–151 (2014)
Pedro, D., Brito, J.D., Evangelista, L.: Performance of concrete made with aggregates recycled from precasting industry waste: influence of the crushing process. Mater. Struct. **48**(12), 3965–3978 (2015)

Poon, C.S., Shui, Z.H., Lam, L., Fok, H., Kou, S.C.: Influence of moisture states of natural and recycled aggregates on the slump and compressive strength of concrete. Cem. Concr. Res. **34**(1), 31–36 (2004)
Poon, C.S., Kou, S.C., Lam, L.: Influence of recycled aggregate on slump and bleeding of fresh concrete. Mater. Struct. **40**(9), 981–988 (2007)
Rahal, K.: Mechanical properties of concrete with recycled coarse aggregate. Build. Environ. **42**(1), 407–415 (2007)
Rao, M.C., Bhattacharyya, S.K., Barai, S.V.: Influence of field recycled coarse aggregate on properties of concrete. Mater. Struct. **44**(1), 205–220 (2011)
Ravindrarajah, S.R., Tam, C.T.: Recycled concrete as fine aggregate in concrete. The International Journal of Cement Composites and Lightweight Concrete **9**(4): 235–241 (1987)
Ruhl, M., Atkinson, G.: The influence of recycled aggregate concrete on the stress-strain relation of concrete. Darmstadt Conc. **26**(14), 36–52 (1999)
Ryu, J.S.: An experimental study on the effect of recycled aggregate on concrete properties. Mag. Concr. Res. **54**(1), 7–12 (2002)
Ryu ,J.S.: An experimental study on the effect of recycled aggregate on concrete properties. Mag. Concr. Res. **54**(1), 7–12 (2002)
Sáez del Bosque, I.F., Zhu, W., Howind, T., Matías, A., Sánchez de Rojas, M.I., Medina, C.: Properties of interfacial transition zones (ITZs) in concrete containing recycled mixed aggregate. Cement Concr. Compos. **81**, 25–34 (2017)
Sagoe-Crentsil, K.K., Brown, T., Taylor, A.H.: Performance of concrete made with commercially produced coarse recycled concrete aggregate. Cem. Concr. Res. **31**(5), 707–712 (2001)
Seara-Paz, S., González-Fonteboa, B., Eiras-López, J., Herrador, M.F.: Bond behavior between steel reinforcement and recycled concrete. Mater. Struct. **47**(1–2), 323–334 (2014)
Seara-Paz, S., González-Fonteboa, B., Martínez-Abella, F., González-Taboada, I.: Time-dependent behavior of structural concrete made with recycled coarse aggregates. Creep and shrinkage. Constr. Build. Mater. **122**, 95–109 (2016)
Silva, R.V., de Brito, J., Dhir, R.K.: Comparative analysis of existing prediction models on the creep behavior of recycled aggregate concrete. Eng. Struct. **100**, 31–42 (2015b)
Silva, R.V., de Brito, J., Dhir, R.K.: Prediction of the shrinkage behavior of recycled aggregate concrete: a review. Constr. Build. Mater. **77**, 327–339 (2015c)
Silva, R.V., de Brito, J., Dhir, R.K.: Tensile strength behavior of recycled aggregate concrete. Constr. Build. Mater. **83**, 108–118 (2015d)
Silva, R.V., de Brito, J., Neves, R., Dhir, R.: Prediction of chloride-ion penetration of recycled aggregate concrete. Mater. Res. **18**(2), 427–440 (2015e)
Silva, R.V., Neves, R., Brito, J.D., Dhir, R.K.: Carbonation behavior of recycled aggregate concrete. Cement Concr. Compos. **62**, 22–32 (2015f)
Silva, R.V., de Brito, J., Dhir, R.K.: The influence of the use of recycled aggregates on the compressive strength of concrete: a review. Revue Franã§aise De Gã©nie Civil **19**(7), 825–849 (2015)
Taerwe, L., Matthys, S.: Fib model code for concrete structures 2010. Wiley, Ernst & Sohn (2013)
Wang, C., Xiao, J., Zhang, G., Li, L.: Interfacial properties of modeled recycled aggregate concrete modified by carbonation. Constr. Build. Mater. **105**, 307–320 (2016)
Xiao, J., Li, J., Zhang, C.: Mechanical properties of recycled aggregate concrete under uniaxial loading. Cem. Concr. Res. **35**(6), 1187–1194 (2005a)
Xiao, J., Li, J., Zhang, C.: On statistical characteristics of the compressive strength of recycled aggregate concrete. Struct. Concr. **6**(4), 149–153 (2005b)
Xiao, J.Z., Li, W.G., Fan, Y.H., Huang, X.: An overview of study on recycled aggregate concrete in China (1996–2011). Constr. Build. Mater. **31**, 364–383 (2012)
Xiao, J., Li, W., Corr, D.J., Shah, S.P.: Effects of interfacial transition zones on the stress-strain behavior of modeled recycled aggregate concrete. Cem. Concr. Res. **52**, 82–99 (2013a)
Xiao, J., Li, W., Sun, Z., Lange, D.A., Shah, S.P.: Properties of interfacial transition zones in recycled aggregate concrete tested by nanoindentation. Cement Concr. Compos. **37**, 276–292 (2013b)

Xiao, J., Li, P., Qin, W.: Study on bond-slip between recycled concrete and rebars **34**(1), 13–16 (2006)

Zhang, H., Ji, T., Liu, H.: Performance evolution of the interfacial transition zone (ITZ) in recycled aggregate concrete under external sulfate attacks and dry-wet cycling. Constr. Build. Mater. **229**, 116938 (2019a)

Zhang, W., Wang, S., Zhao, P., Lu, L., Cheng, X.: Effect of the optimized triple mixing method on the ITZ microstructure and performance of recycled aggregate concrete. Constr. Build. Mater. **203**, 601–607 (2019b)

Zhou, C., Chen, Z.: Mechanical properties of recycled concrete made with different types of coarse aggregate. Constr. Build. Mater. **134**, 497–506 (2017)

Zilch, K., Roos, F.: HAUPTAUFSATZE-An equation to estimate the modulus of elasticity of concrete with recycled aggregates. Bauingenieur, **76**(4), 187–190 (2001)

Long-Term Properties of RAC and Its Product

Jianzhuang Xiao, Luc Courard, Miren Etxeberria, Yuxi Zhao, and Zengfeng Zhao

The relentless exploitation of natural resources for construction purposes, specifically that of non-renewable resources, has led to millions of tons of CDW being produced every year. Since most countries have no specific treatment plan for these materials they are sent to landfill instead of being reused and recycled in new construction. When compared to the impact of the cement industry, aggregate mining and production have little environmental impact as it basically consists of simple extraction, without fundamental alteration (e.g. thermal activation) of the material. However, the mining of aggregates and rock is becoming an ecological issue in several parts of the world as the demand for sand and gravel increases rapidly with new infrastructure activities. To counteract this, there have been several studies worldwide on the use of RA, from processed CDW, as a substitute for NA in the production of concrete.

According to existing specifications (BCSJ 1977, RILEM 1994, BRE 1998, DAfStb 1998, DIN-4226 2002, WBTC-No.12 2002, PTV-406 2003, TFSCCS 2004, NBR-15116 2005, BS-8500 2006, LNEC-E471 2006, OT-70085 2006, EHE-08 2010, JIS-5021 2011, JIS-5022 2012, JIS-5023 2012), there are three main types

J. Xiao · Z. Zhao
Department of Structural Engineering, College of Civil Engineering, Tongji University, Shanghai, China

L. Courard
University of Liège, Urban and Environmental Engineering, GeMMe Building Materials, Liège, Belgium

M. Etxeberria (✉)
Department of Civil and Environmental Engineering, Universitat Politècnica de Catalunya (UPC-BarcelonaTech), Barcelona, Spain
e-mail: miren.etxeberria@upc.edu

Y. Zhao
College of Civil Engineering and Architecture, Zhejiang University, Hangzhou, China

J. Xiao et al. (eds.), *Structural Behaviour and Innovation of Recycled Aggregate Concrete*, RILEM State-of-the-Art Reports 41,
https://doi.org/10.1007/978-3-031-84136-1_5

of RA (RA) resulting from CDW, which, after being subjected to proper beneficiation processes in certified recycling plants, are suitable for the production of structural concrete; these materials are crushed concrete, crushed masonry, and mixed demolition debris. Some of these specifications (DIN-4226 2002, PTV-406 2003, NBR-15116 2005, LNEC-E471 2006) have reached a consensus that, in order to be considered as RCA, a given aggregate must comprise a minimum of 90%, by mass, of Portland cement-based paste and NA. RA sourced from crushed masonry, here named RMA, may include: aerated and lightweight concrete blocks; ceramic bricks; blast-furnace slag bricks and blocks; ceramic roofing tiles and shingles; and sand-lime bricks (Hansen 1992). RMA are composed of a minimum of 90%, by mass, of the summation of the aforementioned materials. Aggregates acquired from mixed demolition debris, or mixed RA (MRA), are a mixture of the two main components obtained from the beneficiation process of CDW: crushed and graded concrete and masonry rubble. Some specifications (NBR-15116 2005, BS-8500 2006) state that they are composed of less than 90%, by mass, of Portland cement-based fragments and NA. In other words, they may contain several other common CDW materials such as masonry based materials.

In view of the necessity of mitigating the environmental impacts of the construction industry, especially those brought by the manufacture of concrete, it is vital that the properties of RAC are further researched in order to facilitate the wider use of RA. This chapter presents a literature review on the influence of using RA on the durability-related properties of structural concrete. The following sections contain analyses of the results of several authors on the deformation of RAC by shrinkage and creep phenomena, its long-term porosity, permeation, and absorption capacity, as well as carbonation and chloride-ion penetration.

1 Shrinkage

While hardening, concrete experiences volume variation due to water loss by evaporation, and hydration and carbonation related reactions (Neville 2011), which depend on a number of factors including the concrete's constituents, age, relative humidity and temperature of the environment, and also the shape and size of the element (Tam and Tam 2007). Among the known shrinkage mechanisms, those that are potentiated by the incorporation of RA are the drying shrinkage and carbonation shrinkage. In the former, water that is not physically bound evaporates from the capillaries. Considering that RAC normally present greater porosity and higher water absorption by capillary action, as it will be discussed in the corresponding section, the greater movement of water induces more internal stresses, which, alongside the lower restraining ability of RA, results in higher shrinkage strain. The carbonation related shrinkage may also be potentiated by the inclusion of RA, as there is a clear increase in the amount of carbonated matrix as a number of RA increases. Since the products of hydration from the carbonation process take up less volume than the original elements, this will lead to a contraction of the element.

Buyle-Bodin and Hadjieva-Zaharieva (2002) manufactured concrete with coarse and fine RA from CDW, randomly collected from a French CDW recycling facility. The authors kept a constant workability in all mixes and observed that complete replacement of coarse NA with RA caused an 80% increase in shrinkage after having been stored in open air for 1 year. After replacing all the NA with fine and coarse RA, the resulting materials exhibited a shrinkage strain three times higher than the control specimens during the same 1-year period.

Kou et al. (2004) incorporated 20, 50 and 100% of coarse RCA by mass of coarse NA fraction. The complete replacement of coarse NA led to a 22% increase after 112 days. Still, since all specimens were either cured in water or with steam, the abovementioned value does not include the drying shrinkage variable, which is the most significant in RAC.

Khatib (2005) produced concrete with fine RCA or RMA at 25, 50, 75 and 100% replacing natural sand and placed the specimens in a chamber with a relative humidity of 60% and at 20 °C for 90 days to measure their shrinkage. As expected, the complete substitution of sand with fine RCA led to a shrinkage increase of around 55% when compared with the control samples. However, specimens with 25.100% of fine RMA resulted in lower shrinkage strains (10–25% higher than those of the control mix) when compared with concrete made with fine RCA. This may have been a result of internal curing provided by the fine RMA throughout the drying period.

Evangelista and De Brito (2004) noticed significant increases in shrinkage strain for concrete mixes with 100% fine RCA (i.e. about 75% higher than that of the control concrete) after being cured for 90 days in a controlled chamber with a relative humidity of 50% and at 20 °C. However, it was also observed that the incorporation of 30% had little effect on this property.

Ferreira et al. (2011) found that concrete made with pre-saturated coarse RCA showed higher shrinkage strain when compared with the control specimens, while a similar performance was observed between the control and RAC mixes when the RCA's additional water absorption was compensated with an equivalent amount of water. The authors thus concluded that pre-saturation of coarse RCA, apart from the concept's impracticality from an industrial viewpoint, was also detrimental to the shrinkage of concrete, when compared with the water compensation method.

Manzi et al. (2013) detected a 60% increase in shrinkage for mixes with 50% of fine and coarse RCA after having been subjected to a relative humidity of 60% and at 20 °C for 400 days. The authors claimed that this was mostly due to the lower modulus of elasticity of RCA relatively to that of NA. Since RCA are less stiff as a result of the high AM content, these are less capable of restraining the cement paste's shrinkage.

Cartuxo et al. (2015) analyzed the shrinkage behavior of concrete with fine RCA with and without the effect of superplasticizers. After being cured for 90 days in a controlled chamber with a relative humidity of 50% and at 20 °C, the authors found that the shrinkage strain of 100% fine RCA-containing mixes and without superplasticizer increased 47% when compared to the control concrete. This value was 57% for the counterpart with 1% of a high water-reducing admixture by mass of cement. These findings suggest that the use of superplasticizers has little effect on this property when compared to the incorporation of RCA.

Silva et al. (2015a, b, c, d, e) carried out a systematic literature review, wherein, by means of a meta-analysis, the authors made a comparative analysis on the using a sample size over 4000 values taken from several experimental campaigns in order to ascertain the applicability of (ACI-209 2008, EN-1992-1-1 2008, CEB Model Code 99, Gardner and Lockman 2001) overestimate the shrinkage strain of concrete, the magnitude of which is greater for NAC. The authors also established that replacing the coarse NA fraction with RA could lead to an increase of 80%. Lye et al. (2016) also carried out a very similar study and obtained comparable findings.

Pedro et al. (2014) evaluated the influence of using RCA subjected to a different crushing procedure in the properties concrete. The studied RCA, one that underwent a primary crushing stage only, while the other was also subjected to a secondary crushing stage, were incorporated in concrete fully replacing NA. Apart from the enhanced strength (i.e. 7–15% higher than specimens with the first RCA) in mixes produced with the more processed RCA, the use of these aggregates also improved the shrinkage performance of concrete. Since the second crushing stage produces RCA with a rounder shape and less content of AM, it becomes stiffer than the less processed RCA and thus capable of restraining the cement paste's shrinkage more effectively.

Andal et al. (2016) also observed comparable findings when assessing the effect of incorporating up to 100% RCA from a controlled sourced (25 MPa concrete) versus those from a CDW recycling plant. The results showed the use of controlled RCA improved the mechanical performance when compared to samples made with RA from the recycling plant. The authors noticed that this also positively influenced the shrinkage of RAC thereby further demonstrating the importance of a high-quality processing to treat the RA prior to their incorporation in concrete.

Over time, there have been several entities and individuals (Hansen 1992; Lamond et al. 2002; TFSCCS 2004) that have established average values for shrinkage increase when 100% coarse RCA is used. These may vary from 20 to 50% greater shrinkage when compared to that of the control NAC. In fact, the Task Force of the Standing Committee of Concrete of Spain (TFSCCS 2004) presented correction factors, determined in previous specifications, for the use of 100% coarse RCA in concrete production (Table 1).

Table 1 Correction factors to calculate shrinkage (TFSCCS 2004)

Source	Shrinkage correction factors	
	100% coarse RCA	20% coarse RCA
Belgium	1.50	1.00
RILEM	1.50	1.00
The Netherlands	1.35.1.55	1.00

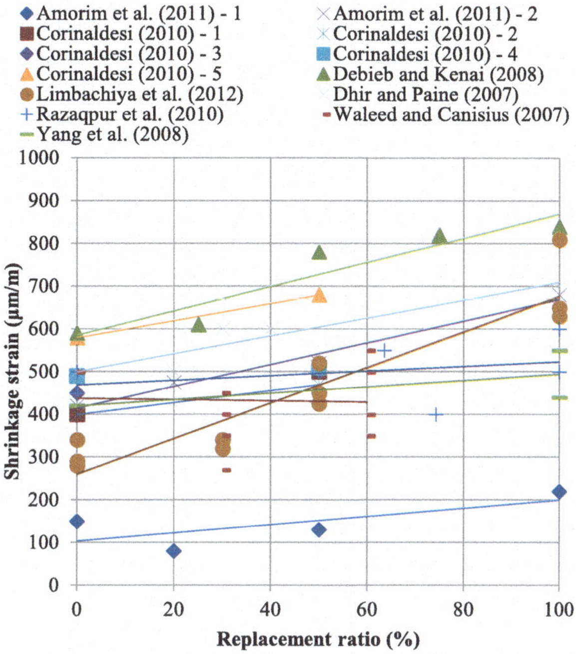

Fig. 1 Shrinkage strain of RAC with increasing coarse RA

Figure 1 presents the linear regressions of a number of studies (Dhir and Paine 2007, Waleed and Canisius 2007, Debieb and Kenai 2008, Yang et al. 2008, Corinaldesi 2010, Razaqpur et al. 2010, Amorim et al. 2012, Limbachiya et al. 2012) concerning the effect of incorporating an increasing amount of coarse RA on the shrinkage strain of RAC. As the replacement ratio increases, the shrinkage strain also tends to increase, which depends on the RA's modulus of elasticity, which is affected by the AM content.

2 Creep

Creep deformation of concrete is a complex phenomenon, influenced by many factors, including mix design, temperature and relative humidity, load conditions, and the element's shape and size (Neville and Brooks 2010). Apart from the natural loss of water prompting drying shrinkage, when a hardened concrete element is loaded, C-S-H will lose a large amount of water with a subsequent increase in creep strain (Tam and Tam 2007). A more recent study (Vandamme and Ulm 2009) showed that the creep of C-S-H, in the presence of sustained stress, may suffer sliding of nanoparticles, causing a local increase of packing density toward the jammed state associated with limit packing densities, beyond which no particle sliding is possible without dilation of the granular media.

Wesche (1982) found that the creep deformation of mixes made with RCA was around 50% higher than that of the control concrete. The authors explained the results with the fact that the RCA mixes contained a more significant amount of cement mortar, directly proportional to the creep of concrete. Ravindrarajah and Tam (1987) studied concrete with coarse RCA and observed average increases of creep deformation of 45% relative to those of the control concrete. De Pauw et al. (1998) evaluated the creep of concrete with coarse RCA having reached values 35% higher than those of the control mix after 560 days of loading. The authors also studied the behavior of concrete with coarse RMA from crushed ceramic bricks and observed improved results when compared to RCA-containing concrete.

Gomez-Soberon (2002a) evaluated the creep of concrete with RCA, having used cylindrical specimens with 150 mm in diameter and 450 mm in height. The cylinders were subjected to a load equal to 35% of their 28-day compressive strength. The test results showed that mixes with 100% fine and coarse RA exhibited a 39% increase in creep coefficient when compared with the control concrete.

Fathifazl et al. (2011) evaluated the creep deformation of concrete with two types of coarse RCA from different recycling plants. The authors followed two approaches for replacing the NA with RCA, wherein one was replacing the material by volume and the other consisted of slightly reducing the cement content so that the extra AM content from the RCA could amount to the same quantity of total mortar of the control samples. Although comparable creep coefficients had been obtained, it should be noted that all mixes were produced using different mix designs and with little equivalent criteria.

Manzi et al. (2013) obtained a 15% increase in creep for mixes with 50% fine and coarse RCA, relative to the control concrete, after being loaded for 400 days. The authors claimed that this increase in creep was caused by the RA's lower Young's modulus when compared to that of the NA. Similar reasoning was given by the authors regarding the shrinkage behavior of RAC, which is less stiff due to the greater content of cement mortar (old and new) thereby leading to a lower capacity to restrict deformation by creep phenomena.

Cartuxo et al. (2015) studied the effect of incorporating an increasing amount of fine RCA in the properties of concrete. The results showed contradictory findings in terms of the RAC's creep behavior. The authors inferred that the greater roughness of RA in relation to NA could contribute to improved adhesiveness between those aggregates and the new cement paste (i.e. enhanced ITZ) thereby reducing creep. However, the reduced modulus of elasticity of RA may lead to an increase in creep deformation, which can be worsened by the addition of extra mixing water to maintain the same workability in all the mixes. They also evaluated the creep of concrete with fine RCA, with and without superplasticizers. The authors observed that the use of 100% fine RCA led to a 154% increase in creep deformation after a loading period of 90 days and that water-reducing admixtures had little effect in offsetting the greater deformation induced by RCA.

Domingo-Cabo et al. (2009) analyzed the creep of concrete with 20%, 50% and 100% coarse RA from CDW, maintaining the cement content and effective water/cement ratio constant in all mixes. The authors observed that the total replacement of coarse NA with coarse RA led to a creep deformation increase of 51%, relative to the control specimens, after a loading period of 180 days.

Silva et al. (2015a, b, c, d, e), similarly to what they did for the shrinkage of RAC, carried out a comparative analysis of the creep behavior using the existing models of prediction. The results showed that the ACI-209 (2008) exhibited the highest accuracy, though slightly overestimated the compliance of concrete. The authors also appraised the main creep influencing factors related to the use of RA, wherein, apart from the replacement ratio, the quality of RCA (i.e. water absorption, density, AM content, the strength of the original concrete) also significantly affected the creep of concrete. The deformation can further increase with increasing quantity of ceramic particles in the RA, as a result of their lower modulus of elasticity when compared with NA and RCA.

More recently, (Geng et al. 2016) studied the effect of incorporating RCA from concrete specimens with different effective w/c ratios on the creep behavior of RAC. After a loading period of over 200 days at a relative humidity of 77% and temperature of 11.5 °C, the authors observed increases in the specific creep of 60, 75 and 125% for RAC mixes with the coarse NA fraction completely replaced with RCA from concrete with effective w/c ratios of 0.30, 0.45, and 0.65, respectively. This further reinforces the notion that the quality, apart from the quantity, of the AM has a significant influence on the creep deformation of concrete.

In view of the potential variation of creep increase, some authorities have suggested correction factors to estimate the creep of RAC mixes with a given replacement ratio and RCA with known properties. The Task Force of the Standing Committee of Concrete of Spain (TFSCCS 2004) reported indicative correction factors derived from some sources to calculate the creep of concrete made with RCA (Table 2). Although the exact figures may be open to debate, since the actual deformation may vary significantly depending on several variables, the use of such factors would be helpful in practice.

Figure 2 presents the relative creep coefficient of RAC with increasing coarse RA content relative to that of the corresponding NAC. As mentioned, the incorporation of RA leads to an increase in the deformation induced by creep phenomena in a similar way as for the shrinkage of concrete. It is also clear that this property shows a linear relationship with increasing coarse RA content in a similar way as for other properties of RAC.

Table 2 Correction factors to calculate creep coefficients (TFSCCS 2004)

Source	Creep coefficient correction factors	
	100% coarse RCA	20% coarse RCA
Belgium	1.25	1.00
RILEM	1.25	1.00
The Netherlands	1.25.1.45	1.00

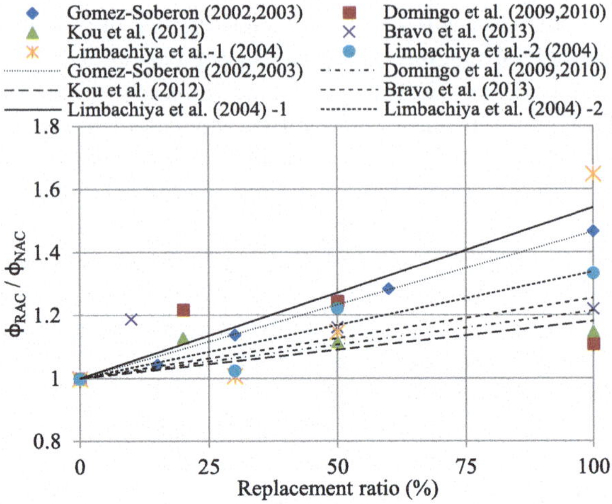

Fig. 2 Relative increase in creep coefficient of RAC with increasing coarse RA

3 Porosity and Absorption

The durability of concrete is generally connected with its water absorption capacity because water transports ions into concrete, which potentiates its degradation and the corrosion of the reinforcement (Eckert and Oliveira 2017). The porosity and pore size distribution are the most important characteristics of the pore system of concrete which influence the ingress of foreign substances to the interior of concrete (Garboczi 1990).

Thus, the durability of reinforced concrete is primarily influenced by its microstructure—the connectivity, continuity, tortuosity, and radius of its pores—as this determines how gases, liquids and other substances penetrate the concrete cover to the reinforcement (Kropp et al. 1995). Water and chloride ions can ingress concrete through a combination of transport mechanisms, namely absorption by capillary action, diffusion and permeation (Tuutti 1982). The transport of liquid in concrete predominantly occurs through the pores of the cement matrix. However aggregates also play an important role. The specific gravity of aggregates (or particle density) can be a good indicator of their water absorption properties (Ryu and Monteiro 2002; Neville 2011). Therefore, they should be protected from moisture in aggressive environments.

3.1 *Porosity*

According to Thomas et al. (2013), it is noted that the accessible porosity of the RAC, compared with a control concrete, increases with the w/c ratio and the degree of substitution at 28 days, 180 days and 365 days. They described a greater dispersion

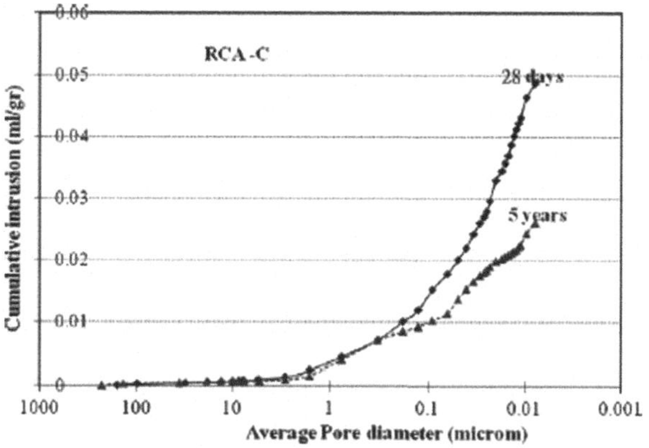

Fig. 3 Development of pore size distribution of an RAC mix at 28 days and 5 years (Kou et al. 2011)

in the values of the fitting curves of concrete mixes produced employing 100% of RCA at advanced ages.

According to Gomez-Soberon (2002a, b), RAC that was prepared with 100% RCA (with natural sand and fine aggregates) had higher porosity than that of NAC. From the curing age of 28 days to 90 days, the total porosity of both the RAC and NAC decreased, but the RAC reduction was higher. A similar conclusion was obtained by González-Corominas and Etxeberria (2016).

Kou et al. (2011) analyzed the pore structure of NAC and RAC were also measured at 28 days, 1 year and 5 years. After 5 years of water curing the concrete made with 100% of RCA had the lowest porosity. The porosity was reduced by 45% between 28 days and 5 years (Fig. 3) after achieving the highest reduction. Thus, the RCA significantly improved the long-term interfacial properties of the new concrete probably due to the long-term self-cementing effects of the old cement mortar and the interaction between the new cement paste and the old cement mortar.

According to (González-Corominas et al. 2016), the use of RCA mitigated the macro-capillary pore increase generated by the use of steam curing at the long-term, which is typically observed in steam-cured conventional concrete. The influence of steam curing was especially beneficial in the reduction of the average pore size of RAC in comparison with those concrete mixes that only underwent air curing. This reduction was higher in mixes produced with lower quality RCA.

3.2 *Absorption Coefficient*

The absorption coefficient increases with the w/c ratio and the degree of RCA incorporation. Figure 4 shows that, with time, the concrete with high percentages of substitution and high w/c ratio reduces the absorption gap further than the standard concrete (Thomas et al. 2013).

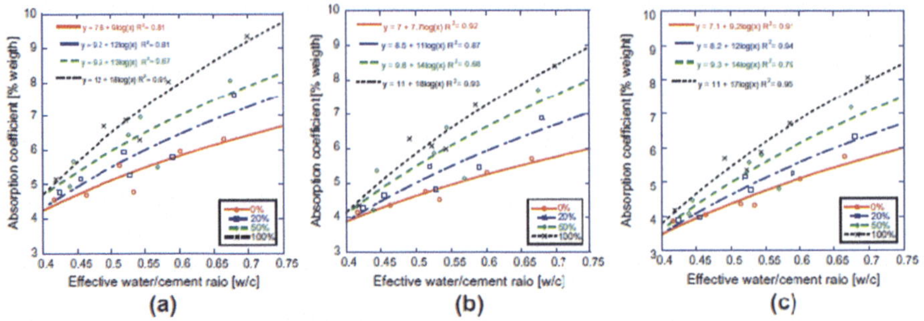

Fig. 4 Absorption coefficient versus water/cement ratio at different ratios of substitution after 28 days (**a**), 180 days (**b**) and 365 days (**c**) (Thomas et al. 2013)

Note that the negative effect of RA is significantly lower for low w/c. In this case, both the porosity and absorption of the RA are partially isolated for the surrounding less absorbent cement paste.

According to (Etxeberria and Vegas 2015), the use of up to 50% of fine ceramic or mixed RA as a substitute for natural sand caused a reduction in concrete density of less than 5% compared with conventional concrete and the absorption increased by more than 40% in comparison with conventional concrete at early ages. In contrast, at long-term, the durability improves, achieving similar or better properties than conventional concrete.

Wainwright and Cabrera (1994) evaluated the water absorption by immersion of RAC with fine and coarse RCA. The authors noticed that the strength of the original concrete that subsequently influence the quality of the resulting RCA had a greater influence on the porosity of the RAC specimens than the target strength class of the resulting concrete.

Evangelista and De Brito (2010) also observed a directly proportional relationship between the replacement level of natural sand with fine RCA and the resulting RAC's water absorption by immersion. After complete replacement of the fine fraction, the authors observed a 45% increase in this property.

3.3 Capillary Water Absorption

Capillary water absorption by a capillary of RAC was significantly greater than that of the NAC concrete; this has also been reported by other researchers (Wirquin et al. 2000; Levy and Helene 2004; Zega and Di Maio 2011). The use of fly ash as a partial replacement of cement and an addition of cement decreased the water absorption of natural and RAC.

According to Kapoor et al. (2016), SCC mixes containing SF and metakaolin (MK) reduced the capillary suction comparing to the SCC mixes produced using Portland cement. Furthermore, the SF and MK were also effective at 50% replacement level

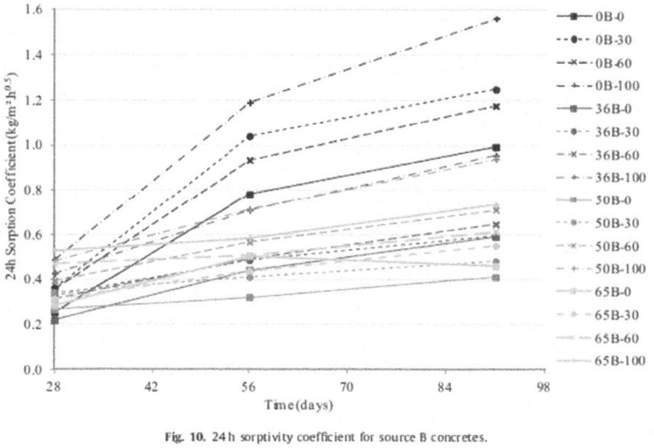

Fig. 5 24 h sorptivity coefficient for RAC (Dodds et al. 2017)

of natural aggregate with RCA, as the capillary suction values of mix made with 50% of RCA and SF, C-SFR50, were 10.4, 14.7 and 25.7% lesser relative to the control mix NAC at the curing ages of 28, 56 and 120 days respectively. It was also observed that SF and MK were relatively ineffective in reducing the capillary suction in the SCC mixes with 100% RCA to a value below than that of the control SCC mix.

According to Dodds et al. (2017), the sorption coefficient generally increases with time for concrete mixes produced with coarse RA employing any type of cement. However, at 56 and 91 days CEM III/A concrete mixes produced with up to 100% RCA had lower sorption coefficients than the CEM I control concrete. CEM III/A concrete mixes produced with up to 100% RCA content had a lower 24 h sorption coefficient by a factor 1.1 to 2.2 than the control CEM I concrete mixes at 91 days (Fig. 5).

The magnitude of difference in the measured results between CEM I and CEM III/A concrete mixes has shown that up to 100% coarse RCA, irrespective of the RCA sources adopted, can be incorporated in structural CEM III/A concrete and have a better durability performance than that of a control CEM I concrete (Dodds et al. 2017), which is higher than the previously reported values of 25–50% (Kou and Poon 2013; Zega et al. 2013; Soares et al. 2014; Lotfy and Al-Fayez 2015) and a positive finding for the wider implementation of coarse RCA to produce sustainable structural concrete.

Zaharieva et al. (2003) studied the water absorption by capillary action of RAC with fine and coarse RA from a recycling plant. The results showed that the complete replacement of coarse NA led to an increase of water absorption of 16%, while the integral substitution of NA with fine and coarse RA caused an increase of 42%.

Correia et al. (2006) evaluated both the water absorption by immersion and by capillarity of RAC with replacement ratios up to 100% of coarse RMA. The results showed increases of 62 and 70%, respectively when compared to the control concrete.

Vieira, Alves et al. (2016) measured the water absorption by immersion and by capillarity of concrete with fine RA from crushed ceramic bricks and sanitary ware. The results showed that both types of RA had a similar effect on the water absorption by immersion (increases of 45% and 47%, for 100% fine RA from ceramic bricks and sanitary ware, respectively). However, capillary action testing showed that the use of 100% finely crushed brick aggregates led to less 31% water absorption when compared to the control concrete, but increased by 38% when 100% fine RA from crushed sanitary ware was used. The authors explained that this may have been a result of the pozzolanic reactions between the high SiO_2 and Al_2O_3 contents of the fine crushed ceramic brick aggregates and the cement's products of hydration, which also improved the mixes' strength development and chloride-ion penetration.

4 Durability

Sustainable development holds a more and more important rank in our society. In this context the re-use of the aggregates coming from deconstruction in new structures is really relevant for the concrete industry. But first of all, it is necessary to study the mechanical and durability performances of the recycled concretes. In France, a national project called "RECYBETON" involving 43 partners for a total cost of 4.8 M€ and a research ANR project named "ECOREB" are in progress to study this thematic. The final results of these two projects will allow an evolution in the standards and in particular will precise the conditions for using of the recycled aggregates into new concretes.

Several studies were performed to evaluate the influence of recycled aggregates as partial or total replacement for NA (Evangelista et al. 2007, Khatib 2005). It has been shown that the mechanical and durability performances of recycled concretes are different from those of the natural concretes. For many researchers (Dhir et al. 1999, Dhir et al. 1998), (Padmini et al. 2009), the differences are due to the attached mortar coming from the old concrete which is very porous (Fig. 6, De Juan et al. 2009).

Fig. 6 Recycled aggregate

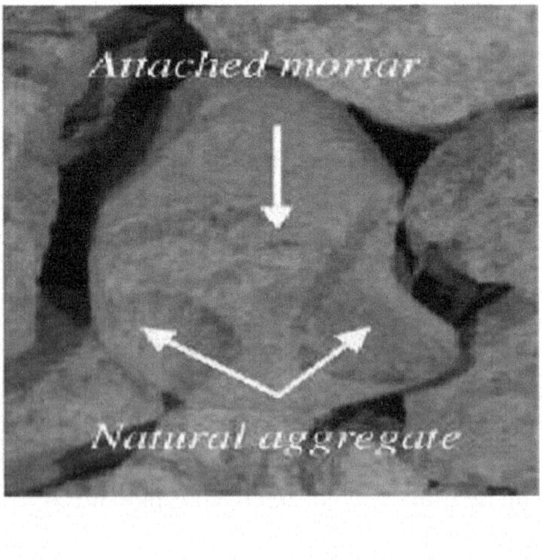

Indeed the properties of the recycled aggregates depend on the properties of the parent concrete. Overall some properties of recycled aggregates linked to the durability mechanisms are not as good as NA, such as water absorption and porosity (Etxeberria et al. 2007).

This paper presents some results about durability properties of recycled concretes and in particular it deals with the performance of concretes in two different aggressive environments for the rebars. Two performance tests are carried out: the accelerated carbonation and the chloride ions migration under electric field. The obtained results allows to conclude on the performance of the recycled concretes and provide enough information to achieve some calculations with durability models.

4.1 Experimental Procedures

4.1.1 Mix Compositions

Three strength classes are studied (C25/30, C35/45 and C45/55 according to NF EN 206/CN specifications) and for each class three different concretes are mixed. So a total of nine mixtures are studied (Table 3).

Table 3 Mix proportions of concretes, per cubic meter

	C25/30			C35/45			C45/45		
	RC	F30C30	C100	RC	F30C30	C100	RC	F30C30	C100
Water (kg/m^3)	184	177	176	178	158	172	168	148	150
CEM II/A-L 42,5 N	267	282	279	299	324	336	390	371	369
Filler	45	49	70	58	44	53	100	65	73
NFA (0/4)	772	492	794	769	495	782	732	483	775
A (0/4)	0	233	0	0	216	0	0	229	0
NCA (4/10)	264	167	0	264	169	0	250	164	0
RCA (4/10)	0	151	161	0	143	158	0	148	157
NCA (6,3/20)	811	539	0	808	546	0	769	529	0
RCA (10/20)	0	167	691	0	164	682	0	162	676
Superplasticizer	0.14	0.13	0.08	0.29	0.66	0.25	1.23	1.28	0.92
Slump (cm)	19.5	20	19.5	20	20	20	19.5	19	19
w/equivalent binder	0.57	0.54	0.53	0.49	0.46	0.47	0.34	0.35	0.36

Table 4 Properties of the aggregates

	Dr (g/cm^3)	A (%)	LA (%)
FRA (0/4)	2.08	8.9	–
RCA (4/10)	2.29	5.6	–
RCA (10/20)	2.26	5.8	37
NFA (0/4)	2.58	0.8	–
NCA (4/10)	2.71	0.51	–

Where Dr, is the relative density of particle (g/cm^3), A the water absorption (%) and LA is the Los-Angeles Index (%)

- 3 reference concretes with NA corresponding to the three strength classes, named RC.
- 3 concretes for which 30% of NFA are replaced by recycled fine aggregate (RFA) and 30% of NCA by RCA, named F30C30.
- 3 concretes for which 100% of NCA are substituted by RCA, named C100.

A CEM II/A-L 42.5 cement and limestone filler are used as binder. The superplasticizer content is adjusted in order to keep the workability constant (slump between 19 and 20 cm) which corresponds to a S4 class according to the norm NF EN 206/CN. The slump measurement is performed following the standard NF EN 12,350–2. Crushed limestone aggregates are used as the NA and the recycled aggregates come from construction demolition wastes. The main properties of the NA and RA are studied in RECYBETON project and are displayed in Table 4. Due to its higher porosity the recycled aggregates present a lower density than NA and that leads to a higher water absorption.

Due to the higher water absorption capacity of RA by comparison with NA (Table 4), water pre- saturation of the RA is adopted before adding it into the mix. In this study, the RA are over saturated at 1% that means that the water content of recycled aggregates for mixing will be equal to the water absorption plus 1%. First the water content of RA is measured before conditioning. Then the quantity of water required to obtain the designed water content is determined as follow:

$$Water\ quantity\ to\ add = \frac{Mass\ of\ RA}{(1 + water\ content\ of\ TA)*(Final\ water\ content - water\ content\ of\ RA)}$$

The RA and the water quantity to add are introduced in a hermetic barrel. Rolling up the barrel during 10 min on the floor leads to increase the homogeneity of the water in the RA. Then the barrel is at rest two hours before mixing.

4.1.2 Specimens Casting and Curing

For each composition, prismatic specimens (70 × 70 × 280 mm) and cubic specimens (150 × 150 × 150 mm) are used to determine the physical properties of concrete. All prismatic samples are casted in steel moulds and all cubic samples in plastic moulds. They are compacted using a vibrating table. Neither segregation nor bleeding is observed. All specimens are demoulded after 24 h from casting. Two types of curing are studied:

- Wet curing: in water (20 °C) until the day of testing (90 days in this case)
- Dry curing: three days in water (20 °C) and 87 days in laboratory conditions (20 °C ± 2, 50% ± 5 relative humidity).

Then some cylindrical specimens of 50 × 100 mm are taken from cubic specimens in order to carry the chloride migration test. Before this test, the samples are vacuum saturated with alkaline solution (NaOH 0.1 M) during 72 h. The prismatic specimens are used to perform the carbonation test.

4.1.3 Transport and Durability Properties

Chloride migration test is performed according to XP P 18–462 standard. Three cylindrical test specimens are tested for each composition. Each specimen is placed between two compartments of a cell where flat silicone circular seals ensure that the system is leak tight. The solutions are made with NaOH (0.025 mol/l) + KOH (0.083 mol/l) in the upstream and the downstream compartment. NaCl (0.513 mol/l) is added in the upstream solution. A tension (30 V) is applied between the two compartments in order to force the chloride ions movement from upstream to downstream. The test is carried out at temperature T = 20 °C ± 5. The test duration is about 24 h for these concrete mixes. At the end of the test, the specimen are broken and pulverized with silver nitrate. The depth reached by the chloride in the specimen at a pre-determined age is measured and then used to calculate the chloride diffusion coefficient (Fig. 7).

The carbonation test is performed according to the European standard project prCEN/TS 12,390–12. After the curing period, the sample are placed in a room at 20 °C ± 2 and 50% ± 5 relative humidity during 15 days. After this preconditioning, the 27 specimens (three for each concrete) are placed in the CO_2 enclosure where the CO_2 concentration is 4% ± 0.5. The relative humidity for this test is 55% ± 5 and the temperature 20 °C ± 2. After 90 days in the enclosure the specimen are broken and pulverized with phenolphthalein. The depth reached by the CO_2 is measured. (Fig. 8).

Fig. 7 Measurement of the chloride ions front

Fig. 8 Measurement of the carbonation front

4.2 Results of the Tests

4.2.1 Chloride Migration

Chloride penetration into cement matrix is one of the main causes of rebar corrosion. This issue concerns concrete in marine environment or concrete subjected to de-icing salts. When the chloride ions reach a certain level (so called critical concentration) close to the rebar, the protecting passive layer is destroyed and the corrosion process can begin.

Figure 9 shows that the diffusion coefficient strongly depends on the W/equivalent binder ratio. Within a mechanic class the diffusion coefficients of chloride ions present the same order. These results confirm previous studies (Gonçalves 2004). However, the penetration of chloride ions seems easier in the concretes containing

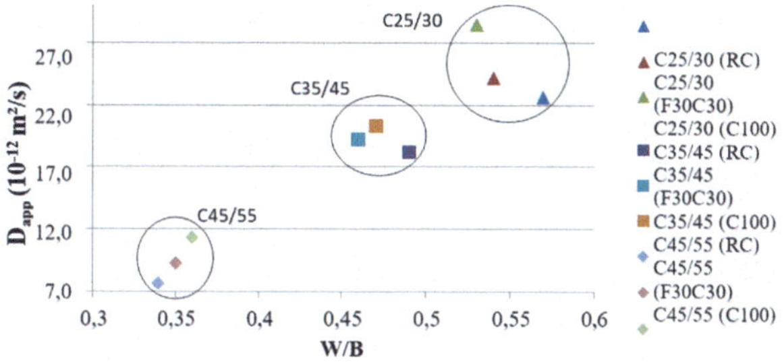

Fig. 9 Apparent diffusion coefficient of chlorides in function of W/equivalent binder ratio

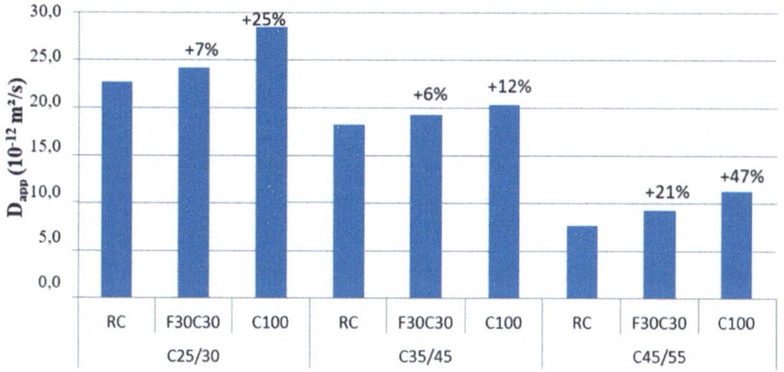

Fig. 10 Apparent diffusion coefficient of chlorides

recycled aggregates (Fig. 10). The relative increase between the reference concretes and the recycled concretes is highlighted on the Fig. 10. A significant increase is observed for the C45/55 class. That could be due to the difference between the paste quality and the RA quality.

4.2.2 Carbonation Phenomenon

Carbonation phenomenon can also induce corrosion. In the atmosphere the percentage of CO_2 is 0.04% in average. When carbon dioxide is penetrating into the cement matrix, the portlandite is consumed and calcite is formed. It induces a pH decrease in the interstitial solution from 13.5 to 8.5. As the pH of the cement matrix decreases with carbonation, the rebar is not protected anymore and corrosion can begin.

In the carbonation test, the results highlight the influence of the W/B ratio on the transfer properties (Fig. 11). The cement matrix is less porous when the W/B ratio is small so the carbonation depth for C45/55 concretes is lower than the other ones. Also the porosity of RA is very high so the porosity of the concretes strongly increases with the quantity of recycled aggregates. But these results also show that paste porosity and aggregates porosity do not have the same influence on the concrete porosity.

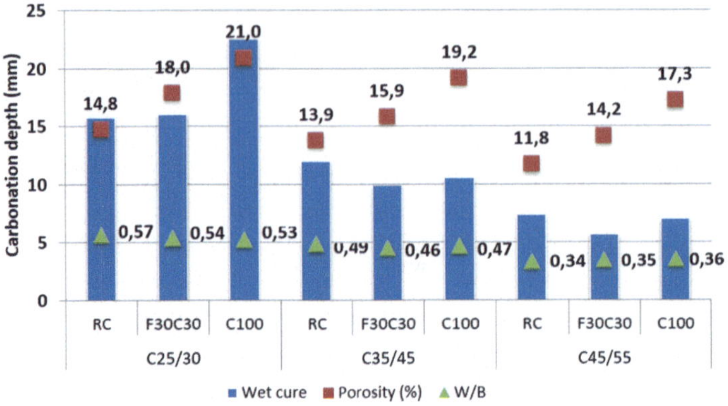

Fig. 11 Carbonation, porosity and W/B ratio of the concretes

Indeed, C25/30 RC and C45/55 F30C30 have almost the same global porosity but their carbonation depths present an important difference.

Moreover, the porosity influence on the transfer properties is more important for the C25/30 concretes (low compactness of the cement matrix) than for the C45/55 concretes (high compactness). Thus, the porosity, generally used as a durability indicator, is not appropriate to characterize the durability of recycled concretes.

For the F30C30 concretes the carbonation depth is lower as those of the RC. The recycled aggregates are pre-saturated before mixing and they have a higher water absorption than the NA. So, after the preconditioning part, RA are probably wetter than the cement matrix what induces a decrease of the transfer properties of the CO_2 and therefore a decrease of the carbonation depth. The dry curing favor the carbonation phenomenon (Fig. 12). Indeed, during the drying period (in the lab or in the oven) the saturation degree of the concrete decreases and that permits a better diffusion of the CO_2 into the cement matrix.

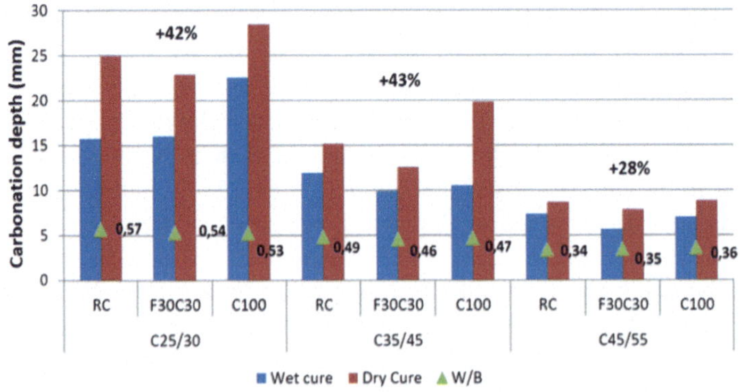

Fig. 12 Carbonation phenomenon, comparison wet curing/dry curing

It is highlighted on the Fig. 12 that the dry curing more influences the C25/30 concretes than the C45/55 concrete. It seems that a compact matrix can limit the influence of the other factors (cure, porosity, quantity of RA) on the concrete and on the transfer properties.

4.3 *Modeling the Porosity of Concrete in Function of the Aggregate's Porosity*

The previous results showed the influence of the aggregates porosity on the concrete porosity. Indeed, for two different concretes leading to the same porosity, the carbonation depth is not the same (Fig. 13, C25/30 RC and C45/55 F30C30). With the Powers formula (Taylor 2004) it is possible to calculate the paste porosity and to confirm that the paste of the C45/55 concretes is less porous than the paste of the C25/30 concretes.

Within a same mechanic class the porosity of the paste is the same. For the three C45/55 concretes the paste porosity is around 40% compared to 52% for the three C25/30 concretes (Table 5). We can calculate the concrete porosity using the paste porosity, the aggregates porosity and the proportions of paste and aggregates in the concrete (Fig. 13).

The results that we obtained with the model are close to the experimental results. That proves the importance of using this kind of model to predict the porosity of concrete containing recycled aggregates and in particular to be sure that the high porosity is due to recycled aggregates and that it will not affect the transfer properties in the cement past.

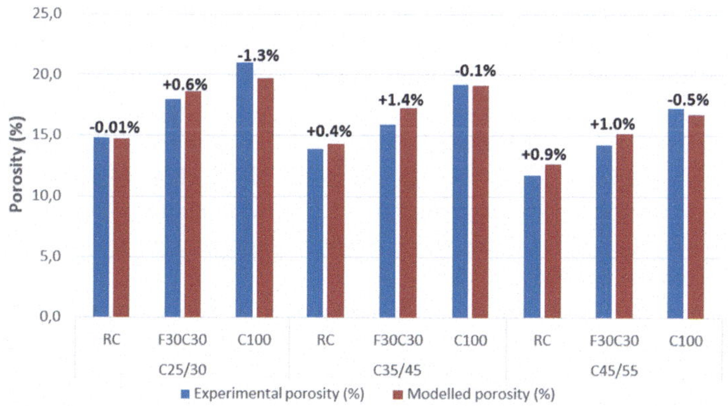

Fig. 13 Comparison between experimental and modeled porosity

Table 5 Porosity of the concretes paste

	C25/30			C35/45			C45/55		
	RC	F30C30	C100	RC	F30C30	C100	RC	F30C30	C100
Hydration rate (%)	89	89	89	85	84	84	75	76	76
Water (kg/m^3)	180	189	185	175	179	185	165	160	160
Cement (kg/m^3)	270	285	280	299	321	336	390	375	375
W/C	0.67	0.66	0.66	0.59	0.56	0.55	0.42	0.43	0.43
Paste porosity (%) (with Power model)	52.3	52.1	52.0	48.7	47.3	46.9	39.8	39.9	39.9

5 Carbonation Behavior

The degradation mechanisms of concrete that deserve the highest attention is the corrosion of steel reinforcements, which is a pathological manifestation in reinforced concrete structures that are very difficult to intervene (Neves et al. 2012). The inherent alkalinity of concrete protects steel reinforcements from corrosion until one of the two main mechanisms initiate the destruction of the passive coating of the reinforcing bars, which are carbonation and chloride-ion ingress. The former involves chemical reactions due to the presence of CO_2, which penetrates the material through a diffusion mechanism depending on the permeability of concrete, the environment's CO_2 content, the material's moisture content and relative humidity. With the decrease of $Ca(OH)_2$ content and the consequent increase of $CaCO_3$ content, concrete suffers a pH reduction, which potentiates steel corrosion.

Evangelista and De Brito (2010) studied the carbonation depth of RAC with 100% fine RCA and observed an increase of 65% relative to the carbonation depths presented by the control specimens. Sim and Park (2011) investigated the influence of using varying contents of fly ash and fine RCA on the carbonation resistance of RAC, with a fixed amount of 100% coarse RCA. As expected, the carbonation depths increased with the inclusion of fly ash. However, the authors observed decreased carbonation in samples with increasing fine RCA, which contradicts the rest of the literature. It was claimed that this was a result of the increased amount of $Ca(OH)_2$ in the mix from the fine RCA's AM.

Medina et al. (2012) studied the carbonation of RAC containing up to 25% coarse RA from crushed sanitary ware. The authors observed that the resulting specimens exhibited a slight decrease in carbonation depth in relation to the control for a replacement ratio of 25%. This was a result of the volume decrease of pores smaller than 0.067 μm, which also yielded higher compressive strength than that of the control.

Ryu (2002) produced RAC mixes with coarse RCA sourced from concrete specimens with different strength classes and varying amounts of AM content. All RAC specimens with 100% coarse RCA showed 20% higher carbonation depths than that of the control concrete. The author inferred that, provided different RCA but with similar water absorption values are used, the strength class of the original concrete and the amount of old AM would have little effect on the carbonation depth of RAC when compared to the control NAC.

Vieira et al. (2016) analyzed the carbonation resistance of RAC with fine RA from crushed ceramic bricks and sanitary ware. The authors observed that a complete replacement of natural sand with fine RA caused increasing carbonation depths of 150 and 250% for 100% fine RA from crushed bricks and sanitary ware, respectively. Concerning the use of crushed sanitary ware, the authors stated that their use led to the formation of clusters, as there was a great cohesive force between the water molecules existing in those aggregates' vitreous surfaces. For this reason, it became necessary to incorporate an additional amount of mixing water to maintain similar workability in all mixes.

Silva et al. (2015e) studied a great deal of the literature on the use of RA in the production of structural concrete and observed that there is a clear consensus in that the carbonation depth of RAC increases with increasing replacement ratios. They observed that, as a result of the greater porosity of RA in comparison with that of NA, the porosity of the RAC also becomes greater, thereby facilitating the diffusion of CO_2 and increasing the surface area within the cement matrix that can become carbonated. They observed that the use of up to 100% coarse RCA can lead to carbonation depths two times greater than that of the control concrete.

The effect of incorporating RA on the carbonation depth of concrete has been studied by numerous researchers. Considering mixes with equivalent composition, the results of several of those publications (Hansen 1992; Kikuchi et al. 1993; Merlet and Pimienta 1993; Dillmann 1998; Teranishi, Dosho et al. 1998; Dhir, Limbachiya et al. 1999; Sagoe-Crentsil et al. 2001; Buyle-Bodin and Hadjieva-Zaharieva 2002; Lamond et al. 2002; Ryu 2002; Rao et al. 2007; Li 2008; Gomes and de Brito 2009; Evangelista and De Brito 2010; Amorim et al. 2012) suggest that increasing the replacement level leads to greater carbonation depth (Fig. 14). The analysis of their results showed that, when 100% coarse RCA are incorporated, the resulting concrete may show carbonation depth increases between 20 and 80%.

6 Chloride-Ion Penetration

Alongside carbonation, chloride-ion penetration is also responsible for destroying the reinforcements' passive layer. The existence of chloride ions around the reinforcements react at anodic sites and form HCl destroying the passive layer. Subsequently, the steel's surface becomes activated locally to form the anode, with the passive surface forming the cathode. The ensuing corrosion is in the form of localized pitting (Neville and Brooks 2010). Therefore, considering the importance of controlling this corrosion mechanism, several researchers have measured RAC's resistance to chloride-ion penetration, when exposed to chloride-rich environments (e.g. pools, bridge decks and sea coasts).

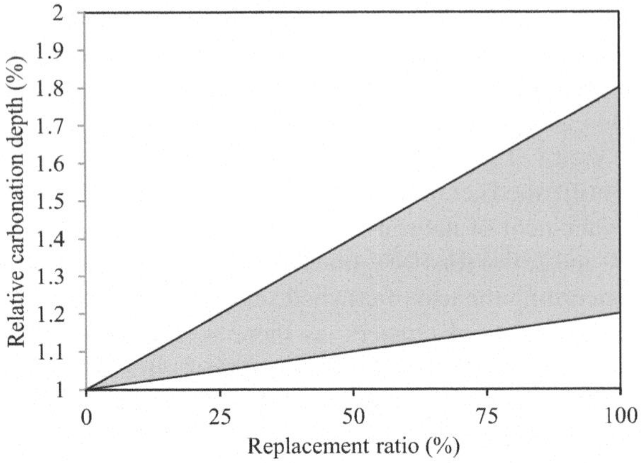

Fig. 14 Relative carbonation depth increase with increasing coarse RCA content

Fraaij et al. (2002) studied the chloride penetration of RAC with increasing replacement levels of fine natural sand with fine RCA. The authors claimed that the inclusion of RA did not affect the concrete's resistance to chloride-ion penetration. Olorunsogo and Padayachee (2002) assessed the conductivity of RAC with coarse RA from CDW recycling plants, after 3, 7, 28 and 56 days of curing. As expected, the increasing incorporation of RA led to higher conductivity. However, the authors observed that, with succeeding hydration of the cement, the concrete specimens exhibited significantly lower conductivity over time, which had a much more relevant role on this property than the inclusion of RA.

Evangelista and De Brito (2010) evaluated the non-steady state migration of chloride ions in RAC with the fine natural fraction completely replaced with fine RCA. The authors observed that there was a clear increase in the migration coefficient with increasing replacement ratios (33% increase when 100% fine RCA were used), due to the resulting porosity increased prompted by the fine RCA.

Kou et al. (2004) evaluated the resistance to chloride-ion penetration of RAC with coarse RCA, at 20, 50 and 100% by mass of coarse NA. The authors observed that the complete substitution of coarse NA with RA led to an increase in the total charge passed by 53% and 47% for curing ages of 28 and 90 days, respectively.

Kou and Poon (2009b, a) evaluated the properties of self-compacting RAC using 100% coarse RCA and varying amounts of fine RCA replacing river sand. Contrary to the findings of others, as the fine RCA content increased, the resistance to chloride-ion penetration also increased. Since the authors used large amounts of fly ash in the mix design, it is possible that the decrease in total charge passed may have been due to the pozzolanic reactions between that addition and the old AM of the fine RCA.

Corinaldesi and Moriconi (2009) studied the properties of concrete containing fly ash and SF, and with the fine and coarse NA fractions completely replaced with MRA (comprised of 70% of concrete and 27% of ceramics). Although the authors stated that the resistance to chloride-ion penetration and carbonation could be improved with the combined use of MRA, fly ash and SF, the comparative analysis was carried out between the mixes exhibiting significantly different cement and water contents, different proportions of aggregates and also water-reducing admixtures.

Berndt (2009) investigated the properties of concrete with the coarse NA fraction completely replaced with RCA and the use of ground granulated blast-furnace slag and fly ash as partial cement substitutes. The author reported that the use of ground granulated blast-furnace slag could compensate for that fact that the incorporation of RCA caused increased permeability. Mixes containing the pozzolanic material exhibited about 50% less chloride diffusion coefficient when compared with the corresponding NA and RCA concrete mixes.

Pacheco-Torgal and Jalali (2010) studied the chloride diffusion coefficient of several concrete mixes containing fine RMA of ceramic origin. The results showed that a complete replacement of the fine NA fraction led to an enhanced resistance to chloride-ion penetration by around 30% when compared to the control concrete. It was explained that the fine RMA may have prompted pozzolanic reactions with the cement's products of hydration to form additional phases capable of binding chloride ions.

Vieira et al. 2016 assessed the resistance to chloride-ion penetration of RAC with increasing fine RA from crushed ceramic bricks and sanitary ware. Similarly to the previous authors (Pacheco-Torgal and Jalali 2010), Vieira et al. 2016 reported that fully replacing the fine NA fraction with fine RA from crushed ceramic bricks led to an enhanced resistance to chloride-ion penetration (i.e. the chloride migration coefficient decreased by 57% when compared to the control concrete). It was explained that this was prompted by the pozzolanic reactions between the fine ceramic RA and the cement. However, the use of 100% fine RA from crushed sanitary ware increased the migration coefficient by 38%. Still, it should be noted that, in the latter case, the authors required notably higher quantities of mixing water to maintain equivalent workability levels, caused by the conglomeration of the lamellar particles of fine sanitary ware aggregates, which led to increased porosity of the resulting concrete mixes.

Silva et al. (2015a, b, c, d, e) carried out a systematic literature review on a great number of studies on the effect of using RA, of different types, sizes, and quality, on the resistance to chloride-ion penetration of concrete. Bearing in mind the use of the material in structural concrete and its practical design, the authors carried out a statistical analysis on the relationship between the chloride migration coefficient and the 28-day compressive strength. It was observed that the varying coarse RCA replacement level, size, type, and quality had little influence on the relationship between the two properties, despite playing a great role on the properties of RAC relative to those of the control NAC.

Figure 15a presents the non-steady state migration coefficient and Fig. 15b plots the total charge passed of RAC mixes with increasing coarse RA content relative to the measurements taken from the corresponding NAC mixes, the results of which were sourced from several experimental investigations on the subject (Moon et al. 2002; Shayan and Xu 2003; Kou et al. 2004, 2007, 2008, 2011, 2012; Poon et al. 2006; Tu et al. 2006; Tam and Tam 2007; Kou and Poon 2009b, a; Kou and Poon 2009b, a; Kong et al. 2010; Corral-Higuera, Arredondo-Rea et al. 2011; Amorim

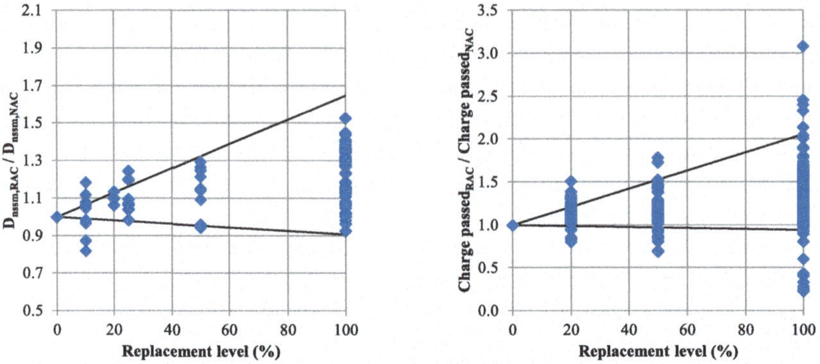

(a) Total charge passed (b) Charge pass of RAC with increasing coarse RA

Fig. 15 Relative non-steady chloride migration coefficient

et al. 2012; Kou and Poon 2012, 2013; Vaishali and Rao 2012; Bravo et al. 2008; Hwang et al. 2013; Kim et al. 2013; Lima, Caggiano et al. 2013; Duan and Poon 2014; González and Etxeberria 2014; Pedro 2014, Pedro et al. 2014). As seen in the trend lines, chloride-ion penetration tends to increase with increasing replacement ratio of coarse RA. Although some values are below 1.0, as mentioned previously, this may have been due to the use of RA with pozzolanic or to the mix design not being equivalent throughout the evaluated mixes (e.g. RAC without water compensation could reduce the effective w/c ratio resulting in a less porous microstructure when compared to that of the control NAC mixes).

7 Bond Between Recycled Aggregate Concrete and Eroded Rebars

Zhao et al. (2014) supplied drying and wetting cycles with chloride-ion to RAC with different RA replacement ratios (0, 33, 67 and 100%). After each drying and wetting cycle, the corrosion current density of the working electrode was measured with an electrochemical workstation.

As shown in Fig. 16a, it can be seen that the corrosion current density of the RAC with higher RA replacement ratio increased, indicating more serious steel corrosion in concrete. The corrosion current density of 0.1 $\mu A/cm^2$ was commonly used to assess depassivation of steel. In Fig. 16b, the initial corrosion time advanced with RA replacement ratio increasing. This is because there were more ITZs in RAC, inducing more penetration paths for chloride-ion ingression.

Dodds et al. (2020) studied the effects of RA on the risk of corrosion initiation, the predicted time to corrosion initiation and the depth of chloride-ion ingress by exposing concrete with the RA replacement ratios (0, 20, 40 and 60%) to aggressive chloride environments. However, they found that the replacement ratio of RA had

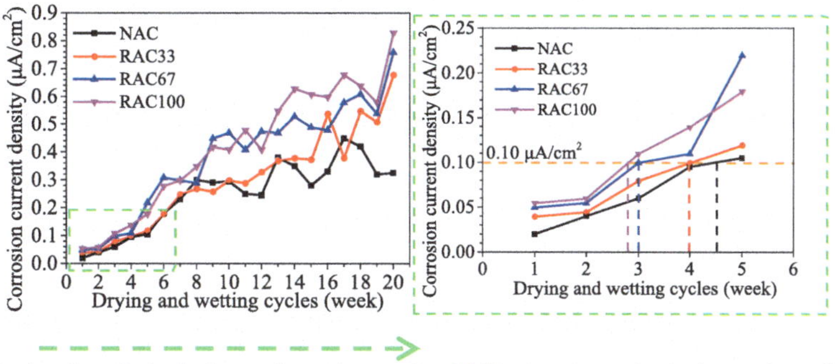

(a) During the whole drying and wetting cycles (b) During the early drying and wetting cycles

Fig. 16 Corrosion current density of RAC with different RA replacement ratios

little difference in the corrosion initiation risk over longer periods of monitoring due to the limited damage level. Therefore, further research was required over longer periods of monitoring-term effects, the degradation of the bond strength between concrete and reinforcement must be considered. This degradation may be caused by steel rebars corrosion due to RAC carbonation or chloride-ion penetration. This part will consider the degradation rules of the bond behavior between RAC and corroded steel rebars through the center pull-out test of RAC and corroded steel rebars of different corrosion rate, which were obtained through the accelerated method of electrochemistry.

7.1 Test and Analysis

7.1.1 Materials

The strength grade of RAC is C30. Common Portland cement (C) type 32.5R conforming to the Chinese standard GB 175.1999, river sand (S), and drinking water (W) were used for the test specimens. Table 6 lists the fundamental physical properties of RCA, and Table 7 lists the mix proportion and compressive strength of RAC. With a diameter of 14 mm, HRB335 steel rebars with yield strength 374 MPa and ultimate strength 571 MPa were used.

Table 6 Physical properties of RCA

Particle radius/mm	Bulk density/ kg/m^3	Apparent density/kg/m^3	Mud content/ %	Water absorption/%	Crush index/ %
5 ~ 31.5	1257	2578	1.39	4.50	16.1

Table 7 The mix proportion and compressive strength of RAC

RCA replacement percentage/%	Amount/kg/m^3					Compressive strength/MPa
	C	S	NCA	RCA	W	Cube compressive strength
100	430	473	0	1058	185	32.5

7.1.2 Preparation of Specimens

Recycled aggregate concrete (RAC) specimens with the dimensions of 200 × 200 × 200mm^3 were used. 3 specimens were contained in each group, and the total amounts of the seven groups are 28. The cube compressive strength was determined on six 150 × 150 × 150mm^3 cubes. Figures 17 and 18 show the sketch of pull-out specimens and pull-out specimens after casting, respectively.

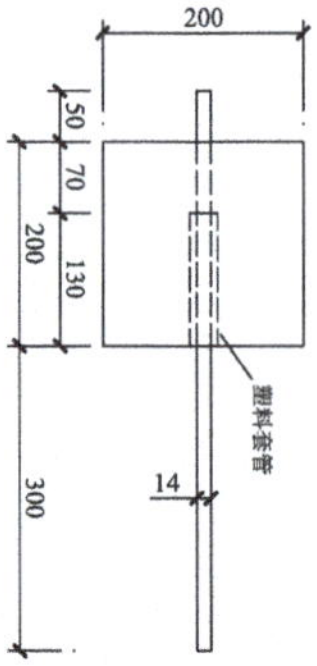

Fig. 17 Sketch of pull-out specimens

Fig. 18 Pull-out specimens after casting

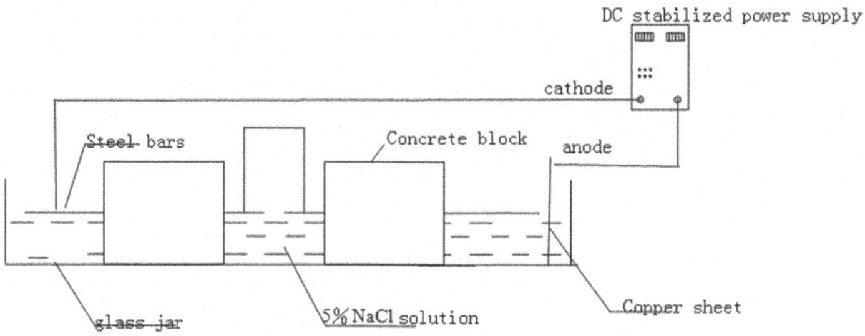

Fig. 19 Setup for accelerated corrosion of steel rebars

7.1.3 Corrosion Setup

After curing, the electrochemical methods were used to accelerate corrosion, and the device is shown in Fig. 19. Put the specimens into approximately 5% NaCl solution, and the surface is no higher than the steel bottom. Link DC stabilized power supply and wires at the steel end and copper used as the cathode, and open power to corrode steel rebars. Ammeter and transformers are contained in the DC stabilized power supply, and the current size (Xiao 2010) of the corroded steel rebars is determined by ammeter, which can estimate corrosion current density. In the corrosion process, it is necessary to check and adjust the current (i.e. record current size) passing the steel rebars regularly (i.e. general interval 6 to 7 h, and 3 to 4 h in the early corrosion) in order to ensure the current stability.

7.1.4 Determination of Steel Corrosion Rate

Remove the steel rebars after the pull-out test. According to relevant provisions of the Chinese standard "Standard for test methods of long-term performance and durability of ordinary concrete" (GB/T50082-2009), it is necessary to scrape the concrete adhered to the steel bars and to carry out pickling with 12% hydrochloric acid solution. After washing steel bars with clean water and limewater, it is necessary to rinse with clean water. After toweling off and drying 4 h in the drying capacity, the weight of steel rebars can be determined by the analytical balance (i.e. accurate to 0.001g). The ratio of the difference of weighing mass and the mass before corrosion to the mass before corrosion is termed as the corrosion percentage of the steel rebars, as is shown in Table 8.

7.1.5 Pull-Out Test Setup and Loading System

Figure 20 shows the test setup. A displacement meter ranging to 30 mm was attached to the top of pull-out steel bars, which measured the displacement of the free end. Two displacement meters ranging to 100 mm were attached to the bottom of steel

Table 8 The corrosion percentage of the steel bar embedded in the pull-out test specimens (unit: %)

No	No. 1	No. 2	No. 3	No. 4	No. 5	No. 6	No. 7
Steel bar 1	0.00	0.28	0.88	1.25	1.51	2.40	8.09
Steel bar 2	0.00	0.32	0.89	1.21	1.14	3.19	7.15
Steel bar 3	0.00	0.31	0.76	1.28	1.57	2.74	–
Average corrosion rate	0.00	0.30	0.84	1.25	1.41	2.78	7.62

Fig. 20 Test setup

bars by special devices, and the average results are the loading end displacement. Deformation control was used during the loading process, the loading rate is: $v_p = 0.5d^2 = 98N/\sec$ (v_p is the loading rate, d is diameter of the steel bar).

7.1.6 Failure Mode

When the corrosion percentage of the steel bars is relatively small, the failure of RAC bond specimens is the steel bar pull-out; with the corrosion rate increasing, due to corrosive cracks on the RAC specimens before loading, a number of smaller sub-cracks also appeared in single or opposite splitting cracks. In the process of loading, cracks mainly develop along corrosive cracks, and split along there. The failure mode turns to splitting failure. The failures of steel bar pull-out and RAC splitting after steel bar pull-out are shown in Figs. 21 and 22, respectively.

Fig. 21 The failure of steel bar pull-out

Fig. 22 The failure of recycled concrete splitting after steel bar pull-out

7.1.7 Bond-Slip Curves

The load-average slip (the loading end and the free end) curves of each group are shown in Fig. 23. Each curve shown in Fig. 23 reflects the bond failure of corrosion steel bars at different stages which are micro-slip, internal cracking and slipping, pull-out, descending and residual. From the bond versus slip curves of the degradation of the bond behavior, the load declines rapidly and the residual strength decreases with the corrosion percentage increasing. As is shown in Fig. 23, the steel bar of one specimen of No.7 was pulled off at the interface, so there are only two curves.

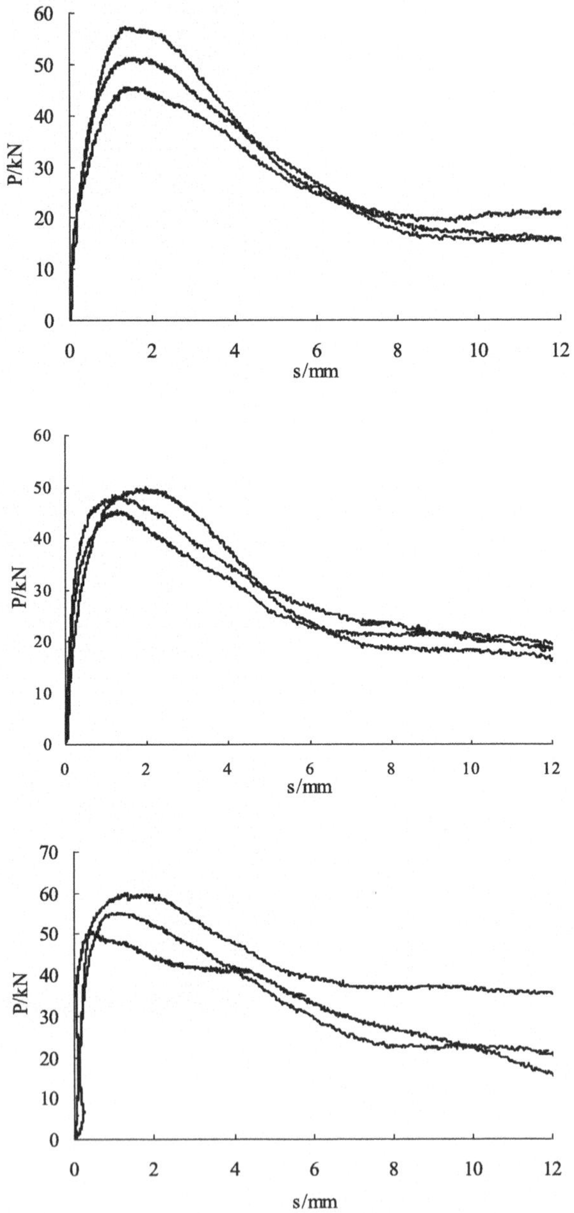

Fig. 23 Test curve of load versus slip

Fig. 23 (continued)

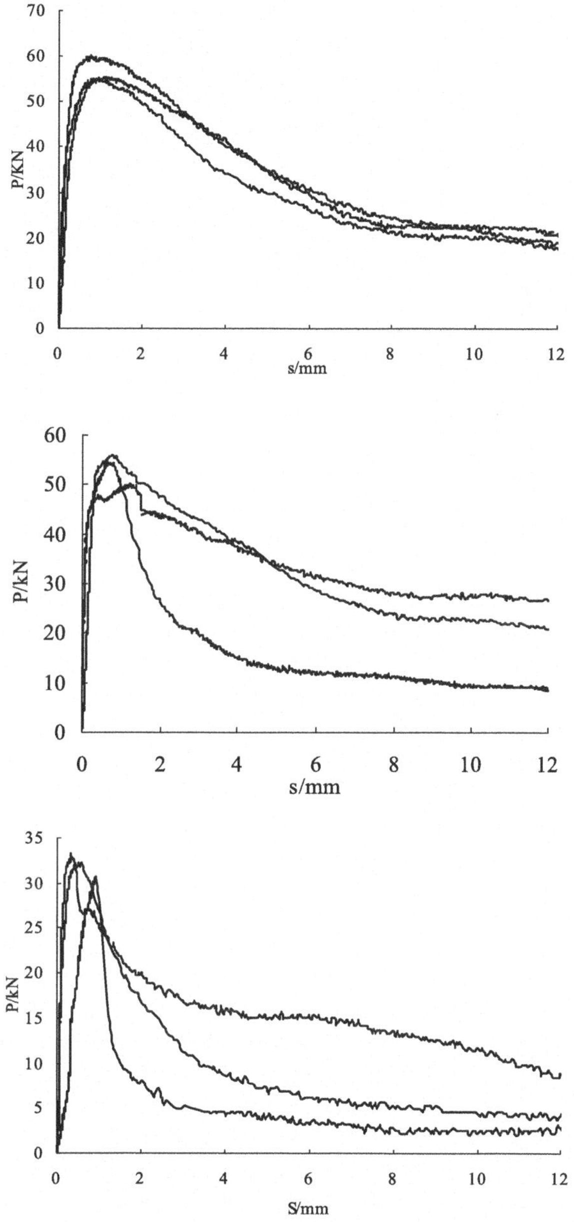

Fig. 23 (continued)

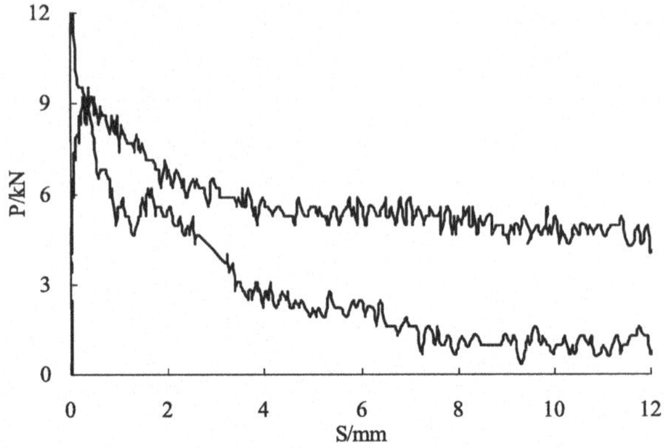

7.1.8 Bond Strength

Table 9 lists pull-out test results of bonding behavior between RAC and corroded steel rebars. From Table 8, it can be concluded that with the steel corrosion ratio increasing, the bond strength between RAC and steel rebars increases first and decreases afterwards. Average peak slip decreases when the steel corrosion ratio increases.

From Fig. 24, it can be seen that the test results and fitting curves of the bond strength between RAC and corrosion steel bars and that of normal concrete and corrosion steel bars with the steel corrosion ratio increasing. The rules of degradation of bond behavior of RAC are similar to that of normal concrete, and the rate of the degradation of RAC is faster than that of normal concrete.

The mix proportion of the center pull-out test referred above. The size of the concrete block is $100 \times 100 \times 100$ mm^3, the diameter of the steel bar is 10 mm, and the bond length of the steel rebar is 50mm. By analyzing the test results, the bond strength of the RAC and normal concrete before degradation is the same, the value of which is 17.39 MPa, and the ratio of the bond strength to the corresponding cube compressive is the relative bond strength: 0.5 and 0.4, respectively. From the test results of group 1 shown in Table 9, the bond strength between RAC and steel rebars is 16.63MPa, and the relative bond strength is 0.51. This further confirms that under the equivalent mix proportion, the bond strength between the RAC and steel rebars is close to that of normal concrete and steel rebars, whereas under the same compressive strength, the bond strength between the RAC and steel rebars is higher than that of normal concrete and steel rebars. Bond constitution relationship.

According to the research results, the average bond stress–strain curve was fitted in this chapter:

$$\text{The ascending part: } \frac{\tau}{\tau_u} = \left(\frac{s}{s_u}\right)^a \tag{1}$$

Table 9 Pull-out test results of bonding behavior between recycled concrete and corroded steel bars

No	Load/kN	Bond strength/MPa	Average bond strength/MPa	Bond strength reduction factor	Relative bond strength (corresponding to compressive strength)	The peak slip of the free end/mm	The peak slip of the loading end/mm	The average peak slip of two ends/mm	The average peak slip/mm
No. 1	45.42	14.76	15.88	1.00	0.49	1.08	1.48	1.28	1.41
	49.97	16.24				1.32	1.78	1.55	
	51.17	16.63				1.11	1.70	1.41	
No. 2	48.16	15.65	15.38	0.97	0.47	1.48	1.52	1.50	1.41
	48.43	15.74				1.21	1.61	1.41	
	45.41	14.76				1.04	1.59	1.32	
No. 3	59.94	19.48	17.97	1.13	0.55	0.88	1.71	1.30	1.31
	50.86	16.53				0.77	1.94	1.36	
	55.11	17.91				0.93	1.62	1.28	
No. 4	55.11	17.91	18.40	1.15	0.57	0.87	1.56	1.22	1.01
	59.94	19.48				0.44	1.13	0.79*	
	45.42	17.81				0.58	1.01	0.80	
No. 5	49.96	16.24	17.38	1.09	0.53	0.89	1.61	1.25*	0.80
	55.72	18.11				0.48	1.08	0.78	
	54.20	17.61				0.11	1.51	0.81	
No. 6	33.19	10.79	10.41	0.66	0.32	0.14	0.55	0.35	0.43
	32.26	10.48				0.25	0.77	0.51	
	30.68	9.97				0.53	2.43	1.48*	
No. 7	11.96	3.89	3.49	0.22	0.11	0.04	0.09	0.07*	0.36
	9.51	3.09				0.34	0.37	0.36	

* It means this value exceeds the mean value of 15%, not included in the calculation of the mean value

$$\text{The descending part: } \frac{\tau}{\tau_u} = \frac{s/s_u}{b(s/s_u - 1)^2 + s/s_u} \tag{2}$$

where τ is the bond stress, s is slip, τ_u is the bond strength, and s_u is the slip corresponding to the τ_u.

By fitting the test results, the value of the parameter a was determined to 0.3 for all test, and the values of the parameter b are given in Table 10. The value of parameter a does not change with variations of the steel corrosion ratio, showing that no changes occur in the ascending part of the dimensionless bond—slip curve. (Fig. 25)

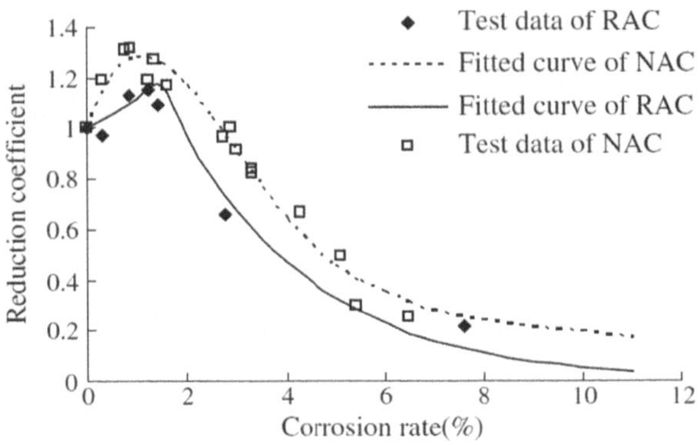

Fig. 24 Reduction coefficient bond strength of recycled concrete and normal concrete changes with increase in corrosion rate of steel bars

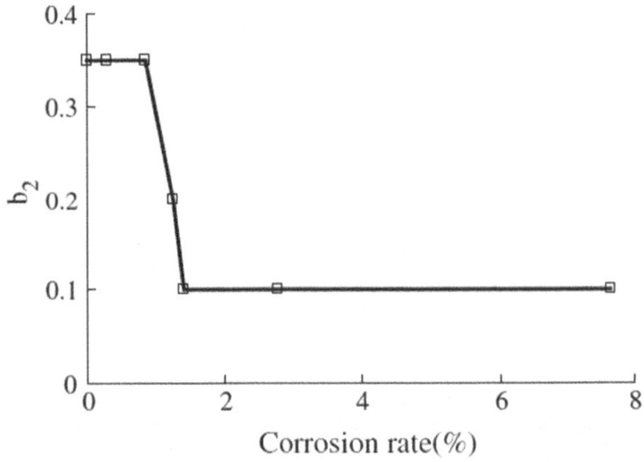

Fig. 25 The value of b changes with increase in corrosion rate of steel bars

Table 10 The value of parameter ***b***

No.	No. 1	No. 2	No. 3	No. 4	No. 5	No. 6	No. 7
b	0.35	0.35	0.35	0.2	0.1	0.1	0.1

Figure 26 lists the variation of the *b*-value referring to the rinsing of steel corrosion ratio. The value of parameter *b* is related to the area under the descending part of the dimensionless bond—slip curve. The area under the curve does not decrease with the increase of the value of *b*, which indicates that the energy absorbing capacity does not decrease from the bond-slip process. From Fig. 26, it can be seen that the value of *b* decreases with the increasing of the steel corrosion ratio, and when the steel corrosion ratio is small (less than 0.84%), there is no change; the value of *b* decreases more with the rate continues to increase, and becomes nearly constant when the rate is higher than 1.41%. It indicates that the energy dissipation capacity of bond-slip process between RAC and steel rebars does not decrease when the steel corrosion ratio is small; however, the energy dissipation capacity decreases much more with the steel corrosion ratio increasing, and after the steel corrosion ratio is higher, the capacity keeps unchanged.

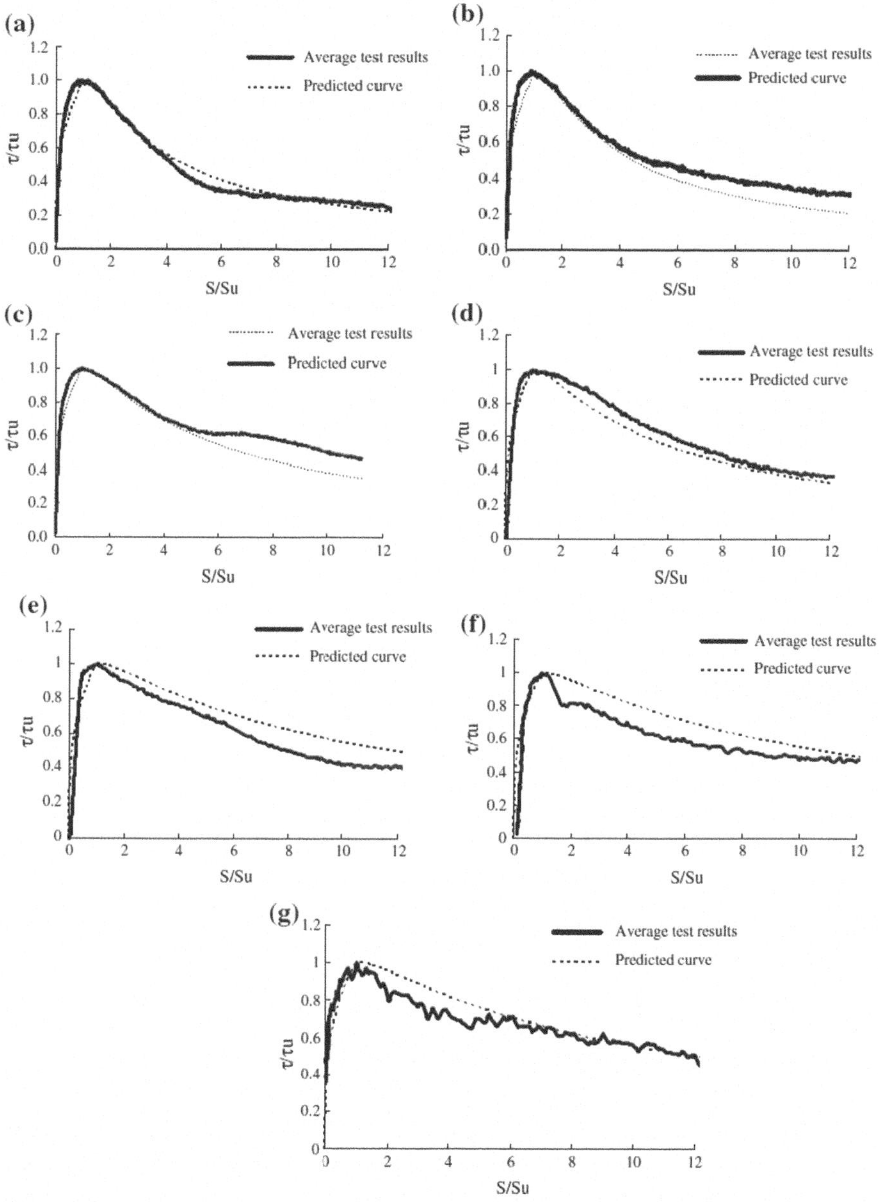

Fig. 26 Comparisons between average test results and predicted bond-slip curve

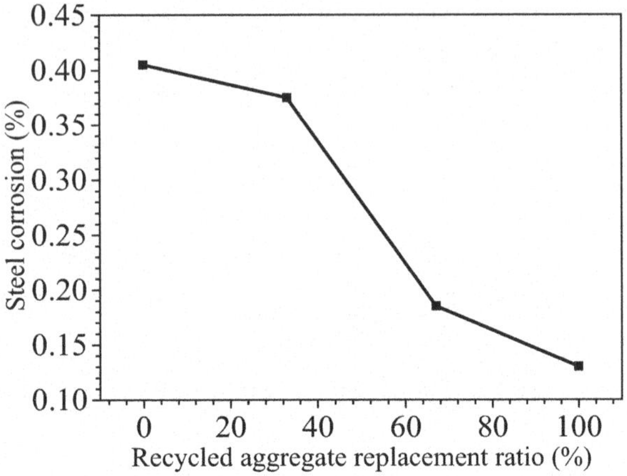

Fig. 27 The steel corrosion at the time of concrete outer surface initial cracking of RAC

The average test curves and the predicted curves are plotted in Fig. 27. It can be seen that the predicted curves are fitting well to the test curves, which demonstrates that Eqs. (1) and (2) can be used to simulate the whole process of bond-slip relationship of RAC and corrosion steel bars.

7.2 Performance of RAC Beams with Corroded Beams

Existing researches almost only considered single impact applied onto reinforced RAC components, e.g., single load, or single chloride-ion ingression. However, concrete structures always suffer the coupled damages induced by both loads and the ingression of harmful materials. Several researchers have studied the performance of RAC under the coupled damage of sustained load and chloride-ion ingression. But direct current supply was used to accelerate the steel corrosion in these literatures, causing uniform corrosion in the steel bars, which was not in accordance with the actual situation. Therefore, it is necessary to discover the performance of corroded RAC components suffering the sustained loads coupled with the natural ingression.

7.3 Cracks Induced by Corroded Bars

Using direct current power coupled with chloride-ion drying and wetting cycles to accelerate the steel corrosion in the reinforced RAC cylinders with different RA replacement ratio, the previous research (Zhao et al. 2014) found that the more RA concrete contained, the earlier corrosion-induced crack appeared at the surface of concrete cover as shown in Fig. 27. This result illustrates that the critical steel corrosion at concrete outer surface cracking decreases with the increasing replacement ratio of RA, meaning that RAC is prone to cracking induced by corrosion than NAC.

In addition, the cracks pattern within the RAC was also discovered by (Zhao et al. 2014). For different RA replacement ratios, a typical internal cracks pattern is shown in Fig. 28. It is found that corrosion-induced cracks tended to propagate through the ITZ, which was known to be a mechanically weak area in concrete. For example, in Fig. 28d the internal cracks in RAC mainly propagated along with the interface 1 between NA and new mortar, labeled as Frame 2, or along with the interface 2 between original NA and old mortar, labeled as Frame 3.While, different from NAC, corrosion-induced cracks in RAC also might propagate into RA due to its internal flaws, labeled as Frame 1, as shown in Fig. 28b.

Moreover, in areas containing several interfaces, corrosion-induced cracks tended to propagate through the interface between the NA and new mortar and the interface between the old coarse aggregate and new mortar but not the interface between the old mortar and new mortar.

The internal corrosion-induced cracks propagation process was discussed (Zhao et al. 2015), as shown in Table 11. They found that before surface cracking, the width of internal crack in both NAC and RAC decreased along with the radial direction. At the concrete surface cracking moment, the internal crack maintained the constant width along the radial direction, and this critical crack width of NAC, W_c^{NAC}, was larger than that of RAC, W_c^{RAC}. Meanwhile, after the surface cracking of concrete cover, RAC exhibited a larger crack angle θ and a wider-opening crack shape than NAC when the steel corrosion level was same.

7.4 Mechanical Performance

Zhang and Zhao (2016) applied sustained loads by anchor bolts to reinforced RAC beams with three types of RA replacement ratios, and supplied drying-wetting cycles to accelerate the ingression of chloride-ion. For the uninjured beams, the cracking moment and the bending stiffness decreased with the increasing RA replacement. The deflections of RAC beams developed faster than those of NAC beams. While, the flexural capacity was comparable for RAC beams with different RA replacement.

After 20-months damage, the steel rebars in three groups of beams were all corroded. The flexural capacity and failure modes of the injured beams was also comparable to those of the uninjured beams. However, the initial flexural stiffness and the ductility of the injured beams were lower than those of the uninjured beams. Moreover, RA replacement seemed to have no significant influence on the cracking moment, the bending stiffness and the flexural capacity.

After 80-months damage, the corrosion level of steel rebars became more serious. The ultimate flexural capacity of three groups of beams at different damage age are

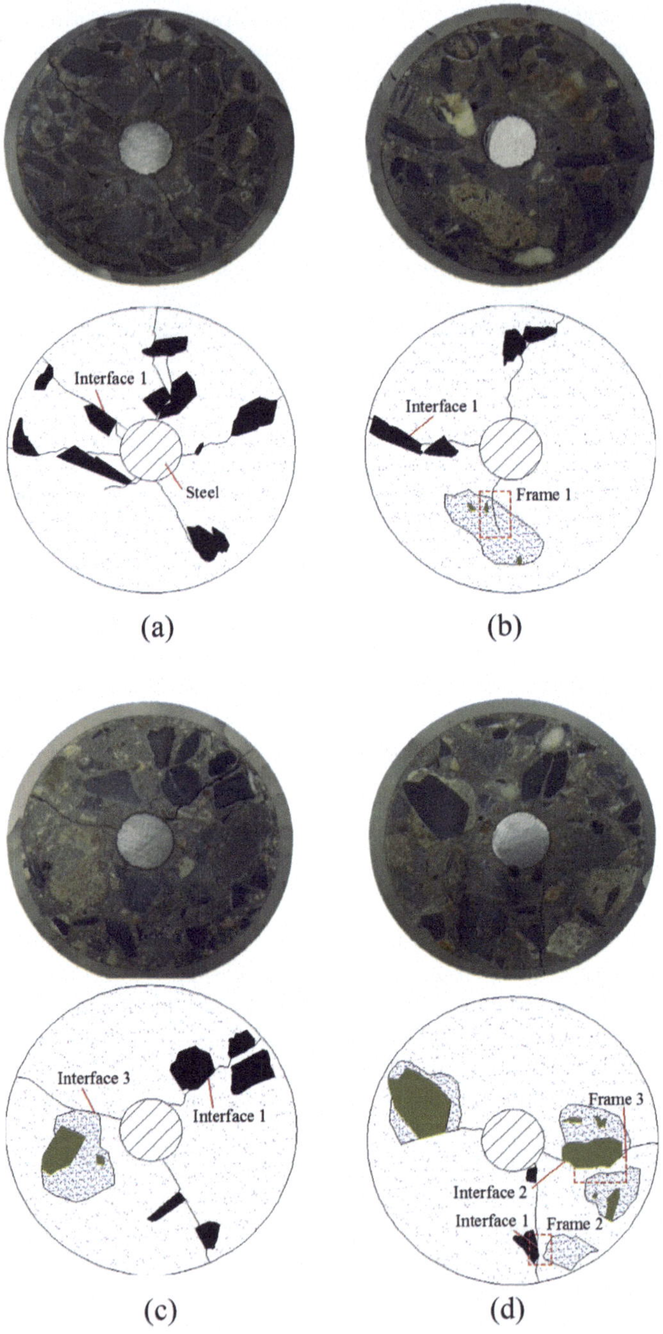

Fig. 28 Optical scanning pictures and schematic diagrams of internal crack propagation path: **a** NAC, **b** RAC33, **c** RAC67, **d** RAC100

Table 11 Crack shapes of NAC and RAC during the corrosion-induced cracks propagation process

Cracking state	NAC	RAC	
Before surface cracking	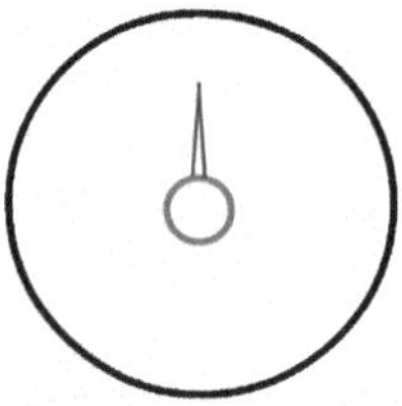	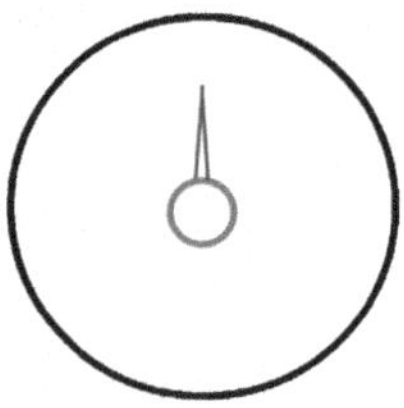	Shapes of inner crack are mostly same
Surface cracking moment	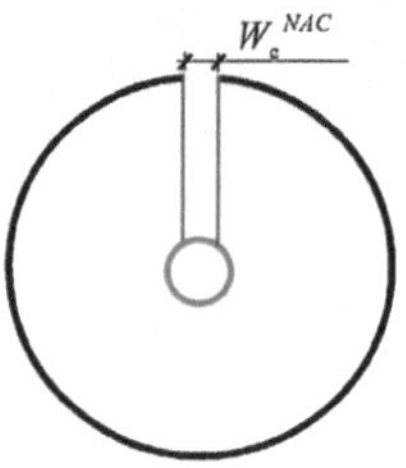	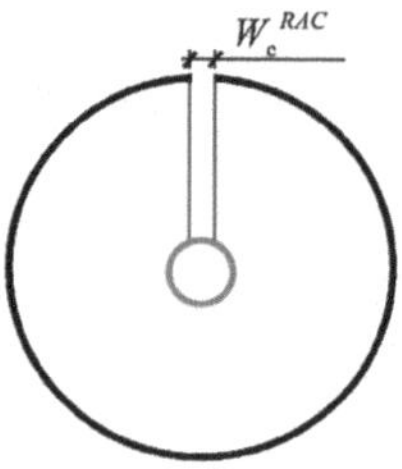	$W_c^{NAC} > W_c^{RAC}$
After surface cracking	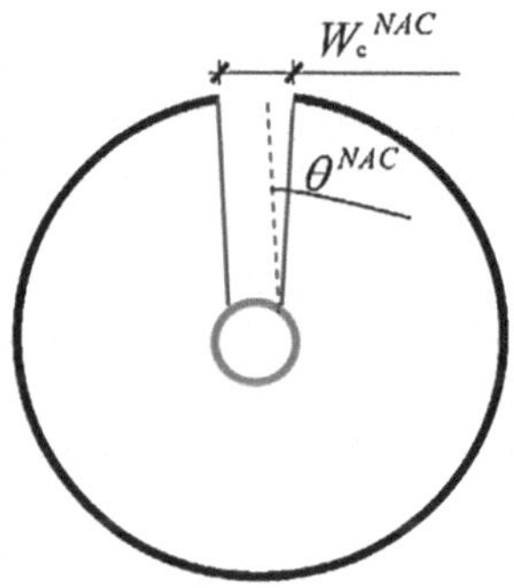	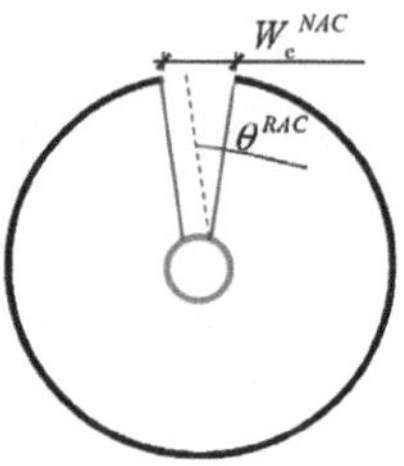	$\theta^{NAC} < \theta^{RAC}$ $W_c^{NAC} < W_c^{RAC}$

show in Fig. 29. It can be seen that at the damaged age of 80 months, the ultimate flexural capacity of RAC100 obviously decreased, compared with the uninjured beam. While the ultimate flexural capacity of NAC slightly increased after 80-months damage. As for RAC50, the ultimate flexural kept stable during the damage age. Therefore, it is believed that the mechanical performance of RAC beams degraded faster than NAC beams under the sustained loads coupled with chloride-ion ingression.

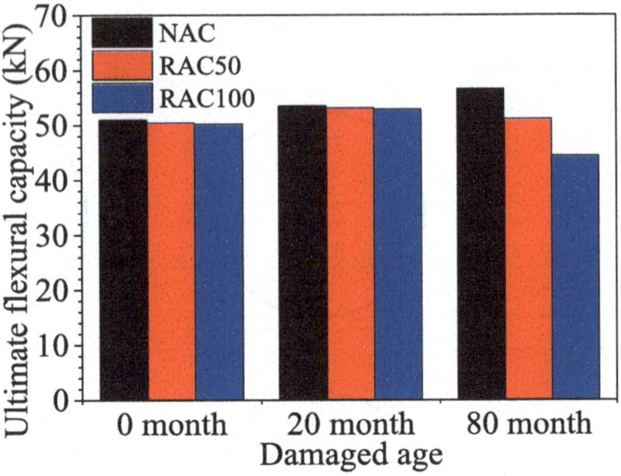

Fig. 29 The ultimate flexural capacity of three groups of beams at different damaged age

7.5 *Crack Propagation*

Except for the mechanical performance, it is also important to investigate the durability of reinforced RAC beams. Zheng (2020) observed the crack propagations on the external surface of RAC beams during the 64-months damaged age.

The results show that when subjected to the coupled damages of sustained loads and chloride-ion ingress, the beam with 100% RA replacement ratio generated more cracks. The spacing, the average width, as well as the propagating height of the transverse cracks mainly induced by sustained loads, were found smaller for beams with 100% RA.

In order to analyze the damage level of injured beams, the areas of external cracks were calculated as shown in Fig. 30. It is found that the area of external cracks of RAC100 increased slowly during the beginning damaged age, then exhibited fast increase during the middle damaged age, and finally slowed down during the later damaged age. While, the area of external cracks of NAC increased linearly. As for RAC50, the development of cracks area was between RAC100 and NAC. Besides, the area of external cracks increased significantly over the RA replacements because of the worse resistance to chloride-ion penetration. Hence, RAC beams demonstrated worse durability against corrosion-induced cracking, compared with NAC beams.

7.6 *Summary*

From the experimental results of bond between RAC and eroded rebars, the following observations and conclusions can be drawn.

The failure of the RAC specimens is steel bar pull-out when the steel corrosion ratio is small. With the rate increasing, the bond failure modes turn into RAC splitting failure.

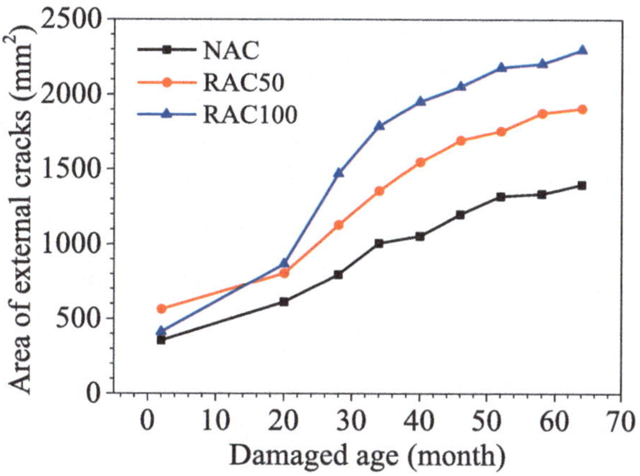

Fig. 30 The areas of external cracks of three groups of beams during the damaged age

The bond failure of corrosion steel rebars contains five stages, which are micro-slip, internal cracking and slipping, pull-out, descending and residual.

The bond strength between the RAC and deformed rebars increases with corrosion up to a certain amount, and then the bond strength decreases as the corrosion ratio further increases, which is similar to that of normal concrete.

Under the equivalent mix proportion condition, the bond strength between RAC and steel rebars is close to that of normal concrete; under the equivalent compressive strength, the bond strength between RAC and steel rebars is higher than that between normal concrete and steel rebars.

The normalized bond versus slip relationship for the RAC and corrosion steel rebars can be fitted by Eqs. (1) and (2).

8 RAC Blocks

All over Europe, more than 5500 companies with around 8000 production plants are producing concrete pre-cast products. It is estimated that the sector generated 24 billion euros in 2015 according to European Federation of the Pre-cast Concrete Industry (Factbook). The European Federation of the Pre-cast Concrete Industry estimates that about 25% of concrete production is represented by concrete pre-cast products (Delvoie, Zhao et al. 2018). Concrete pre-cast producers consume large quantities of aggregates and generate voluminous amounts of concrete waste, generally about 1–2% of total production. Concrete building blocks are a commodity product and its profit margin is low (Soutsos et al. 2011) with a local distribution area. The raw materials used to manufacture blocks could be virgin aggregate, lightweight or recycled materials from CDW. The feasible use of RAs from CDW in the production of concrete blocks has recently attracted more research interest (Table 12) Poon and Chan (2006), Poon and Chan (2007), Poon et al. (2009), Soutsos et al. (2011), Xiao et al. (2012), Guo et al. (2018).

Poon and Lam (2008) evaluated the effects of aggregate to cement ratio and the influences of the combinations of aggregates on the properties of blocks. RCA was mainly obtained by concrete rubble sources from a CDW recycling facility in Hong Kong. The maximum size of all aggregates was less than 5 mm. RCA (water absorption = 10.3%) was used to replace 25, 50, 75 and 100% natural crushed aggregate, respectively, in the production of concrete blocks (200 × 100 × 60 mm); an aggregate to cement ratio of four was used. The blocks were manufactured in the laboratory with three layers. The authors found that the compressive strength of the paving blocks decreased as the aggregate to cement ratio increased. The use of RCA as a replacement of natural crushed aggregate in the production of concrete blocks reduced the density and strength but increased the water absorption of the blocks. The compressive strengths of the blocks after 28 days were 79.9, 67.4, 65.8, 63.5 and 64.8 MPa, respectively, with 0, 25, 50, 75 and 100% replacement of natural crushed aggregate by RCA.

Poon et al. (2009) reported the influence of low-grade RAs on the properties of concrete blocks. The low-grade RAs were obtained from a construction waste sorting facility and they were contaminated with higher percentages of no concrete components (e.g. > 10% soil, brick, tiles etc.). The blocks were prepared using coarse RA and an aggregate to cement ratio of 10:1. Fine RA replaced the crushed fine sand at differing levels of 25, 50, 75 and 100%. They discovered that the mechanical strength of blocks decreased with the increasing low-grade recycled fine aggregate content (the compressive strengths of the blocks were 37 and 25 MPa respectively for the reference block and 100% fine RA block). The drying shrinkage of the blocks increased with an increase in fine RA content.

Soutsos et al. (2011) investigated the effect of partially replacing limestone aggregate by recycled demolition aggregates in the manufacture of pre-cast concrete building blocks. A specifically modified electric hammer drill was used to compact the blocks in the laboratory. The maximum replacement levels were 60% for coarse fraction RCA and 20% for fine fraction RCA, respectively: this had no significant detrimental effect on the mechanical properties of blocks. For all the mixes with 100 kg/m^3 of cement below the maximum replacement level, the compressive strength of blocks is around 7.5.8.5 MPa after 28 days, which is higher than the target strength (7 MPa). They found that the maximum replacement levels were 20% for coarse fraction recycled masonry aggregate and 20% for fine fraction recycled masonry aggregate. Factory trials showed that there were no practical problems with the use of recycled demolition aggregate in the manufacture of building blocks.

Guo et al. (2018) explored the possible use of RCA to produce concrete building blocks by using 75% of RCA. The results indicated that the incorporation of RCA slightly declined the compressive strength and impaired the durability of concrete blocks. However, concrete blocks with 75% RCAs satisfied the strength, drying shrinkage and freeze–thaw resistance requirements for concrete blocks specified by Chinese standards. The compressive and shear performances of masonry prisms made with RA blocks were similar to those of conventional concrete masonry.

Table 12 Summary of recent studies on the feasible use of RAs in the production of concrete blocks from Zhao et al. (2020)

Reference	Origin of recycled aggregates	Block type	Test conditions	Replacement levels	Main results
Poon and Chan (2006)	RCA from recycling facility and crushed clay brick	Paving blocks	Laboratory test	Two-series on 0, 25, 50 and 75%	Feasible to produce paving blocks prepared with 25% crushed clay brick that satisfied the compressive strength requirement
Poon and Chan (2007)	RCA from recycling facility and other contaminants	Paving blocks	Laboratory test	Two-series on 100% RCA and 10% other contaminants	An allowable contamination level in the RCA can be increased from 1% to a maximum of 10% in the production of paving blocks
Poon and Lam (2008)	Recycled crushed glass and RCA (Concrete rubble from CDW recycling facility)	Paving blocks	Laboratory test	Six series (two-series on 0, 25, 50, 75 and 100%)	Recommend to produce the blocks with 50% RCA and 50% recycled crushed glass and with A/C ratio of 4 or below
Poon and Kou (2009)	Low-grade RCA from CDW recycling facility with high content of no concrete components	Non-structural concrete blocks	Laboratory test	Three series (one series on 0, 25, 50, 75 and 100%)	Potential to use low-grade RCA for making non-structural concrete blocks; optimal percentage of recycle fine aggregate is 50%
Soutsos et al. (2011)	RCA and masonry derived aggregate from local demolition company	Concrete building blocks	Laboratory test and factory trial	Four series laboratory tests and factory trial	The maximum replacement levels were 60% for coarse fraction RCA and 20% for fine fraction RCA; 20% for coarse fraction RMA and 20% for fine fraction RMA

(continued)

Table 12 (continued)

Reference	Origin of recycled aggregates	Block type	Test conditions	Replacement levels	Main results
Guo et al. (2018)	RCA from 30 MPa waste concrete	Concrete building blocks	Laboratory test and factory trial	Two mixtures (0 and 75%) laboratory tests and a factory trial	Concrete blocks made with 75% RCAs exhibit favorable mechanical and durability performances and satisfy Chinese standard
Xiao et al. (2011)	RCA (concrete rubble waste from earthquakes) and crushed clay brick	Concrete masonry partition wall blocks	Laboratory test	Four series on 0, 25, 50, 75 and 100%	The amount of crushed clay brick should be controlled at less than 25% for coarse aggregates and within 50–75% for fine aggregates

8.1 Formulation and Construction

Three concrete building blocks with different substitution rates (0, 30 and 100%; these substitution rates were determined according to preliminary laboratory tests) of NA by the same fraction of RCA (only fraction 2/6.3 mm) were manufactured on the production site (in real industrial conditions). Table 13 shows the composition of concrete building blocks (dimension 39 cm × 14 cm × 19 cm with two holes, see Fig. 31 right) for masonry unit. European standard EN 206:2013 + A1 was applied in the concrete blocks mix design. CEM III/A 42.5 cement and a water/cement ratio of 0.5 were used for block production. The air-dried RAs were used for the concrete blocks production. The absorbed water of natural and RAs was adjusted according to the water content of the aggregates and their water absorption in the mixer (Table 13). More water was used for the concrete blocks made with RAs (an additional 28.5 kg of water was used for the concrete blocks B_RCA100 compared with the reference concrete block—see Table 13) due to the higher porosity and water absorption of RAs compared to the NA. After mixing, the fresh concrete blocks were placed by mechanical vibrations (Fig. 31, left photo). The block was stored in a wet chamber at 20 °C for three days (Fig. 31, right photo). After that, the blocks were stored outside for two weeks. They were then stored in the laboratory at a temperature of 20 ± 2 °C and relative humidity of 60% ± 2% until the tests were conducted.

Table 13 Compositions of concrete building blocks—production of a wall unit (1 m^3) (Zhao et al. 2020)

	B_RCA0	B_RCA30	B_RCA100
NA 2/7 (kg)	1010	707	0
RCA 2/6.3 (kg)	0	282	940
NS 0/2 (kg)	822	822	822
Yellow sand 0/1 (kg)	63	63	63
Cement CEM III/A 42.5 (kg)	175	175	175
Efficient water (kg)	87.5	87.5	87.5
Absorbed water (kg)	9.49	18.04	38
W_{eff}/C	0.50	0.50	0.50

Fig. 31 Production of block with RCA on the production site, under real industrial conditions (left: installation for mechanical vibration; right: marking and storage) Zhao et al. (2020)

8.2 Behavior of Blocks

The 28-day and 360-day compressive strength of the blocks are given in Fig. 32. The values are an average from three measurements. It can be seen that the compressive strengths of concretes with RCA were lower than those of concrete with NA. The compressive strength of the block specimens slightly decreased with an increase of RCA content. The compressive strength of concrete block made with 100% RCA at 28 days deceased 16.5% compared to the reference concrete, while the concrete block made with 30% RCA at 28 days decreased 6.0%. The lower mechanical strengths are caused by the poorer physical properties of RCA in comparison to NA used, i.e. the presence of adherent cement paste led to higher porosity and worse mechanical and physical properties of RCA (less resistant) compared to the NA (Xiao et al. 2007, 2013; Zhao et al. 2015). It could also be associated with an increase in the water/cement ratio, from the higher initial free water content in the concrete mixture (since

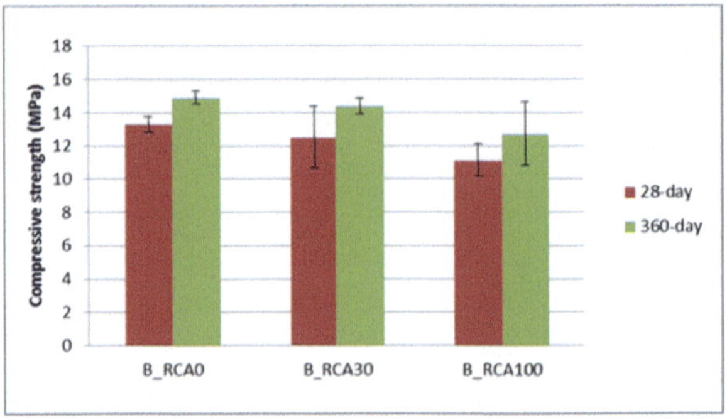

Fig. 32 Compressive strengths of concrete blocks at 28-day and 360-day from (Zhao et al. 2020)

the RCA was used at the air-dried condition with moisture content of aggregate much lower than the water absorption, about 28.5 kg more water was used for the concrete blocks B_RCA100 compared to the reference concrete block), due to the existence of a second zone of transition.

The compressive strength of concrete block made with 100% RCA could reach 11.1 MPa after 28 days and 12.7 MPa after 360 days. The normalized compressive strength of masonry unit (f_{bm}, $f_{bm} = d \times f_b$) can be determined by the shape factor d and mean compressive strength of masonry specimens (f_b) according to EN 772–1. The shape factor d is defined as being a multiplying factor used to convert the mean compressive strength of the masonry specimens to the normalized compressive strength of masonry unit f_{bm} ($d = 1.18$ for all blocks: the shape factor is equal to 1 for specimen with height and width each equaling 100 mm according to the EN 772–1). The normalized compressive strength of concrete block made with 100% RCA was 13 MPa, which is within the requirement for this type of block according to PTV 21–001 (class f10, $f_{bm} \geq 10$ MPa). Moreover, all types of concrete block made with RCA or without RCA were in the quality category of class "10/2.2" according to PTV 21–001 (category of class f10 with $f_{bm} \geq 10$ MPa and category ρ 2.2 with 1900 kg/m^3 < $\rho \leq$ 2200 kg/m^3).

The drying shrinkage of the concrete blocks is shown in Fig. 33. The values are an average from three measurements. The drying shrinkage of the blocks increased with an increase in RCA content. The hardened cement paste attached to the RCA contributed to an increase in the volume of the paste (old hardened cement pastes in RCA + new), thus increasing the drying shrinkage of the resulting concrete. This is consistent with the results of previous studies (Guo et al. 2018; Poon et al. 2009). In addition, the shrinkage of blocks with 100% of RCA in this test (i.e., 0.052%) is consistent with the result obtained by Poon et al. (2009) ($\leq$0.06%). Moreover, the drying shrinkage values of the blocks were beneath the limit ($\leq$0.06%) prescribed by BS 6073 regardless of the type of block.

After 14 freeze–thaw cycles, a visual inspection of the specimen did not reveal any significant deterioration in all the blocks (this is in line with the requirement for this type of block according to PTV 21–001). The loss of mass in all the concrete blocks did not exceed 1% in any case (the mass loss of concrete blocks was 0.25,

Fig. 33 Drying shrinkage of concrete blocks from (Zhao et al. 2020)

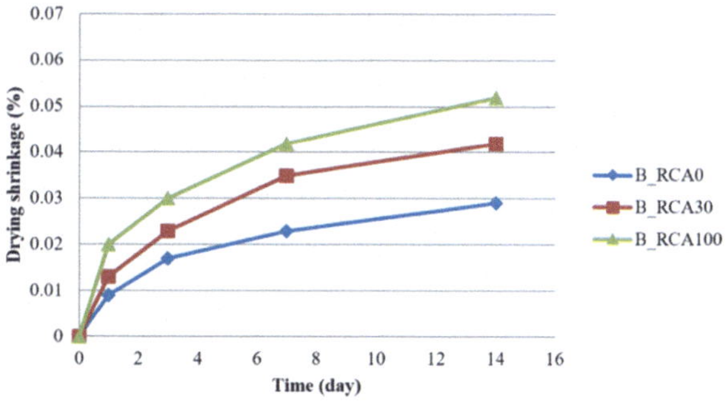

Fig. 34 Residual resonant frequencies of concrete blocks after freeze–thaw action from (Zhao et al. 2020)

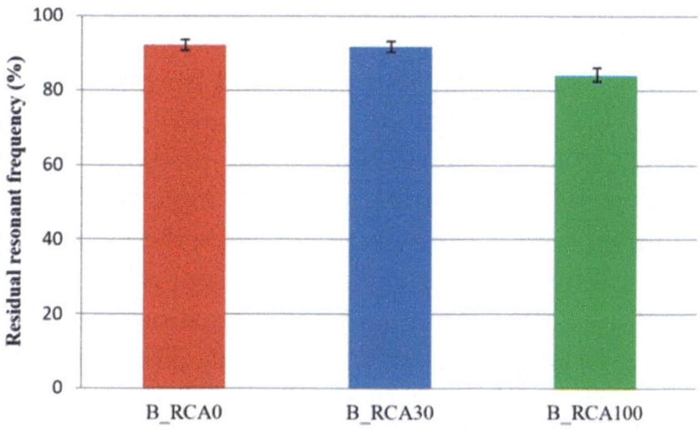

0.27 and 0.47% for B_RCA0, B_RCA30 and B_RCA100, respectively). The mass loss of concrete blocks made with RCA was higher than those of the reference concrete blocks. The residual resonant frequency of concrete blocks is shown in Fig. 34. The residual resonant frequency of blocks made with RCA after freeze–thaw cycles was lower than that of the reference concrete blocks (the scope of reduction in resonant frequency of the blocks after freeze–thaw cycles were 7.91, 8.32 and 15.84% respectively for B_RCA0, B_RCA30 and B_RCA100). As can be seen, the specimens containing RCA were less durable in freeze–thaw action, which is due to their higher initial water/cement ratio and consequent higher capillary porosity. This is a direct consequence of the higher porosity of RCA and their lower stiffness (Bogas et al. 2015) and is consistent with the results of a previous study (Guo et al. 2018).

8.3 Life Cycle Assessment of Concrete Blocks

The comparative life cycle assessment (LCA) of concrete blocks without RCA (B_RCA0) and with 30% and 100% of RCA in substitution of NA (B_RCA30 and B_RCA100) do not show significant differences across almost all impact categories

(Fig. 35, Table 14) with the exception of land use, especially for the 100% substitution case (in green in Fig. 35). The benefits are respectively 16.1% and 53.1% for the B_RCA30 and the B_RCA100 compared to the NA scenario. Although these findings appear obvious without even doing an LCA, it is worthwhile to be able to quantify objectively the potential gains from the substitution of NA by RCA.

Usually cement is the main influential element when evaluating the environmental burden of concrete (as it is for instance for a generic concrete entry in Ecoinvent database) (Wernet et al. 2016). The NA (*gravel* in the inventory) only represent a small part of the impact in a classical formulation of a concrete. This has been verified in the test.

Figure 36 presents the detailed impact of the different elements of the inventory of B_RCA0 concrete blocks. The cement (in violet in Fig. 36) is clearly the element with the most impact in all categories except land use. This category is mainly affected by NA and sand. NA or sand have no impact in the mineral resource depletion category because there is no characterization factor associated with the "gravel" in the ILCD method (nor in any other method). This is due to the fact that gravel is not considered a potentially missing mineral resource by any method. However, NA and sand have a rather large impact in the land use category since they are crushed stones coming from quarries, with occupation of land for the mineral extraction site. Land uses (occupation and transformation) are generic data included in the entries of Ecoinvent database (Wernet et al. 2016). This fact explains why even a small substitution of NA can bring gains in this category (see Fig. 35).

The internal production of RCA 2/6.3 mm from old block waste for the production of B_RCA30 blocks can only substitute 3% of the global production of concrete

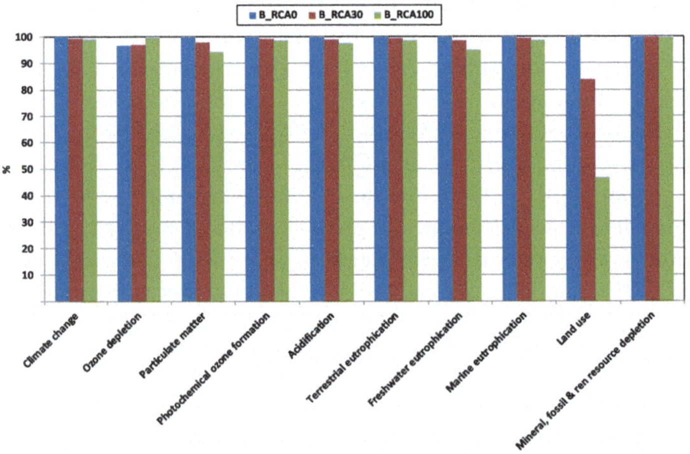

Fig. 35 Comparison of the impacts of the production of 1 m3 of concrete blocks with NA only (B_RCA0—reference scenario) and the production of 1 m^3 of blocks with the substitution of 30% (B_RCA30) and 100% (B_RCA100) of NA by RCA—Characterization ILCD 2011 Midpoint + (1.10) from (Zhao et al. 2020)

Table 14 Comparison of the impact of the production of 1 m^3 of concrete blocks with NA only (B_RCA0—reference scenario) and the production of 1 m^3 of blocks with the substitution of 30% (B_RCA30) and 100% (B_RCA100) of NA by RCA—Characterization ILCD 2011 Midpoint + (1.10) from (Zhao et al. 2020)

	Units	B_RCA0	B_RCA30	B_RCA100
Climate change	kg CO_2 eq	1.09E + 02	1.08E + 02	1.08E + 02
Ozone depletion	kg CFC-11 eq	6.30E-06	6.32E-06	6.51E-06
Particulate matter	kg PM2.5 eq	3.36E-02	3.29E-02	3.17E-02
Photochemical ozone formation	kg NMVOC eq	3.01E-01	2.99E-01	2.97E-01
Acidification	molc H^+ eq	3.82E-01	3.79E-01	3.74E-01
Terrestrial eutrophication	molc N eq	1.15E + 00	1.14E + 00	1.14E + 00
Freshwater eutrophication	kg P eq	2.90E-03	2.85E-03	2.76E-03
Marine eutrophication	kg N eq	1.00E-01	9.98E-02	9.93E-02
Land use	kg C deficit	1.86E + 02	1.56E + 02	8.72E + 01
Mineral, fossil and ren resource depletion	kg Sb eq	7.81E-02	7.81E-02	7.81E-02

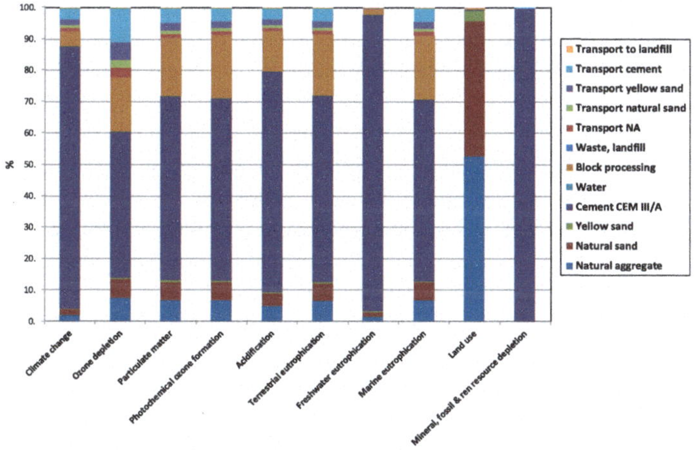

Fig. 36 Impact of the production of 1 m^3 of concrete blocks with NA only (B_RCA0—reference scenario)—characterization ILCD 2011 midpoint + (1.10) from (Zhao et al. 2020)

blocks at Prefer because a large quantity of materials is needed for the global production of concrete blocks. As the block waste in Prefer is limited, recycled materials from a nearby CDW sorting site could be another solution (that is why the scenario of LCA of B_RCA30 and B_RCA100 by using the RCA 2/6.3 mm from old block waste plus RCA from a nearby CDW recycling site was included). Further investigation is needed for the production of blocks based on recycled materials obtained from the unknown source.

The use of RCA in the production of concrete blocks is interesting. It can decrease the quantity of NA used in production and protect the environment. Meanwhile, block waste can be reused in the production of new blocks and therefore the amount of waste sent to landfills can be reduced. From an economic point of view, waste recycling is worthwhile when the recycled product is competitive with natural resources in terms of cost and quality. This study has demonstrated that RCA can be successfully used for the production of concrete blocks on an industrial scale. Due to a reduction in transportation costs, recycled materials will be more competitive in regions with scarce raw materials.

9 Conclusions

For some time now, several researchers have been investigating a way to improve the mechanical performance of RAC, but its durability has generally been neglected. Indeed, there have been a series of investigations concerning the effects of incorporating RA on the mechanical properties of concrete, most of which have proven its technical feasibility. Nevertheless, though acceptable levels have been obtained from a structural behavior point of view, durability-related properties appear to be much more affected by the incorporation of RA and must be considered when designing RAC structure.

Even though concrete is a very versatile material, in which one could incorporate almost anything, not all materials are suitable for structural concrete production. Although the literature review showed that it is possible to produce good quality concrete using only RMA or MRA, the durability performance of the resulting materials was significantly worse than that of a corresponding NAC and also concrete made with RCA. Therefore, specifications that allow the use of these materials in structural concrete must impose strict limits on both their incorporation level and their physical properties (i.e. water absorption and oven-dried density).

RCA is by far the most studied, successful and compatible RA used in the production of concrete. This is only natural considering that they are sourced from a material similar to that in which they are going to be used in. However, RCA varies widely, depending on the original materials and the recycling procedure used to obtain them. Proper certification and quality control must be applied at all stages of construction, demolition, transportation, and processing, to ensure that the best possible RA is produced. This will help maximize the use of these materials in the production of structural concrete, with minimum detrimental effects to its performance.

Some authors have established relationships and developed methodologies that can predict to some extent the durability of RAC with certain characteristics in order to meet the recommended limiting values of conventional concrete mixes and accomplish a given target service life. This may facilitate the wider use of RA in the production of structural concrete since it shows compatibility with conventional NAC and it is reasonably easy to design for a given environmental class, target service life, and reinforcement cover.

References

ACI-209.: Guide for modeling and calculating shrinkage and creep in hardened concrete, vol. 48. American Concrete Institute, Farmington Hills, Michigan, USA (2008)

Amorim, P., de Brito, J., Evangelista, L.: Concrete made with coarse concrete aggregate: Influence of curing on durability. ACI Mater. J. **109**(2), 195–204 (2012)

Andal, J., Shehata, M., Zacarias, P.: Properties of concrete containing recycled concrete aggregate of preserved quality. Constr. Build. Mater. **125**, 842–855 (2016)

BCSJ.: Proposed standard for the use of recycled aggregate and recycled aggregate concrete (in Japanese). Committee on Disposal and Reuse of Construction Waste, Building Contractors Society of Japan, Tokyo, Japan (1977)

Berndt, M.L.: Properties of sustainable concrete containing fly ash, slag and recycled concrete aggregate. Constr. Build. Mater. **23**(7), 2606–2613 (2009)

BRE.: Recycled aggregates. BRE Digest 433, CI/SfB p(T6), vol. 6. Building Research Establishment, Watford, UK (1998)

Bravo, M., de Brito, J., Evangelista, L., Pontes, J.: Mechanical and durability-related performance of concrete made with recycled aggregates sourced from construction and demolition wastes of Portuguese recycling plants (in Portuguese). Project FCT-PTDC/ECM/108682/:Instituto Superior Técnico, University of Lisbon. Portugal. Task **4**(4), 55 (2008)

BS-8500.: Concrete—complementary British Standard to BS EN 206–1–Part 2: Specification for constituent materials and concrete, vol. 52. British Standards Institution (BSI), United Kingdom

Buyle-Bodin, F., Hadjieva-Zaharieva, R.: Influence of industrially produced recycled aggregates on flow properties of concrete. Mater. Struct. **35**(252), 504–509 (2002)

Cartuxo, F., de Brito, J., Evangelista, L., Jiménez, J.R., Ledesma, E.F.: Rheological behavior of concrete made with fine recycled concrete aggregates—Influence of the superplasticizer. Constr. Build. Mater. **89**, 36–47 (2015)

Corral-Higuera, R., Arredondo-Rea, S.P., Neri-Flores, M.A., Gomez-Soberon, J.M., Almaral-Sanchez, J.L., Castorena-Gonzalez, J.H., Martinez-Villafane, A., Almeraya-Calderon, F.: Chloride-ion penetrability and corrosion behavior of steel in concrete with sustainability characteristics. Int. J. Electrochem. Sci. **6**(4), 958–970 (2011)

Corinaldesi, V., Moriconi, G.: Influence of mineral additions on the performance of 100% recycled aggregate concrete. Constr. Build. Mater. **23**(8), 2869–2876 (2009)

Corinaldesi, V.: Mechanical and elastic behavior of concretes made of recycled-concrete coarse aggregates. Constr. Build. Mater. **24**(9), 1616–1620 (2010)

Correia, J.R., de Brito, J., Pereira, A.S.: Effects on concrete durability of using recycled ceramic aggregates. Mater. Struct. **39**(2), 169–177 (2006)

DAfStb: DAfStb: Richtlinie - Beton mit rezykliertem zuschlag (in German). Germany, Deutscher Ausschuss Für Stahlbeton (German Committee for Reinforced Concrete), Berlin (1998)

De Juan, M.S., Gutiérrez, P.A.: Study on the influence of attached mortar content on the properties of recycled concrete aggregate. Constr. Build. Mater. **23**(2), 872–877 (2009)

Debieb, F., Kenai, S.: The use of coarse and fine crushed bricks as aggregate in concrete. Constr. Build. Mater. **22**(5), 886–893 (2008)

De Pauw, P., Vyncke, J., Thomas, P., Desmyter, J.: Shrinkage and creep of concrete with recycled materials as coarse aggregates. In: Dhir, R.K., Henderson, N.A., Limbachiya, M.C. (eds.), Proceedings of the international symposium on sustainable construction: Use of recycled concrete aggregate, pp. 213–225. Thomas Telford, London, UK (1998)

DIN-4226.: Aggregates for mortar and concrete, Part 100: Recycled aggregates, vol. 29. Deutsches Institut für Normungswesen (DIN), Berlin, Germany (2002)

Dillmann, R.: Concrete with recycled concrete aggregates. In: Dhir, R.K., Henderson, N.A., Limbachiya, M.C. (eds.), Proceedings of the international symposium on sustainable construction: Use of recycled concrete aggregate, pp. 239–254 . Thomas Telford, London, UK (1988)

Dhir, R.K., Paine, K.A.: Performance related approach to the use of recycled aggregates. Waste and Resources Action Programme (WRAP) Aggregates Research Programme, Banbury, Oxon, UK (2007). STBF 13/14C, ISBN: 1–84405–302–4: 77

Dhir, R.K., Limbachiya, M.C., Leelawat, T.: Suitability of recycled concrete aggregate for use in BS 5328 designated mixes. Proc. Inst. Civ. Engineers–Structures Build., **134**(3), 257–274 (1999)

Dodds, W., Goodier, C., Christodoulou, C., Austin, S., Dunne, D.: Corrosion risk assessment of structural concrete with coarse crushed concrete aggregate. Proc. Inst. Civ. Eng.-Constr. Mater., **173**(2), 57–69 (2020)

Dodds, W., Goodier, C., Christodoulou, C., Austin, S., Dunne, D.: Durability performance of sustainable structural concrete: Effect of coarse crushed concrete aggregate on microstructure and water ingress. Constr. Build. Mater. **145**, 183–195 (2017)

Domingo-Cabo, A., Lázaro, C., López-Gayarre, F., Serrano-López, M.A., Serna, P., Castaño-Tabares, J.O.: Creep and shrinkage of recycled aggregate concrete. Constr. Build. Mater. **23**(7), 2545–2553 (2009)

Duan, Z.H., Poon, C.S.: Properties of recycled aggregate concrete made with recycled aggregates with different amounts of old adhered mortars. Mater. des. **58**, 19–29 (2014)

Eckert, M., Oliveira, M.: Mitigation of the negative effects of recycled aggregate water absorption in concrete technology. Constr. Build. Mater. **133**, 416–424 (2017)

EHE-08.: Code on structural concrete, vol. 556. Publication Center, General Technical Secretariat, Ministry of Development, Madrid, Spain

EN-1992–1–1.: Eurocode 2—Design of concrete structures: Part 1–1: General rules and rules for buildings, vol. 259. Comité Européen de Normalisation (CEN), Brussels, Belgium (2008)

Etxeberria, M., Vegas, I.: Effect of fine ceramic recycled aggregate (RA) and mixed fine RA on hardened properties of concrete. Mag. Concr. Res. **67**(12), 645–655 (2015)

Etxeberria, M., Vázquez, E., Marí, A., Barra, M.: Influence of amount of recycled coarse aggregates and production process on properties of recycled aggregate concrete. Cem. Concr. Res. **37**(5), 735–742 (2007)

Evangelista, L., De Brito, J.: Criteria for the use of fine recycled concrete aggregates in concrete production. In: Vázquez, E., Hendriks, C., Janssen, G.M.T. (eds.), International RILEM conference on the use of recycled materials in buildings and structures, pp. 503–510. RILEM Publications SARL, Barcelona, Spain (2004)

Evangelista, L., De Brito, J.: Durability performance of concrete made with fine recycled concrete aggregates. Cement Concr. Compos. **32**(1), 9–14 (2010)

Evangelista, J.R.E., Oliveira, J.A., Botelho, F.J.E., Oliveira, R.M.E., Pereira, C.E.: Desempenho de sementes de soja peliculizadas em solo com diferentes teores de água. Ciência e Agrotecnologia, **31**, 994–999 (2007)

Ferreira, L., de Brito, J., Barra, M.: Influence of the pre-saturation of recycled coarse concrete aggregates on concrete properties. Mag. Concr. Res. **63**(8), 617–627 (2011)

Fathifazl, G., Razaqpur, A.G., Isgor, O.B., Abbas, A., Fournier, B., Foo, S.: Creep and drying shrinkage characteristics of concrete produced with coarse recycled concrete aggregate. Cement Concr. Compos. **33**(10), 1026–1037 (2011)

Fraaij, A. L., Pieterson, H. S., Vries, J.: Performance of concrete with recycled aggregate. In: Proceedings of the international conference: sustainable concrete construction. Dundee, Scotland (2002)

Gardner, N.J., Lockman, M.J.: Design provisions for drying shrinkage and creep of normal-strength concrete. ACI Mater. J. **98**(2), 159–167 (2001)

Garboczi, E.J.: Permeability, diffusivity, and microstructural parameters: a critical review. Cem. Concr. Res. **20**(4), 591–601 (1990)

Geng, Y., Wang, Y.Y., Chen, J.: Creep behavior of concrete using recycled coarse aggregates obtained from source concrete with different strengths. Constr. Build. Mater. **128**, 199–213 (2016)

González, A., Etxeberria, M.: Experimental analysis of properties of high performance recycled aggregate concrete. Constr. Build. Mater. **52**, 227–235 (2014)

Gonzalez-Corominas, A., Etxeberria, M.: Effects of using recycled concrete aggregates on the shrinkage of high performance concrete. Constr. Build. Mater. **115**, 32–41 (2016)

Gomez-Soberon, J.M.V.: Porosity of recycled concrete with substitution of recycled concrete aggregate - An experimental study. Cem. Concr. Res. **32**(8), 1301–1311 (2002a)

Gomez-Soberon, J.M.V.: Creep of concrete with substitution of normal aggregate by recycled concrete aggregate. In: Malhotra, V.M. (ed.), Innovations in design with emphasis on seismic, wind, and environmental loading: Quality control and innovations in materials/hot-weather concreting, vol. 209, pp. 461–474. (2002b)

Gomes, M., de Brito, J.: Structural concrete with incorporation of coarse recycled concrete and ceramic aggregates: durability performance. Mater. Struct. **42**(5), 663–675 (2009)

Guo, Z., Tu, A., Chen, C., Lehman, D.E.: Mechanical properties, durability, and life-cycle assessment of concrete building blocks incorporating recycled concrete aggregates. J. Clean. Prod. **199**, 136–149 (2018)

Hansen, T.C.: Recycling of demolished concrete and masonry. E & FN Spon, London, UK (1992)

Hwang, J.P., Shim, H.B., Lim, S., Ann, K.Y.: Enhancing the durability properties of concrete containing recycled aggregate by the use of pozzolanic materials. KSCE J. Civ. Eng. **17**(1), 155–163 (2013)

JIS-5021.: Recycled aggregate for concrete-class H (in Japanese), vol. 30. Japan Standards Association, Tokyo, Japan (2011)

JIS-5022.: Recycled aggregate for concrete-class M (in Japanese), vol. 74. Japan Standards Association, Tokyo, Japan (2012)

JIS-5023.: Recycled aggregate for concrete-class L (in Japanese), vol. 68. Japan Standards Association, Tokyo, Japan (2012)

Kapoor, K., Singh, S.P., Singh, B.: Durability of self-compacting concrete made with recycled concrete aggregates and mineral admixtures. Constr. Build. Mater. **128**, 67–76 (2016)

Khatib, J.M.: Properties of concrete incorporating fine recycled aggregate. Cem. Concr. Res. **35**(4), 763–769 (2005)

Kim, K., Shin, M., Cha, S.: Combined effects of recycled aggregate and fly ash towards concrete sustainability. Constr. Build. Mater. **48**, 499–507 (2013)

Kikuchi, M., Yasunaga, A., Ehara, E.: The total evaluation of recycled aggregate and recycled concrete. In: Lauritzen, E.K. (ed.), Proceedings of the third international RILEM symposium on demolition and reuse of concrete and masonry, pp. 425–436. CRC Press, Odense, Denmark (1993)

Kong, D., Lei, T., Zheng, J., Ma, C., Jiang, J., Jiang, J.: Effect and mechanism of surface-coating pozzalanics materials around aggregate on properties and ITZ microstructure of recycled aggregate concrete. Constr. Build. Mater. **24**(5), 701–708 (2010)

Kropp, J., Hilsdorf, H., Grube, H., Andrade, C., Nilsson, L.O.: Transport mechanisms and definitions. In: Hilsdorf, H., Kropp, J. (eds.), RILEM Report 12—performance criteria for concrete durability, vol. 327. CRC Press, (1995)

Kou, S.C., Poon, C.S., Chan, D.: Properties of steam cured recycled aggregate fly ash concrete. In: Vázquez, E., Hendriks, C., Janssen, G.M.T. (eds.), International RILEM conference on the use of recycled materials in buildings and structures, pp. 590–599. RILEM Publications SARL, Barcelona, Spain (2004)

Kou, S.C., Poon, C.S., Chan, D.: Influence of fly ash as cement replacement on the properties of recycled aggregate concrete. J. Mater. Civ. Eng. **19**(9), 709–717 (2007)

Kou, S.C., Poon, C.S., Chan, D.: Influence of fly ash as a cement addition on the hardened properties of recycled aggregate concrete. Mater. Struct. **41**(7), 1191–1201 (2008)

Kou, S.C., Poon, C.S., Agrela, F.: Comparisons of natural and recycled aggregate concretes prepared with the addition of different mineral admixtures. Cement Concr. Compos. **33**(8), 788–795 (2011)

Kou, S.C., Poon, C.S.: Enhancing the durability properties of concrete prepared with coarse recycled aggregate. Constr. Build. Mater. **35**, 69–76 (2012)

Kou, S.C., Poon, C.S., Wan, H.W.: Properties of concrete prepared with low-grade recycled aggregates. Constr. Build. Mater. **36**, 881–889 (2012)

Kou, S.-C., Poon, C.-S.: Long-term mechanical and durability properties of recycled aggregate concrete prepared with the incorporation of fly ash. Cement Concr. Compos. **37**, 12–19 (2013)

Kou, S.C., Zhan, B.J., Poon, C.-S.: Use of a CO_2 curing step to improve the properties of concrete prepared with recycled aggregates. Cement Concr. Compos. **45**, 22–28 (2014)

Kou, S.C., Poon, C.S.: Properties of concrete prepared with crushed fine stone, furnace bottom ash and fine recycled aggregate as fine aggregates. Constr. Build. Mater. **23**(8), 2877–2886 (2009a)

Kou, S.C., Poon, C.S.: Properties of self-compacting concrete prepared with coarse and fine recycled concrete aggregates. Cement Concr. Compos. **31**(9), 622–627 (2009b)

Lamond, J.F., Campbell, R.L., Giraldi, A., Jenkins, N.J.T., Campbell, T.R., Halczak, W., Miller, R., Cazares, J.A., Hale, H.C., Seabrook, P.T.: Removal and reuse of hardened concrete. ACI Mater. J., 300–325 (2002)

Levy, S.M., Helene, P.: Durability of recycled aggregates concrete: a safe way to sustainable development. Cem. Concr. Res. **34**(11), 1975–1980 (2004)

Limbachiya, M., Meddah, M.S., Ouchagour, Y.: Use of recycled concrete aggregate in fly-ash concrete. Constr. Build. Mater. **27**(1), 439–449 (2012)

Lima, C., Caggiano, A., Faella, C., Martinelli, E., Pepe, M., Realfonzo, R.: Physical properties and mechanical behavior of concrete made with recycled aggregates and fly ash. Constr. Build. Mater. **47**, 547–559 (2013)

Li, X.: Recycling and reuse of waste concrete in China : Part I. Material behavior of recycled aggregate concrete. Resour. Conserv. Recycl. **53**(1), 36–44 (2008)

LNEC-E471.: Guide for the use of coarse recycled aggregates in concrete (in Portuguese), vol. 6. National Laboratory of Civil Engineering (Laboratório Nacional de Engenharia Civil—LNEC), Portugal (2006)

Lotfy, A., Al-Fayez, M.: Performance evaluation of structural concrete using controlled quality coarse and fine recycled concrete aggregate. Cement Concr. Compos. **61**, 36–43 (2015)

Lye, C.Q., Dhir, R.K., Ghataora, G.S.: Shrinkage of recycled aggregate concrete. Proceedings of the Institution of Civil Engineers-Structures and Buildings **169**(12), 867–891 (2016)

Manzi, S., Mazzotti, C., Bignozzi, M.C.: Short and long-term behavior of structural concrete with recycled concrete aggregate. Cement Concr. Compos. **37**(37), 312–318 (2013)

Medina, C., Frias, M., Sanchez de Rojas, M.I., Thomas, C., Polanco, J.A.: Gas permeability in concrete containing recycled ceramic sanitary ware aggregate. Constr. Build. Mater. **37**, 597–605 (2012)

Merlet, J.D., Pimienta, P.: Mechanical and physico-chemical properties of concrete produced with coarse and fine recycled concrete aggregates. In: Lauritzen, E.K., Proceedings of the third international RILEM symposium on demolition and reuse of concrete and masonry. CRC Press, Odense, Denmark, pp. 400–411

Moon, D. J., Moon, H. Y., Nagataki, S., Hisada, M., Saeki, T.: Improvement on the qualities of recycled aggregate concrete containing super fine mineral admixtures. In: Kyokai, P.K., Kyokai, N.K. (eds.), Proceedings of the 1st fib congress, pp. 113–118. Osaka, Japan (2002)

Neville, A.M., Brooks, J.J.: Concrete technology. Harlow, United Kingdom, Pearson (2010)

Neves, R., Branco, F.A., de Brito, J.: A method for the use of accelerated carbonation tests in durability design. Constr. Build. Mater. **36**, 585–591 (2012)

Neville, A.M.: Properties of concrete. UK, Longman, London (2011)

NBR-15116.: Recycled aggregate of solid residue of building constructions—Requirements and methodologies (in Portuguese), vol. 18. Brazilian Association for Technical Norms (Associação Brasileira de Normas Técnicas—ABNT), Rio de Janeiro, Brasil

Olorunsogo, F.T., Padayachee, N.: Performance of recycled aggregate concrete monitored by durability indexes. Cem. Concr. Res. **32**(2), 179–185 (2002)

OT-70085.: Use of secondary mineral construction materials in the construction of shelters (in French), vol. 16. Berne, Switzerland, Objectif Technique, Instruction Technique, Schweizerische Eidgenossenschaft, (2006)

Pacheco-Torgal, F., Jalali, S.: Reusing ceramic wastes in concrete. Constr. Build. Mater. **24**(5), 832–838 (2010)

Padmini, A.K., Ramamurthy, K., Mathews, M.S.: Influence of parent concrete on the properties of recycled aggregate concrete. Constr. Build. Mater. **23**(2), 829–836 (2009)

Pedro, D.: Mechanical and durability-related performance of concrete produced with recycled aggregates from high performance concrete for use in pre-cast industry (in Portuguese), p. 47. Project FCT-PTDC/ECM/118372/2010. Instituto Superior Técnico, University of Lisbon, Lisbon, Portugal (2014)

Pedro, D., Brito, J.D., Evangelista, L.: Influence of the use of recycled concrete aggregates from different sources on structural concrete. Constr. Build. Mater. **71**, 141–151 (2014)

Poon, C.S., Kou, S.C., Chan, D.: Influence of steam curing on hardened properties of recycled aggregate concrete. Mag. Concr. Res. **58**(5), 289–299 (2006)

Poon, C.-S., Chan, D.: Effects of contaminants on the properties of concrete paving blocks prepared with recycled concrete aggregates. Constr. Build. Mater. **21**(1), 164–175 (2007)

Poon, C.S., Lam, C.S.: The effect of aggregate to cement ratio and types of aggregates on the properties of pre-cast concrete blocks. Cement Concr. Compos. **30**(4), 283–289 (2008)

Poon, C.S., Kou, S.C., Wan, H.W., Etxeberria, M.: Properties of concrete blocks prepared with low-grade recycled aggregates. Waste Manag. **29**(8), 2369–2377 (2009)

PTV-406.: Recycled aggregates from construction and demolition wastes (in French), vol. 16. Prescriptions Techniques, Brussels, Belgium

Razaqpur, A.G., Fathifazl, G., Isgor, G., Abbas, A., Fournier, B., Foo, S.: How to produce high-quality concrete mixes with recycled concrete aggregate. In: Xiao, J.Z., Zhang, Y., Cheung, M.S., Chu, R.P.K. (eds.), 2nd International Conference on Waste Engineering Management, ICWEM 2010, vol. 73, pp. 11–35. Shanghai, China (2010)

Rao, A., Jha, K.N., Misra, S.: Use of aggregates from recycled construction and demolition waste in concrete. Resour. Conserv. Recycl. **50**(1), 71–81 (2007)

Ravindrarajah, S.R., Tam, C.T.: Recycled concrete as fine aggregate in concrete. Int. J. Cem. Compos. Light. Concr. **9**(4), 235–241 (1987)

RILEM: Specifications for concrete with recycled aggregates. Mater. Struct. **27**(173), 557–559 (1994)

Ryu, J.S., Monteiro, P.J.M.: Effect of specific gravity of coarse aggregate on interfacial transition zone, permeability, and strength of concrete. Br. Ceram. Trans. **101**(1), 30–34 (2002)

Ryu, J.S.: An experimental study on the effect of recycled aggregate on concrete properties. Mag. Concr. Res. **54**(1), 7–12 (2002)

Sagoe-Crentsil, K.K., Brown, T., Taylor, A.H.: Performance of concrete made with commercially produced coarse recycled concrete aggregate. Cem. Concr. Res. **31**(5), 707–712 (2001)

Shayan, A., Xu, A.M.: Performance and properties of structural concrete made with recycled concrete aggregate. ACI Mater. J. **100**(5), 371–380 (2003)

Sim, J., Park, C.: Compressive strength and resistance to chloride-ion penetration and carbonation of recycled aggregate concrete with varying amount of fly ash and fine recycled aggregate. Waste Manage. **31**(11), 2352–2360 (2011)

Silva, R.V., Neves, R., Brito, J.D., Dhir, R.K.: Carbonation behavior of recycled aggregate concrete. Cement Concr. Compos. **62**, 22–32 (2015e)

Silva, R.V., de Brito, J., Dhir, R.K.: Comparative analysis of existing prediction models on the creep behavior of recycled aggregate concrete. Eng. Struct. **100**, 31–42 (2015a)

Silva, R.V., de Brito, J., Dhir, R.K.: Prediction of the shrinkage behavior of recycled aggregate concrete: A review. Constr. Build. Mater. **77**, 327–339 (2015b)

Silva, R.V., de Brito, J., Dhir, R.K.: Tensile strength behavior of recycled aggregate concrete. Constr. Build. Mater. **83**, 108–118 (2015c)

Silva, R.V., de Brito, J., Neves, R., Dhir, R.: Prediction of chloride-ion penetration of recycled aggregate concrete. Mater. Res. **18**(2), 427–440 (2015d)

Soares, D., de Brito, J., Ferreira, J., Pacheco, J.: Use of coarse recycled aggregates from pre-cast concrete rejects: Mechanical and durability performance. Constr. Build. Mater. **71**, 263–272 (2014)

Soutsos, M.N., Tang, K., Millard, S.G.: Concrete building blocks made with recycled demolition aggregate. Constr. Build. Mater. **25**(2), 726–735 (2011)

Tu, T.-Y., Chen, Y.-Y., Hwang, C.-L.: Properties of HPC with recycled aggregates. Cem. Concr. Res. **36**(5), 943–950 (2006)

TFSCCS.: Draft of Spanish regulations for the use of recycled aggregate in the production of structural concrete (Task force of the standing committee of concrete of Spain). In: Vázquez, E., Hendriks, C., Janssen, G.M.T. (eds.), International RILEM conference on the use of recycled materials in building and structures, pp. 511–525. RILEM Publications SARL, Barcelona, Spain (2004)

Tam, V.W.Y., Tam, C.M.: Assessment of durability of recycled aggregate concrete produced by two-stage mixing approach. J. Mater. Sci. **42**(10), 3592–3602 (2007)

Tuutti, K.: Corrosion of steel in concrete, pp. 4–82. Swedish Cement and Concrete Research Institute at the Institute of Technology, Stockholm, Sweden (1982)

Thomas, C., Setién, J., Polanco, J.A., Alaejos, P., Sánchez de Juan, M.: Durability of recycled aggregate concrete. Constr. Build. Mater. **40**, 1054–1065 (2013)

Teranishi, K., Dosho, Y., Narikawa, M., Kikuchi, M.: Application of recycled aggregate concrete for structural concrete: Part 3—Production of recycled aggregate by real-scale plant and quality of recycled aggregate concrete. In: Dhir, R.K., Henderson, N.A., Limbachiya, M.C. (eds.), Proceedings of the international symposium on sustainable construction: Use of recycled concrete aggregate. London, UK, Thomas Telford, pp. 143–156

Vaishali, G.G., Rao, H.S.: Strength and permeability characteristics of fiber reinforced high performance concrete with recycled aggregates. Asian J Civ Eng (Build Hous) **13**(1), 55–77 (2012)

Vandamme, M., Ulm, F.-J.: Nanogranular origin of concrete creep. Proc. Natl. Acad. Sci. **106**(26), 10552–10557 (2009)

Vieira, T., Alves, A., de Brito, J., Correia, J.R., Silva, R.V.: Durability-related performance of concrete containing fine recycled aggregates from crushed bricks and sanitary ware. Mater. des. **90**, 767–776 (2016)

Wainwright, P.J., Cabrera, J.G.: Use of demolition concrete to produce durable structural concrete. In: Goumans, J.J.J., vanderSloot, H.A., Aalbers, T.G. (eds.), Environmental aspects of construction with waste materials, vol. 60, pp. 553–562. Amsterdam, The Netherlands, Elsevier (1994)

Waleed, N., Canisius, T.D.G.: Engineering properties of concrete containing recycled aggregates, vol. 104. Waste & Resources Action Programme, Banbury, Oxon, UK (2007)

Wernet, G., Bauer, C., Steubing, B., Reinhard, J., Moreno-Ruiz, E., Weidema, B.: The ecoinvent database version 3 (part I): overview and methodology. The International Journal of Life Cycle Assessment, **21**, 1218–1230 (2016)

Wesche, K.A.S., R.R.: Beton aus aufbereitetem Altbeton, Technologie und. Eigenschaften (Technology and properties of recycled concrete processed from concrete wastes). Beton **32**(2/3), 64–68/108–112 (1982)

Wirquin, E., Hadjieva-Zaharieva, R., Buyle-Bodin, F.: Utilisation de l'absorption d'eau des bétons comme critères de leur durabilité—application aux bétons de granulats recyclés. Mater. Struct. **33**(6), 403 (2000)

WBTC-No.12.: Specifications facilitating the use of recycled aggregates, vol. 16. Works Bureau Technical Circular, Hong-Kong (2002)

Xiao, J.Z., Xie, H., Yang, Z.: Shear transfer across a crack in recycled aggregate concrete. Cem. Concr. Res. **42**(5), 700–709 (2012)

Xiao, Z., Ling, T.-C., Kou, S.-C., Wang, Q., Poon, C.-S.: Use of wastes derived from earthquakes for the production of concrete masonry partition wall blocks. Waste Manage. **31**(8), 1859–1866 (2011)

Xiao, J.Z., Liu, Q., Du, J.T., Zhang, C.Z.: Micro-damage mechanisms and property fluctuation of recycled aggregate concrete. Key Eng. Mater. **348–349**, 61–64 (2007)

Xiao, J., Li, W., Corr, D.J., Shah, S.P.: Effects of interfacial transition zones on the stress-strain behavior of modeled recycled aggregate concrete. Cem. Concr. Res. **52**, 82–99 (2013)

Yang, K., Chung, H., Ashour, A.: Influence of type and replacement level of recycled aggregates on concrete properties. ACI Mater. J. **105**(3), 289–296 (2008)

Zaharieva, R., Buyle-Bodin, F., Skoczylas, F., Wirquin, E.: Assessment of the surface permeation properties of recycled aggregate concrete. Cement Concr. Compos. **25**(2), 223–232 (2003)

Zega, C.J., Di Maio, Á.A.: Use of recycled fine aggregate in concretes with durable requirements. Waste Manage. **31**(11), 2336–2340 (2011)

Zega, C.J., Di Maio, Á.A., Zerbino, R.L.: Influence of natural coarse aggregate type on the transport properties of recycled concrete. J. Mater. Civ. Eng. **26**(6), 1–9 (2013)

Zhao, Z., Courard, L., Groslambert, S., Jehin, T., Leonard, A., Xiao, J.: Use of recycled concrete aggregates from pre-cast block for the production of new building blocks: An industrial scale study. Resour. Conserv. Recycl. **157**, 104786 (2020)

Zhao, Y., Dong, J., Ding, H., Jin, W.: Shape of corrosion-induced cracks in recycled aggregate concrete. Corros. Sci. **98**, 310–317 (2015)

Zhang, H., Zhao, Y.: Performance of recycled concrete beams under sustained loads coupled with chloride-ion (Cl−) ingress. Constr. Build. Mater. **128**, 96–107 (2016)

Zheng W (2020) Degradation of recycled aggregate concrete structural components and modification of recycled aggregate Doctoral Thesis, Zhejiang University

Zhao, Z., Remond, S., Damidot, D., Courard, L., Michel, F.: Improving properties of recycled concrete aggregates by accelerated carbonation. Proc. Inst. Civ. Engineers—Construction Mater. **171**(3), 126–132 (2018)

Zhao, Y., Dong, J., Wu, Y., Wang, H., Li, X., Xu, Q.: Steel corrosion and corrosion-induced cracking in recycled aggregate concrete. Corros. Sci. **85**, 241–250 (2014)

The manufacturer's authorised representative in the EU is Springer Nature Customer Service Centre GmbH, Europaplatz 3, 69115 Heidelberg, Germany. If you have any concerns regarding our products, please contact ProductSafety@springernature.com

Printed and bound by CPI Group (UK) Ltd, Croydon, CR0 4YY
01/10/2026
02225949-0007